Progress in Molecular and Subcellular Biology

Series Editors: W.E.G. Müller (Managing Editor), Ph. Jeanteur, I. Kostovic, Y. Kuchino, A. Macieira-Coelho, R.E. Rhoads

31

Springer-Verlag Berlin Heidelberg GmbH

Progress in Molecular and Subcellular Biology

Volumes Published in the Series

Philippe Jeanteur (Ed.)

Regulation of Alternative Splicing

With 26 Figures

Springer

Professor Dr. PHILIPPE JEANTEUR
Institute of Molecular Genetics
of Montpellier
CNRS, BP 5051
1919 Route de Mende
34293 Montpellier Cedex 05
France

ISSN 0079-6484

Library of Congress Cataloging-in-Publication Data

Regulation of alternative splicing / Philippe Jeanteur, ed.
p. cm. – (Progress in molecular and subcellular biologys; 31)
Includes bibliographical references and index.
ISBN 978-3-642-07858-3 ISBN 978-3-662-09728-1 (eBook)
DOI 10.1007/978-3-662-09728-1

1. Genetic recombination. 2. Exons (Genetics) 3. Split genes. I. Jeanteur, Ph. (Philippe)
II. Series.

QH506 .P76 no. 31 [QH443]
572.8 s–dc21 [572'.645]

http.//www.springer.de

Originally published by Springer-Verlag Berlin Heidelberg New York in 2003
Softcover reprint of the hardcover 1st edition 2003

Cover design: Meta Design, Berlin
Typesetting: SNP Best-set Typesetter Ltd., Hong Kong
SPIN 10858714 39/3130 – 5 4 3 2 1 0 – Printed on acid-free paper

Preface

The discovery in 1977 that genes are split into exons and introns has done away with the one gene – one protein dogma. Indeed, the removal of introns from the primary RNA transcript is not necessarily straightforward since there may be optional pathways leading to different messenger RNAs and consequently to different proteins. Examples of such an alternative splicing mechanism cover all fields of biology. Moreover, there are plenty of occurrences where deviant splicing can have pathological effects.

Despite the high number of specific cases of alternative splicing, it was not until recently that the generality and extent of this phenomenon was fully appreciated. A superficial reading of the preliminary sequence of the human genome published in 2001 led to the surprising, and even deceiving to many scientists, low number of genes (around 32,000) which contrasted with the much higher figure around 150,000 which was previously envisioned.

Attempts to make a global assessment of the use of alternative splicing are recent and rely essentially on the comparison of genomic mRNA and EST sequences as reviewed by Thanaraj and Stamm in the first chapter of this volume. Most recent estimates suggest that 40–60% of human genes might be alternatively spliced, as opposed to about 22% for *C. elegans*. These figures indicate that alternative splicing is an essential element to expand the capacity of metazoan genomes to generate the functional complexity of their proteomes, the more so as a majority of splicing variants affect the function of the protein itself. In terms of pathologies, about 15% of hereditary diseases might result from anomalies in splicing.

The next three chapters focus on the mechanistic aspects of the splicing reaction itself with special emphasis on the role of SR proteins, a highly conserved family of arginine and serine-rich proteins. SR proteins are able to engage in both RNA binding and protein–protein interactions and can, therefore, mediate interactions between the pre-mRNA and the assembling spliceosome. In addition to being essential factors of the basic splicing machinery, they play key roles in the regulation of alternative splice site selection. The chapter by Sanford, Longman and Cáceres reviews the numerous roles of individual SR proteins, or combinations thereof, in both constitutive and alternative splicing.

Dosage of SR proteins is the critical factor which influences the choice of splice sites and, therefore, orients towards alternate splicing pathways, and so is the case for the A1 protein of hnRNP, which behaves antagonistically to SR proteins. The role of hnRNP A1 protein in both constitutive and alternative splicing is the main focus of the chapter by Chabot and collaborators. The ability of hnRNP A1 protein to both dimerize and bind with high affinity to sites on pre-mRNA leads to the straightforward speculation that it could function by remodeling the pre-mRNA structure. This model has obvious implications in alternative splicing, but it is also relevant to the generic splicing of long introns simply by bringing distant splicing partners in closer proximity. hnRNP A1 protein can bind to a wide range of RNA sequences. Therefore, despite their high abundance, the steady-state concentration of free A1 protein may become limiting and lead to a situation of discrimination between sequence elements and, therefore, the choice of splicing patterns.

Phosphorylation/dephosphorylation events have been implicated in the assembly of spliceosomes, the regulation of splice site selection and the subcellular localization of splicing factors. SR proteins, hnRNP and snRNP proteins are obvious targets for these events which might modulate their interactions with RNA or other proteins. In the next chapter, Soret and Tazi review, on the one hand, the molecular components of the splicing machinery which are affected by phosphorylation and, on the other hand, the known SR protein-specific kinases including the DNA-topoisomerase I. Given that alternative splicing is regulated by SR proteins in a dose-dependent fashion, they discuss the hypothesis that the activity of these enzymes might be a key element in the regulation of alternative splicing, as well as in its dysregulation, as is now observed in a growing number of diseases.

The next three chapters describe the role of alternative splicing in different in vivo physiological situations.

The determination of sexual identity in *Drosophila* is a well-known example where sex-lethal protein, exclusively expressed in female flies, acts as a master switch inducing female-specific patterns of alternative splicing in target genes. Förch and Valcarcel's paper reviews the molecular mechanisms which control somatic and germ-line sexual differentiation, sexual behavior and X chromosome dosage compensation.

Apoptosis or programmed cell death (PCD) is another example of a situation regulated by alternative splicing events which may eventually result in an as clear-cut a decision as life or death. Wu, Tang and Havlioglu review the molecular mechanisms underlying alternative splicing of PCD genes leading to functionally distinct or even antagonistic protein isoforms.

As amply discussed in all chapters of this book, it is clear that alternative splicing is common to most genes and tissues in metazoans. In this general context, the nervous system stands apart by the particularly high level of molecular diversity created by alternative splicing especially for proteins involved in two processes: the formation of neuronal connections during development and cell excitability. From hundreds of examples, the paper by Black and Grabowski focuses on a few systems where the regulatory mechanisms or the effects of alternative splicing on protein function are better understood.

Most papers of this book have touched upon the effects of splicing dysfunction in human diseases. The last one by Sazani and Kole not only reviews the use of antisense oligonucleotides to probe the interactions of splicing factors with splice sites, but also addresses the possibility of manipulating alternative splicing pathways with a goal to clinical applications.

PHILIPPE JEANTEUR

Contents

Multiple Roles of the SR Protein Family in Splicing Regulation
J.R. Sanford, D. Longman, and J.F. Cáceres

Heterogeneous Nuclear Ribonucleoprotein Particle A/B Proteins and the Control of Alternative Splicing of the Mammalian Heterogeneous Nuclear Ribonucleoprotein Particle A1 Pre-mRNA
B. Chabot, C. LeBel, S. Hutchison, F.H. Nasim, and M.J. Simard

Prediction and Statistical Analysis of Alternatively Spliced Exons

T.A. Thanaraj[1] and S. Stamm[2]

The completion of large genomic sequencing projects revealed that metazoan organisms abundantly use alternative splicing. Alternatively spliced exons can be found in these sequences by sequence comparison of genomic, mRNA and EST sequences. Furthermore, a large number of alternative exons have been described in the literature. Here, we review computer and manually curated databases of alternative exons and discuss the various approaches used to generate them. Sequence analysis shows that alternative exons often have unusual lengths, suboptimal splice sites and characteristic nucleotide patterns. Despite this progress alternative exons cannot be predicted *ab initio* from genomic data, which is due to the degenerate nature of splicing signals.

1 Overview

The first draft of the human genome has demonstrated that an average human gene contains a mean of 8.8 exons with an average size of 145 nt. The mean intron length is 3365 nt and the 5′ and 3′ UTR are 770 and 300 nt, respectively. As a result, a "standard" gene spans about 27 kbp. After pre-mRNA splicing, the mature message consists of 1340 nt coding sequence and 1070 nt untranslated regions and a poly (A) tail (Lander et al. 2001). The vertebrate splicing machinery is not only capable of accurately recognizing the small exons within the larger intron context, but is also able to recognize exons alternatively. In this process, an exon is either incorporated into the mRNA, or is excised as an intron. This process of alternative splicing is abundantly used in higher eukaryotes. In humans, a detailed analysis was performed for chromosome 22 and 19 (Lander et al. 2001). Of the 245 genes present on chromosome 22, 59% are alternatively spliced and the 544 genes of chromosome 19 result in 1859 different messages. The comparison of expressed sequence tags (ESTs) with the human genome sequence indicates that 47% of human genes might be alternatively

[1] European Bioinformatics Institute, Wellcome Trust Genome Campus, Hinxton, Cambridge, CB10 1SD, UK
[2] Institute for Biochemistry, University Erlangen-Nürnberg, Fahrstrasse 17, 91054 Erlangen, Germany; e-mail: stefan@stamms-lab.net

Progress in Molecular and Subcellular Biology, Vol. 31
Philippe Jeanteur (Ed.)

spliced (Modrek et al. 2001). This is in contrast to data obtained from *C. elegans*, where about 22% of the genes are alternatively spliced. Several databases have been generated that contain a wealth of information about sequences that are involved in alternative splicing. Here, we shall discuss the major findings and the computer programs used to generate them.

2 Mechanism of Splicing

2.1 General Splicing Mechanism

Three major *cis*-elements of the pre-mRNA define an exon, the 5′ splice site, the 3′ splice site and the branch point. All these elements are short (7–14 nt) and can only be described by degenerate consensus sequences. A given *cis*-element in a human gene will follow the consensus sequence only to a certain degree (Burset et al. 2000, 2001). This divergence from consensus sequences in higher eukaryotes is in contrast to the situation in yeast, where splice sites follow a strict consensus. To allow for exon recognition, in higher eukaryotes, additional elements known as exonic or intronic enhancers, depending on their location, are present (Cooper and Mattox 1997; Smith and Valcárcel 2000). Through recognition of these regulatory elements, sequence-specific RNA binding proteins regulate spliceosome assembly. The spliceosome is a 60S complex containing small nuclear RNPs (U1, U2, U4, U5 and U6) and over 50 different proteins. In this complex, U1 snRNP binds to the 5′splice site. SF1 and U2 snRNP bind to the 3′ splice site and the branch point. The consensus sequences of the 5′ splice site and the branch point reflect their binding to U1 and U2 snRNA, and the polypyrimidine tract of the 3′ splice site is reminiscent of the systematic evolution of ligands by exponential enrichment (SELEX) sequence of U2AF (TTTYYYYTNTAGG; Wu et al. 1999).

2.2 Exon Definition

Since the splice sites in higher eukaryotes are less conserved than the ones in yeast, the question arises how these exons are recognized. It was proposed that in vertebrates with their larger introns, the splicing machinery searches for a pair of closely spaced 5′ and 3′ splice sites (Berget 1995). The exon is then defined by binding of U1 and U2 snRNAs, as well as associated splicing factors to the exon. After an exon has been defined, neighboring exons must be juxtaposed. Due to the degenerate nature of splice sites in higher eukaryotes, it is difficult to predict exons from genomic DNA sequences and current computer programs cannot accurately predict exons from genomic DNA (Thanaraj

2000). This finding *in silico* contrasts with the high accuracy and fidelity characteristic for splice sites in vivo.

One reason for the specificity observed in vertebrate cells are additional regulatory elements known as silencers or enhancers. Based on their location they can be intronic or exonic. These sequence elements are again characterized by loose consensus sequences. They can be subdivided into purine-rich (GAR-type) and AC-rich (ACE-type) enhancers (Cooper and Mattox 1997). Enhancers bind to proteins that are able to recruit spliceosomal components, which results in the recognition of splice sites that are located near an enhancer (Hertel and Maniatis 1998). Since enhancers are often exonic, their loose consensus sequences, e.g., their degeneracy, is most likely necessary to allow for the amino acid usage needed in a given protein.

Proteins binding to sequence elements on the pre-mRNA can be subdivided into two major groups: members of the serine/arginine-rich (SR) family of proteins (Fu 1995; Manley and Tacke 1996; Graveley 2000) and heterogenous ribonuclear proteins (hnRNPs) (Weighardt et al. 1996). The binding of individual proteins to enhancer sequences is intrinsically weak and not highly specific. However, in most cases studied several such sequence elements are present. Furthermore, the proteins binding to *cis*-elements often bind to other RNA binding proteins. As a result, a protein:RNA complex is formed and the exon is recognized with high specificity. The composition of the protein:RNA complex is dependent on the concentration of various regulatory proteins, their phosphorylation status and the sequences of the regulatory elements on the pre-mRNA (Fig. 1).

2.3 Splice-Site Recognition Is Influenced by the Relative Concentration of Regulatory Proteins

The regulation of alternative splicing is still under intense investigation. The relative concentration of splicing-associated proteins can regulate alternative splice site selection (Hastings and Krainer 2001). Experiments both in vivo and in vitro show that the relative concentration of SR proteins and hnRNPs can dictate splice site selection (Mayeda and Krainer 1992; Mayeda et al. 1993; Caceres et al. 1994). Furthermore, the expression levels of various SR proteins (Ayane et al. 1991; Zahler et al. 1993; Screaton et al. 1995) and hnRNPs (Kamma et al. 1995) vary amongst tissues and could therefore account for differences in splice site selection. Several examples of antagonistic splicing factors have been described (Mayeda et al. 1993; Caceres et al. 1994; Gallego et al. 1997; Jumaa and Nielsen 1997; Polydorides et al. 2000). Here, one factor promotes inclusion of an exon and the other factor promotes its skipping. In most of these cases, it remains to be determined whether this antagonistic effect is achieved by (1) an actual competition of the factors for an RNA binding site, (2) through sequestration of the factors by protein:protein interaction and (3) by changes

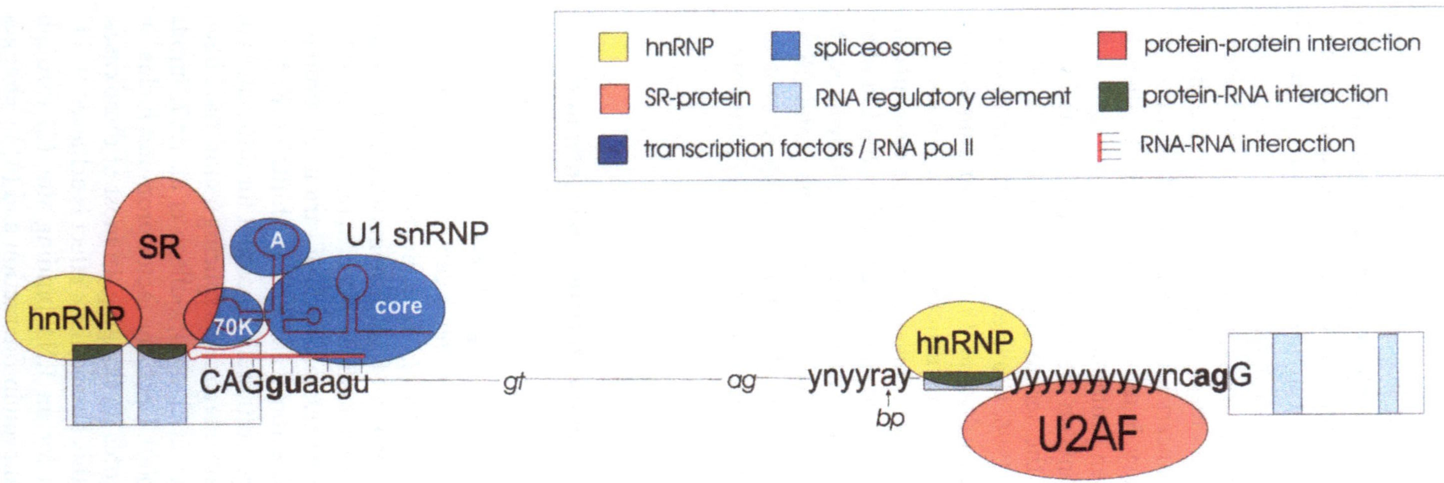

Fig. 1. Elements involved in alternative splicing of pre-mRNA. Exons are indicated as *boxes*, introns as *thin lines*. Splicing regulator elements (enhancers or silencers) are shown as *gray boxes* in exons or as *thin boxes* in introns. The 5′ splice site (CAGguaagu) and 3′ splice site $(y)_{10}$ncagG, as well as the branch point (ynyyray), are indicated (y = c or u, n = a, g, c or u). *Uppercase letters* refer to nucleotides that remain in the mature mRNA. Two major groups of proteins, hnRNPs (*yellow*) and SR or SR related proteins (*orange*), bind to splicing regulator elements; the protein: RNA interaction is shown in *green*. This protein complex assembling around an exon enhancer stabilizes binding of the U1 snRNP close to the 5′ splice site, for example, due to protein:protein interaction between an SR protein and the RS domain of U1 70 K (shown in *red*). This allows hybridization (*thick red line with stripes*) of the U1 snRNA (*red*) with the 5′ splice site. The formation of the multiprotein:RNA complex allows discrimination between proper splice sites (*bold letters*) and cryptic splice sites (small *gt ag*) that are frequent in pre-mRNA sequences. Factors at the 3′ splice site include U2AF, which recognizes pyrimidine-rich regions of the 3′ splice sites, and is antagonized by binding of several hnRNPs (e.g., hnRNP I) to elements of the 3′ splice site. *Orange*: SR and SR related proteins; *yellow*: hnRNPs; *green*: protein:RNA interaction; *red*: protein:protein interaction; *thick red line with stripes*: RNA:RNA interaction

in the composition of protein complexes recognizing the splicing enhancer. In addition, cell-type-specific splicing factors have been detected. In *Drosophila*, for example, the expression of the SR protein transformer is female-specific (Boggs et al. 1987) and determines the sex by directing alternative splicing decisions. Other tissue-specific factors include the male germline-specific transformer-2 variant in *D. melanogaster* (Mattox et al. 1990) and *D. virilis* (Chandler et al. 1997), an isoform of its mammalian homologue htra2-beta3 that is expressed only in some tissues (Nayler et al. 1998), the neuron-specific factor NOVA-1 (Jensen et al. 2000) as well as testis and brain-enriched factor rSLM-2 (Stoss et al. 2001) and NSSR (Komatsu et al. 1999). For most of these factors, the tissue-specific target genes remain to be determined. However, a combination of knockout experiments and biochemical analysis allowed the identification of doublesex, fruitless, and transformer-2 as a target of the transformer-2/transformer complex in *Drosophila* (Hoshijima et al. 1991; Mattox and Baker 1991; Heinrichs et al. 1998) and glycine receptor alpha2 and $GABA_A$ pre-mRNA as a target for NOVA-1 (Jensen et al. 2000). Although this analysis is currently limited, it is likely that a given splicing factor will influence several pre-mRNAs. SR-proteins (Fu 1995; Graveley 2000) from all species and splicing regulatory proteins for *Drosophila* (Mount and Salz 2000) have been compiled.

3
Properties of Alternative Spliced Exons

3.1
Several Types of Splicing

The analysis of human intron sequences demonstrates the existence of several intron types. Out of 53,295 human confirmed exons, 98.12% use the canonical GT and AG dinucleotides at the 5′ and 3′ site respectively. Another 0.76% of introns contain GC-AG dinucleotides at this position (Lander et al. 2001); which is comparable to the estimated 1.1% obtained by modeling ESTs on the human genome (Clark and Thanaraj 2002).These introns are processed similarly to the GT-AG introns. They have been compiled in a database and it was found that one in every twenty alternative introns is a GC-AG intron. About 60% of all GC-AG introns are alternatively spliced (Burge et al. 1998; Thanaraj and Clark 2001). Finally, a different class of introns exists that are flanked by AT-AC dinucleotides at the 5′ and 3′ position. These introns are processed by a variant U12 splicing system (Burge et al. 1998). Further, Clark and Thanaraj (2002) observed that 0.4% of the observed introns can be of the type U12-spliceosome GT-AG. Levine and Durbin could identify 404 EST-confirmed U12-type introns in the whole genome, 20 of which had termini dinucleotides different from GT-AG and AT-AC (Levine and Durbin 2001). A systematic survey of mammalian splice sites revealed even more intron sequence diversity, as introns flanked by GT-AC, GT-CG, GT-TG, AT-AG and GA-AG have been detected (Burset et al. 2000, 2001). However, only a total of 39 such introns have been detected and the exact splicing mechanism remains to be determined. The results of these studies of mammalian introns correlate well with the data from human DNA databases. It was found that 99.24% of all human introns are flanked by GT-AG, 0.69% by GC-AG, 0.05% by AT-AC and 0.02% by other nucleotides (Burset et al. 2000, 2001). Weight matrices for some of the noncanonical splice sites and different categories of GT-AG introns have been compiled in AltExtron database (Clark and Thanaraj 2002; Table 2).

3.2 Types of Exons

Alternative exons can be subdivided into different groups. Depending on their splicing mode, they can be classified as cassettes, mutually exclusive, retained intron, and alternative 3′ and 5′ splice sites (Breitbart et al. 1987; Fig. 2). Database analysis from human curated sets revealed that cassette exons are the most common type, representing about half (53%) of alternative exons (Stamm

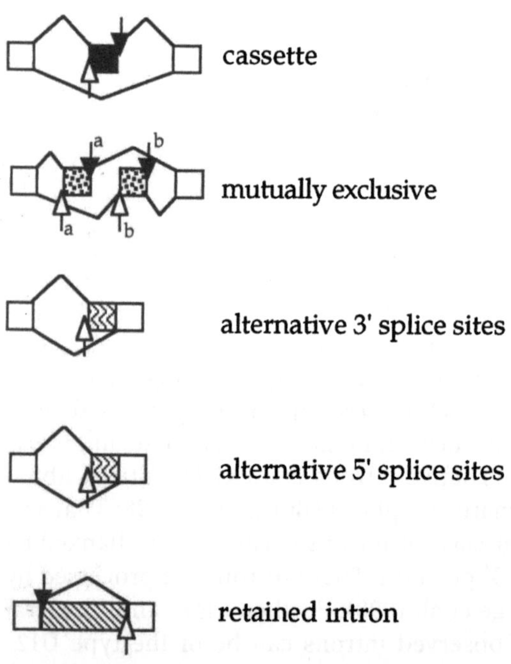

Fig. 2. Type of alternative splicing events: alternative exons are shown as *boxes with different shading*. Flanking constitutive exons are shown as *white boxes*. The *open arrow* indicates the position of the alternative 3′ splice site analyzed; a *closed arrow* indicates the position of the 5′ splice sites analyzed

et al. 2000). This is in agreement with an EST based study, where roughly 61% of alternative exons were found to be either cassette or mutually exclusive exons (Mironov et al. 1999). The next most common splicing modes in this sample of data are alternative 3′ splice sites (10%), mutually exclusive cassette exons (9%), alternative 5′ splice sites (7%) and retained introns (6%). Details of the molecular mechanism that regulates alternative splicing of the various subgroups will most likely be different: in alternative 3′ and 5′ splice sites, there is often a competition between different splice sites. Intronic sequences between mutually exclusive exons are often short and there is evidence for unusual branch point locations in some of these systems studied (Helfman and Rici 1989; Southby et al. 1999). Often, alternatively spliced exons show tissue- or developmental-specific expression. A survey of the methods used to obtain information about the expression patterns shows that most data are obtained by RT-PCR from whole tissue RNA (Stamm et al. 2000). As a result, expression data will be averaged over a vast number of different cell types. When organs are compared, it is striking that most alternative exons are present in multiple tissues, which is in agreement with EST-based studies (Modrek et al. 2001). This could indicate that an intrinsic balance of regulatory factors expressed in all tissues (Hanamura et al. 1998) could be responsible for alternative splice site selection of numerous exons. The organs that express alternative exons devoted only to one tissue were predominantly brain, then muscle, blood and liver (Stamm et al. 2000). Since these data are obtained from the published literature, they could also reflect the current focus on research. However, high expression of the genetic information in the nervous system has been found earlier in clonal and kinetic analysis (Chaudhari and Hahn 1983; Milner and Sutcliffe 1983) and in a recent EST-based study (Modrek et al. 2001). The analysis of tissue specificity is further complicated by the fact that not all researchers use the same tissue to determine alternative splicing patterns. A systematic and unified approach is urgently needed here.

3.3 Regulatory Elements

Since the regulatory elements reside on the pre-mRNA, the comparison of alternative with constitutive exons should reveal overall regulatory features. Several studies therefore compared the nucleotide usage of alternative exons at the 5′ and 3′ splice sites. A major result from this comparison is that the majority of alternative exons contain splice sites that strongly deviate from the consensus sequence. This deviation is more pronounced in exons that are specific for a certain tissue or developmental stage.

3.3.1
5′ Splice Site

The nucleotide composition at the 5′ splice site reflects binding to U1 snRNA (Zhuang and Weiner 1986) as well as an interaction with U6 snRNA (Wise 1993). Overall, vertebrate alternative exons deviate most strongly at the +4 and +5 position. In constitutive cassette exons, the +3 position is occupied by adenosine, while in the case of alternative exons it is more often occupied by guanosine (Clark and Thanaraj 2002). In exons that are exclusively expressed in neurons (Fig. 3) and muscle cells, there is an additional deviation at the −3 position where splice sites from both categories use more adenosine than the consensus (Stamm et al. 1994, 2000; Fig. 3). The corresponding bases U5 and U6 of U1 snRNA that would bind to the +4 and +3 positions are modified to pseudouracils (Reddy 1989). It is interesting that alternative exons deviate from the consensus at positions of the 5′ splice site that are modified in the complementary base position of U1 and U6 snRNA. None of the 16 human and 6 *Drosophila* U1 snRNA genes show a sequence diversion at the positions binding to the 5′ splice site, but the posttranscriptional modifications have not been systematically studied.

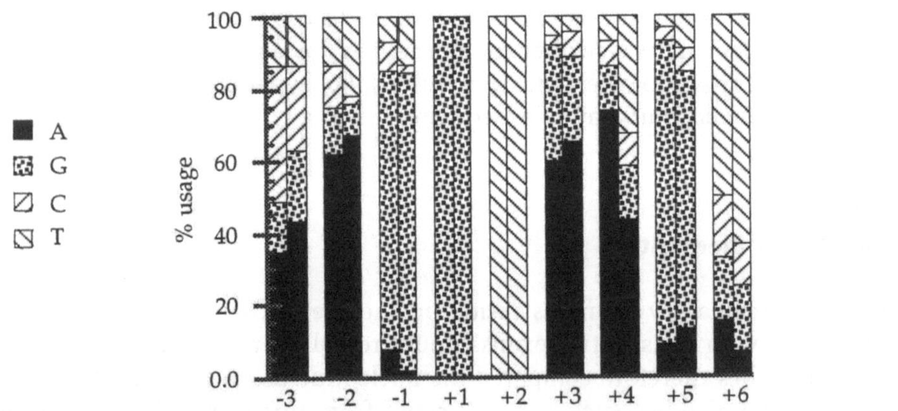

5' splice site consensus C A G g t a a g t

U1 snRNA 3' G U C C A U U C A U ApppG 5'

Fig. 3. Nucleotide usage at the 5′ splice site of neuron-specific exons. The percent nucleotide usage of constitutively (*left*) and neuron-specific alternatively spliced exons (*right*) is compared pairwise. The U1 sequence and the vertebrate consensus splice site sequence are shown at the *bottom*. The different nucleotides are indicated by different patterns as indicated

3.3.2
3′ Splice Site

In general, alternative exons have a more purine-rich polypyrimidine tract. The reduced pyrimidine content in the polypyrimidine tract of alternative exons will most likely reduce its affinity towards U2AF, which was shown to bind to the consensus TTTYYYYTNTAGG (Wu et al. 1999). When subclasses of exons are analyzed, the purine content is strongest in retained introns. In contrast, polypyrimidine tracts of mutually exclusive exons adhere well to the consensus sequence. Inspection of subgroups of cassette exons with different biological regulation reveals some remarkable features. Exons that are expressed only in brain, neurons or muscle, as well as all exons that are developmentally regulated, deviate more from the consensus than any other classes of exons. This deviation appears not to be random, as it is most pronounced at the –3 and –10 position of developmentally regulated exons, at the +1, –3, –9 and –11 position of brain-specific exons, at the +1, –3 and –9 position of neuron-specific exons and at the +1, –7 and –12 position of muscle-specific exons. In all these classes, the single most divergent nucleotide usage is the presence of an adenosine at the –3 position of the 3′ splice site, (Fig. 4; Stamm et al. 1994, 2000). The mechanistical implication for this usage is not clear, but in the Holliday-like structure proposed for the spliceosome (Steitz 1992) the nucleotide at the –3 position would base pair with the G9 of U1 snRNA, which would favor a cytosine at this position. Mutations of this nucleotide at the –3 position influence the usage of exons (Epstein et al. 1994; Stamm et al. 1999). Interestingly, the SELEX consensus sequence of U2AF (Wu et al. 1999) has a T at this position and does not reflect the mammalian splice site consensus.

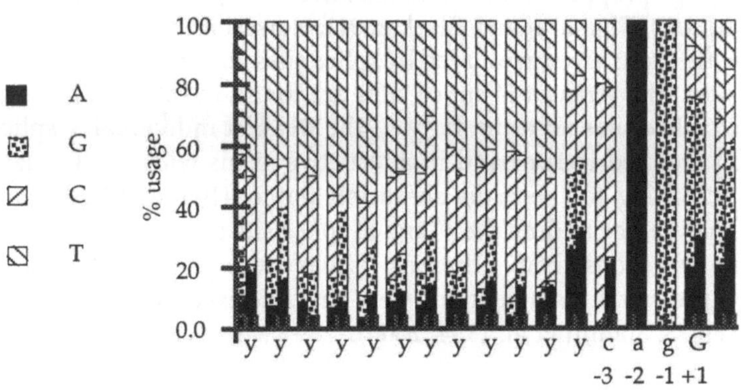

Fig. 4. Nucleotide usage at the 3′ splice site of neuron-specific exons. The percent nucleotide usage of constitutively (*left*) and neuron-specific alternatively spliced exons (*right*) is compared pairwise. The vertebrate consensus sequence is indicated at the *bottom*, exon sequences are in *capital letters*. Note the use of A in alternatively spliced exons at the –3 position. The different nucleotides are indicated using the same patterns as in Fig. 3

3.3.3
Branch Point

The determination of the exact location of the branch point is difficult. In cases where branch points in alternative exons have been studied, the distance between them and the 3′ splice site was found to vary significantly. Cases of alternative 3′ splice site usage were studied systematically. It was found that 15% of the alternative isoforms show strong polypyrimidine tract sequences compared to 59% of constitutive exons. In an equal fraction (40%) of cases both the normal and alternative isoforms show strong branch point signals (Thanaraj and Clark 2002). This implied that the isoforms from alternative 3′ splice site usage probably use the same branch points and it is the variation and the composition in distance that plays a dominant role in determining the competitiveness of the alternative AGs.

3.4
Overall Splice Site Quality

Calculating a score for the splice site can assess the overall quality of a splice site. The score expresses how well the splice site adheres to the consensus sequence. Several methods have been used to calculate splice site scores (Shapiro and Senapathy 1987; Zhang and Marr 1993; Stamm et al. 1994). In general, the score expresses the coincidence of a splice site with the consensus sequence. The higher the score, the better the splice site follows the consensus. When the distribution of splice site scores of alternative exons is analyzed, several features emerge. Alternative splice sites have a broader distribution and a large proportion of the exons have sub optimal splice site scores (Stamm et al. 2000; Thanaraj and Clark 2002). However, about half of the alternative cassette exons are surrounded by splice sites within the range of constitutive splice sites. The weakest 3′ splice sites are found in retained introns and alternative 3′ splice sites. In a similar way, the weakest individual 5′ splice sites are found in retained introns and in alternative exons with an alternative 5′ splice site. According to the exon definition model (Berget 1995), cassette exons were analyzed by adding the 3′ and 5′ splice site scores of a single exon. When alternative exons from different tissues were compared, exons that were only expressed in a single tissue were characterized by the weakest splice site scores (Fig. 5; Stamm et al. 1994, 2000).

3.5
Exonic Elements

Since a considerable number of alternative exons are flanked by suboptimal splice sites, the question arises, why and how these exons are recognized.

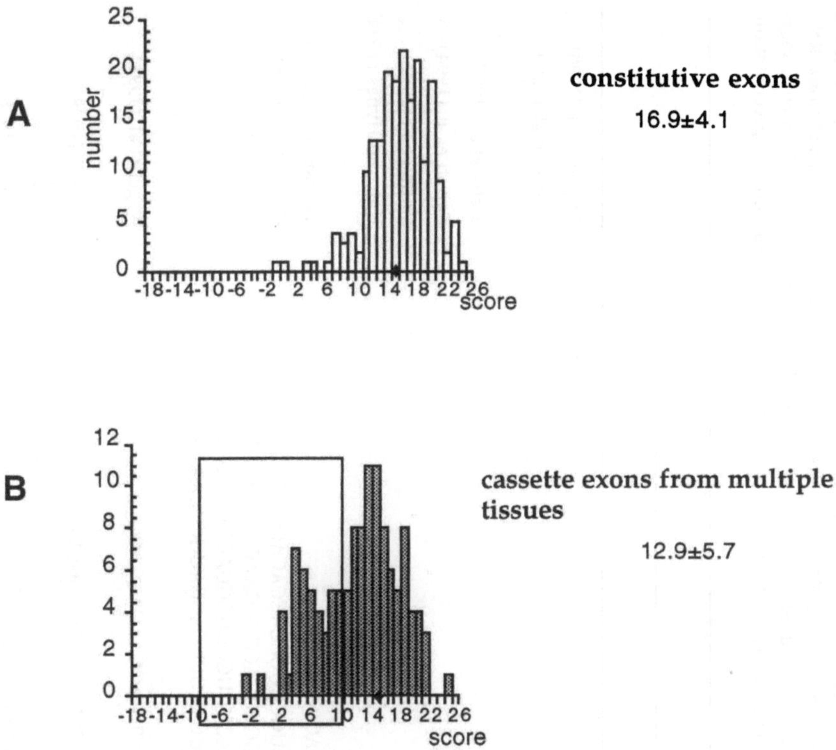

Fig. 5A, B. Distribution of splice site scores in constitutive and alternative exons, illustrating the location of the splice sites. **A** Splice site scores of constitutive exons. The combined scores for an exon are plotted on the *x*-axis, the number of exons on the *y*-axis. The *black diamond* indicates the mean of the distribution in human cassette exons. An exon surrounded by "perfect" splice sites would have a score of 27.8. The mean of the distribution is 16.9 for constitutively spliced exons. **B** Splice site scores of human cassette exons that are expressed in multiple tissues (more than two tissues, but not detectable in all tissues analyzed). The mean of the distribution is 12.9. The *box* indicates exons surrounded with suboptimal splice sites that were analyzed by sequence comparison to find common motifs (Fig. 6)

Experimental data show that purine-rich and AC-rich sequences can act as enhancers. These sequences were either identified by mutagenesis in model substrates, found to be a consensus sequence that binds to regulatory proteins, or were shown to cause human disease when mutated (Table 1). In order to find common motifs in groups of exons with suboptimal splice sites, these exons were compared using a Gibbs sampler. This algorithm allows the determination of redundant sequence motifs that are least likely to occur by chance in a given nucleotide composition (Neuwald et al. 1995). When exons with suboptimal splice sites were tested, the motif GNCNTCCAA was found to be highly abundant. Using the same algorithm, subgroups of alternative exons that were only expressed in neurons, brain, blood, liver, muscle, testis and thymus were tested. Only short motifs from 4 to 9 nucleotides could be found (Fig. 6).

Table 1. Compilation of RNA elements that influence splice-site selection. The first column shows the gene and exon that contain the RNA element, which is characterized in the next columns according to its type (ESE: exonic sequence element, ISE: intronic sequence element) and sequence. *trans*-acting factors are indicated in *bold* under the sequence if they were identified experimentally. Most RNA elements will work in combination with additional RNA regulatory sequences that are not shown. Method indicates the experimental method used: 1, deletion analysis: 2, in vivo splicing assay: 3, in vitro splicing assay: 4, gel mobility shifts; 5, autogenesis; 6, in vitro binding; 7, UV-cross-link; 8, competition experiment; 9, SELEX; 10, immunoprecipitation; 11, spliceosomal complex formation; 12, nuclease protection

	Element			
Gene/Exon	Type	Sequence	Method	Reference
Human CFTR exon 9	ESE	GAUGAC	1	Pagani et al. (2000)
Human CFTR exon 9	ESS	UUAUUUCAAGA **SF2/ASF, SRp55, SRp75**	1, 2	Pagani et al. (2000)
Human fibronectin EDA exon	ESS	CAAGG	1, 2	Staffa et al. (1997)
Human fibronectin EDA exon	ESE	GAAGAAGA	2	Caputi et al. (1994)
Human fibronectin EDA exon	ESE	AGGGUGACC	1, 2	Staffa et al. (1997)
Rat fibronectin intron downstream of exon EIIIB	ISE	UGCAUG repeat		Huh and Hynes (1994)
Mouse IgM exon M2	ESE	GGAAGGACAGCAGAGACCAAGAG	3	Tanaka et al. (1994)
Cardiac troponin T exon 17	ISE	GUAAGUCAGGCUGCAUGCCUCCCACCACACCUGUGCU GCAUGACACCUGGGGCUGACCUGCAACAGAAGUGG GGCUGAGGGAAGGACUGUCCUGGGGACUGGUGUC AGAGCGGGGUUGGUGACUCUCAGGAUGCCCAAAAU GCCCA	2	Carlo et al. (1996)
Chicken cTNT	ESE	AAGAGGAAGAAUGGCUUGAGGAAGACGACG **SRp55**	2, 3, 4, 5	Nagel et al. (1998)
Chicken cTNT intron 4	ISE	GCGCUUCUUCCCUUCCCUCCUCCCUGGCUCAG	5, 2	Cooper (1998)
Chicken cTNT intron 5	ISE	CUCCACCUUUCUU	5, 2	Cooper (1998)
Chicken cTNT intron 5	ISE	UGUGUCCUGUGCCUUUCCCUGCU	5, 2	Cooper (1998)
Chicken cTNT intron 5	ISE	AUUCUUUCACUUCUCUGCU	5, 2	Cooper (1998)
Human GH-1 intron 3	ISE	GGXXXXGGG	5	McCarthy and Phillips (1998)

Human dystrophin exon 27		UAAAGAAGAGGCCCAACAAAAAGAAGCGAAAG	5, 3	Shiga et al. (1997)
Human calcitonin (CT) exon 4	ESE	GGAAAGAAAAGGGA	5, 2	Zandberg et al. (1995)
Human calcitonin (CT) exon 4	ESE	ACUUCAACAAGUU	5, 2	Zandberg et al. (1995)
Drosophila dsx exon 4	ESE	CUUCAAUCAACAU or CAACAAUCAACAU **SR proteins, Tra, Tra2**	3	Tian and Maniatis (1993)
Drosophila dsx exon 4	ESE	UCWWCRAUCAACA **Tra/Tra2**	3	Bruzik and Maniatis (1995)
Drosophila dsx exon 4	ESE	AAAGGACAAAGGACAAAA **Tra/Tra2**	6	Lynch and Maniatis (1995)
Mouse c-src intron B downstream of exon N1	ISE	AGGCUGGGGGCUGCUCUCUGCAUGUGCUUCCU **KSRP, p53/hnRNP** F	4, 6, 7, 3	Min et al. (1997)
Mouse c-src intron B downstream of exon N1	ESE	TGCATG	2, 5, 1	Modafferi and Black (1997)
Human a-tropomyosin gene exon 5/SK	ESS	UAAGUGUUCUGAGCUGGAGGAGGAGCUGAAGAAUGU CACCAACAACCUCAAGUCUCUUGAGGCUCAGGCGG AGAAG	2, 12	Graham et al. (1992)
HIV-1 tat-rev exon 3	ESS	AGAUCCAUUCGAUUAGUGAA	3, 11	Amendt et al. (1995)
HIV-1 exon 6D	ESE	GAGAAAGGAGAGA	5, 2	Wentz et al. (1997)
HIV-1 tat exon 3	ESS	AGAUCC and UUAG	3, 8	Si et al. (1998)
HIV-1 tat-rev exon 3	ESS	GAAGAAGAAGGUGGAGAGAGAGAGACAGAG	3, 11	Amendt et al. (1995)
Mouse CGRP intron 3	ISE	CAGGUAAGAC **PTB, U1 snRNA**	8, 7	Lou et al. (1996)
Rat GABA(A) receptor gamma2 intron	ISE	UUCUCU (within pyrimidine context) **PTB**	7, 8	Ashiya and Grabowski (1997)
Rat beta-TM exon 5, 6	ESE	GARGARGAR	3	Tsukahara et al. (1994)
Human caldesmon exon 5	ESE	(GAGGAAGAGAAAAGGGCAGCAGAGGAGAGGCA)×4	2	Humphrey et al. (1995); Elrick et al. (1998)
Chicken β tropomyosin intron IVS AB	ISS	CCCUCUCUCUAUCGCUGUCUCUUGAGCCACGCC	5, 2	Libri et al. (1990, 1991)
Chicken β-tropomyosin intron 6	ISE	(A/U)GGG repeat	3, 7	Sirand-Pugnet et al. (1995)

Table 1. *Continued*

Gene/Exon	Element Type	Element Sequence	Element Method	Reference
B3, artificial	ESE	UCGACAUUGGGAGCAGUCGGCUCGUGC **SRp40**	9	Tacke et al. (1997)
Human FGFR-2 K-SAM(IIIb)	ESS	UAGGGCAGGC	2	Del Gatto and Breathnach (1995)
Rat clathrin light chain B exon EN	ESE	ACAAAGCGUUCUACCAGCAGCCAGAUGCUG	4	Stamm et al. (1999)
MHC-B exon N30 upstream intron	ISE	UGCAUGUCGUACUGCAUGU	1, 2	Kawamoto (1996)
Human insulin receptor exon B upstream intron	ISE	GUAGUGGGACCCAGAGACGGCAGAAGGGUGGGUGGA GUCUGAAUGGAG	1, 2	Kosaki et al. (1998)
Human insulin receptor exon B upstream intron	ISS	UUACUCGGACACAUGUGGCCUCCAAGUGUCAGAGCC CAGUGGU	1, 2	Kosaki et al. (1998)
Human cd44 exon v5	ESE	GAGUAUCAGGAUGAAGAGGAGACCCCACAUGCUACAA GCA	1, 2	König et al. (1998)
CT/CGRP intron downstream from exon 4	ISE	CUCCGCUCCUCUUC **PTB**	5, 2, 4, 7, 10,12	Lou and Gagel (1999)
CT/CGRP intron downstream from exon 4	ESE	UUAUUUUUCCC **PTB**	5, 7, 4, 8	Lou and Gagel (1999)
BPV-1	ESS	CUGUCUCUUCUUUGCUCGGCUCCCCCGCCUGCGGUC CCAUCAGAGCAG **U2AF(65), PTB, and SR proteins**	3, 11, 7, 10	Zheng et al. (1999)
BPV-1	ESE	GAAGGACCUGAAGGACCCUGCAGGAAAAGCCGAGCCA GCCCAGCC	1, 3	Zheng et al. (1999)
Rous sarcoma virus gag gene	ESS	CGAAUCGACAAAGGGGAGGAAGUGGGAGAAA **HnRNP H**	7,10, 4, 6	Fogel and McNally (2000)
Rous sarcoma virus gag gene	ESS	AUCGAGAAACCAGCAACGGAGCGGCGAAUCGAC **SF2/ASF**	3, 2, 5, 7, 4	McNally and McNally (1996, 1998)

Exons with splice site score < 10

motif	incidence
GNCNTCCAA	(53/100/100/100/100/100/80/60/60)

Neurons

CCYCCA	(70/94/76/89/94/94/74)
GRGGAGG	(93/95/93/80/80/93/93)
(C/G)CNGNCNNCCC	(87/94/n/93/94/n/n/87/94/87)
TCCCAG(G/C)	(81/88/94/94/87/93/79)
CCTCCAG	(81/81/78/81/81/79/80)
TCCCAGN(C/G)	(90/91/91/91/91/90/91/n/94)

Brain

AAA(A/T)AAA	(93/86/93/95/93/86/79)
AGCCCCA	(53/93/93/93/83/93/93)
CNCTCCTC	(93/77/85/93/93/93/60)
GAGRG(T/A)G	(93/93/87/90/51/96/93)
CCARYCC	(93/93/85/87/96/93/69)
CCACYCC	(93/93/93/93/95/93/93)

Blood

TGGT(G/T)(A/T)C	(83/73/83/83/95/96/93)
(A/C)AGAGRA	(96/93/93/93/68/95/60)

Liver

TGA(C/G)TGG	(94/80/80/95/94/93/93)
TTTTATT	(94/84/94/94/53/78/94)
CCAG(ACG)Y	(85/93/86/93/98/70/96)

Muscle

CT(C/G)NNCC(A/T)G	(88/41/97/NN/93/88/95/77)
GAGGRAG	(94/87/82/71/91/76/94)

TESTIS

R(G/T)GACTG	(94/95/70/93/94/93/82)
TGCAGCC	(88/77/94/88/93/88/88)

THYMUS

AGRAGYY	(80/93/95/93/67/95/85)
TCA(A/T)GTT	(93/70/71/96/93/93/93)

Fig. 6. Nucleotide motifs found in exons specific for certain tissues. The motifs were identified using a Gibbs algorithm on the subset of weak exons present in multiple tissues and on exons that are specifically expressed in one tissue

Common to all motifs is a high degree of degeneracy, which has also been found when SELEX procedures were employed in vivo or in vitro (Lui et al. 1998, 1999; Tacke and Manley 1999). Furthermore, the occurrence of a given motif was always restricted only to certain exons in a given subclass. When all motifs were considered together, the nucleotide composition was A = 22%, G = 28%, C = 33% and T = 17%, indicating that overall, these motifs are not purine-rich. Another interesting feature of these motifs is that they do not contain many GAR-enhancer motifs that have been described for a variety of systems. Similarly, AAC and ACC triplets could not be detected, which marks the absence of A/C-rich enhancer (ACE) motifs (Coulter et al. 1997). Together, these data argue that in addition to the previously described GAR and ACE enhancer sequences, other motifs exist in alternatively spliced exons. Often, these motifs are pyrimidine-rich. Interestingly, such pyrimidine-rich motifs have been previously identified as activating motifs in in vitro SELEX procedures (Tian and Kole 1995). However, the *trans*-acting factors binding to these motifs remain to be determined.

In the future, comparison of the complete mouse and human genomes will allow the identification of intronic regulatory sequences. The comparison between the genomes of the related nematodes *C. elegans* and *C. briggsae* has been performed using a newly developed three pass algorithm (Kent and Zahler 2000a). It was found that several alternatively spliced genes, such as the let-2 and bli-4 genes, contain conserved intronic GT-repeat regions that could have regulatory functions (Kent and Zahler 2000b). A similar genome-wide analysis in the vertebrate system is expected to yield intronic motifs, since the comparison of model systems already revealed conserved regions (Zhang et al. 1999). However, due to the larger size of vertebrate introns, these analyses will be more difficult.

3.6 Length of Alternative Exons

The length of exons is conserved among species. Most internal cassette exons have a length between 50 to 200 nt in humans, *Drosophila* and *C.elegans*. However, in the length distribution in *C. elegans* and *Drosophila*, more exons can be found in the classes larger than 250 nt, which results in a larger mean size for internal exons (218 nt for *C. elegans*, versus 145 nt for humans). The difference in size distribution is even more striking, when introns are analyzed. Here, the mean size of introns is 267 nt for *C. elegans*, 487 nt for *Drosophila*, but 3365 nt for humans (Lander et al. 2001).

Experimental evidence suggests that the length of alternative exons (Black 1991), as well as the length of its flanking introns (Bell et al. 1998) are involved in alternative splicing regulation. A computer study showed that human introns associated with alternative 3′ splice sites seem to be shorter on average, with a median length of 625 nt (Clark and Thanaraj 2002). When all types of vertebrate alternative exons are considered, they have a mean length of 174 ± 288 nt (mean of the distribution ± SD), which is comparable to the distribution of human cassette exons (137 ± 123).

However, when cassette exons that are expressed in a single tissue, namely brain, neurons or muscle were analyzed, they were found to be shorter on average (78 ± 58; 101 ± 125 and 58 ± 55) than constitutive exons (Fig. 7). Exons with alternative 5′ and 3′ splice sites still have means of 136 ± 187 and 144 ± 199, respectively, which is comparable to constitutive exons. Inspection of the length distribution shows that the mode, indicating the highest frequency of exons, is smaller in alternative exons than in constitutive ones. This is apparent with alternative 3′ and 5′ splice sites, for which the mode of the distribution is within 1–25 and therefore significantly smaller than the mode for constitutive exons, which is 100–125. In contrast, exons arising from intron retention are on average larger than constitutive cassette exons (308 ± 320) and smaller than constitutive introns that average around 3365 in humans (Lander et al. 2001). However, the mode of their distribution is in the range of 75–100 nt. Common to all distributions is that alternative exon length does not follow a normal distribution, but is skewed toward the smaller size exons, which underlines a bias towards smaller exons in alternative splicing.

The decreased size of alternative exons expressed only in a single tissue could indicate the necessity of a certain amount of RNA binding factors to be

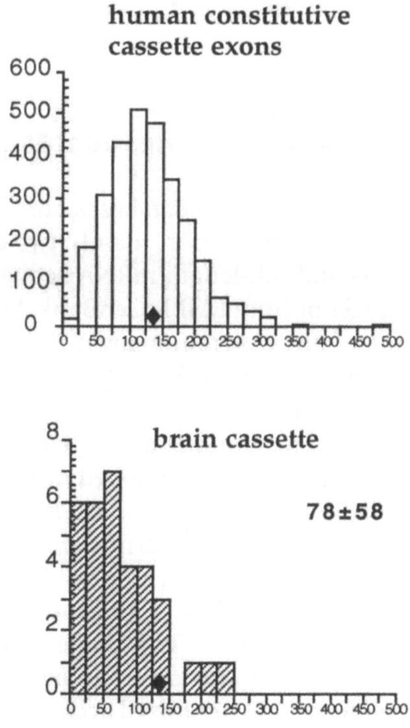

Fig. 7. Length distribution of alternative exons. *Above* Histogram showing the length distribution of human constitutive cassette exons. The *black diamond* indicates the mean of the distribution in human constitutive cassette exons. *Below* For comparison, the length distribution of brain cassette exons are shown. The *y*-axis of each histogram represents the numbers of exons; the *x*-axis represents the nucleotide length

assembled on constitutive exons prior to their recognition. This has been observed in constitutive splicing as well: the 42 detected human exons that are smaller than 19 nt contain 72% purines, which are most likely part of exonic GAR-type enhancers (Lander et al. 2001).

3.7 A Major Function of Alternative Exons Are the Introduction of Premature Stop Codons

From sequence comparison, a general function of alternative exons is not readily visible. The coding frame is maintained in only 55–60% of alternative splicing events (Clark and Thanaraj 2002). In a sample of 1000 alternative spliced exons compiled from the literature, about 22% of the exons contained a stop codon or introduced a frameshift, resulting in a premature stop codon (Stamm et al. 2000), which is in agreement with EST based studies that found 19% frameshifts (Modrek et al. 2001). Therefore, the introduction of premature stop codons seems to be an important biological feature of alternative exons. These stop codons have been compiled and were found to often disrupt purine-rich or AC-rich enhancer sequences (Valentine 1998). Most of these stop codons appear in coding exons and fulfill the requirements for nonsense-mediated decay. In conclusion, the introduction of stop codons and the following nonsense-mediated decay seem to be an important biological role for alternatively spliced exons.

4 Bioinformatics Resources to Identify Alternative Exons

Publicly available databases contain a large amount of transcript sequences, which are present in mainly two forms: partial transcript sequences such as ESTs and full-length mRNA sequences. The EST sequences are available from dbEST or from EMBL/GenBank databases. Since partial and complete sequencing of mRNAs from many different tissues, developmental stages, and pathologies creates these collections they contain diverse transcripts. There is a high amount of redundancy in EST sequences. Related EST sequences that correspond to a unique mRNA transcript are available as clusters and consensus sequences and are compiled in TIGR Gene Indices, NCBI Unigene clusters, and SANBI STACK clusters. TIGR gene indices are created by comparing EST sequences with transcripts and clustering them if there is a minimum of a 40 base pair match with at least 95% of similarity and if there is only a maximum unmatched region of 20 base pairs. These clusters are assembled into consensus sequences. The Unigene database clusters ESTs in a similar manner, but does not create a consensus sequence. In the STACK database the gene sequences are organized according to tissue and each gene is represented with alignments of its expressed fragments; consensus sequences are presented (see Table 3).

4.1 Detection of Alternative Splice Events Using Sequence Comparison

To date, data on alternative splicing have been derived by either examining annotated gene forms in nucleotide and protein sequence databases (such as EMBL/GenBank/DDBJ, SwissProt) and bibliography databases such as MedLine or through examining alignments of nucleic acid sequences. The major sequences alignments are between (1) genomic DNA and ESTs, (2) mRNA and ESTs, as well as (3) mRNA and mRNA. In these approaches, BLAST is often used to identify regions of identical nucleotide bases denoted as high-scoring segments. A comparison that includes genomic DNA gives the most information, as the gene structure and splice sites can be delineated.

1. When genomic *DNA and ESTs* are aligned, the gaps on the DNA sequence correspond to putative introns. They are further analyzed by comparing their terminal bases with splice site consensus sequences. The satisfying gaps are denoted as introns and those that are not compatible are ignored. The region between two such confirmed introns is denoted as a confirmed exon. If these confirmed exons overlap with one another, they are called alternative exons. Specialized versions of BLAST such as Sim4 and SplicedAlignment are used for this computation (see Table 3).
2. Gaps that are produced in the alignment of *mRNA with EST* sequences indicate different isoforms. Databases that used this approach present the data in terms of insertions/deletions in the transcripts rather than in terms of exon-intron structures (Brett et al. 2002) and do not discriminate between the different alternative exon types.
3. *mRNA-mRNA* alignments were mainly performed from mouse sequences by the RIKEN research group (Kawai et al. 2001).This program attempts to detect alternative splice events through clustering cDNAs from redundant clone sets. Redundant cDNAs showing more than two portions of alignments longer than 20 bases are clustered. Similar to that of mRNA-EST alignments, the gaps/inserts are deduced as alternative splice events. Recent databases based on the above approaches are listed in Table 2.

4.2 Computer Programs for Detecting Alternative Splicing

Using alignments between different nucleic acids, alternative splicing events can be detected in databases. Different programs are used, depending whether the gene structure is known or needs to be determined.

Table 2. Overview of existing databases that compile alternative exons, regulatory proteins and diseases

Database	Entry mode	Species covered	Reference	URL
Alternative exon database	Manual	All animals	Stamm et al. (2000)	Cgsigma.cshl.org/new_alt_exon_db2/
AltExtron	Computer-generated	Human	Clark and Thanaraj (2002)	www.ebi.ac.uk/~thanaraj/altExtron
AltRefSeq	Computer-generated	Human	Kan et al. (2001)	Sapiens.wustl.edu/~zkan/TAP/ALTSEQ.htm
ASDB: database of alternatively spliced genes	Computer generated	All animals	Gelfand et al. (1999); Dralyuk et al. (2000)	Cbcg.nersc.gov/asdb
Asforms	Computer-generated	Human and 6 other species	Brett et al. (2002)	www.bioinf.mdc-berlin.de/asforms
AsMAMDB	Computer-generated	Mammalian	Ji et al. (2001)	http://166.111.30.65/ASMAMDB.html
Drosophila splicing protein database	Manual	*Drosophila*	Mount and Salz (2000)	http://www.wam.umd.edu/~smount/DmRNAfactors/table.html
GC-AG introns	Computer-generated	Human	Thanaraj and Clark (2001)	www.ebi.ac.uk/~thanaraj/gcag
HASDB	Computer-generated	Human	Modrek et al. (2001)	www.bioinformatics.ucla.edu/HASDB
Intronerator	Computer-generated	*C. elegans*	Kent and Zahler (2000b)	~kent/intronerator
ISIS	Computer-generated	All species	Croft et al. (1999)	Isis.bit.uq.edu.au/front.html
Neuron-specific exons	Manual	Animals, neurons	Stamm et al. (2000)	ftp://phage.cshl.org/pub/science/alt_exon
RRM-containing proteins	Computer-generated	Metazoan splicing factors	Birney et al. (1993)	www.sanger.ac.uk/cgi-bin/Pfam/getacc?PF00076
Splice-site mutations	Manual	Human	Nakai and Sakamoto (1994)	www.cookie.imcb.osaka-u.ac.jp/nakai/asdb.html

Table 3. Tools to identify predict and analyze alternative splicing events

Databases/tools	URL	Description
AltExtron	www.ebi.ac.uk/~thanaraj/altExtron/	Methods to identify alternative introns and exons
BLAST	www.ebi.ac.uk/blast2/; www.ncbi.nlm.nih.gov/BLAST	Similarity between biomolecular sequences
EMBL	www.ebi.ac.uk/embl/	Nucleotide sequence database
EST-GENOME	www.well.ox.ac.uk/rmott/est_genome.shtml	cDNA-genome alignment
GenBank	www.ncbi.nlm.nih.gov/Genbank/	Nucleotide sequence database
GenScan	genes.mit.edu/GENSCAN.html	Gene identification
HMMGENE	www.cbs.dtu.dk/services/HMMgene/	Gene identification
j_explorer	www.sanbi.ac.za/exon_skipping	Identify exon skipping events
Modrek et al.	www.bioinformatics.ucla.edu/~splice/ HASDB	Derive gene structures and alternative introns/exons
NCBI unigene clusters	www.ncbi.nlm.nih.gov/UniGene/	EST clusters
SANBI STACK	www.sanbi.ac.za/Dbases.html	EST clusters and consensus sequences
SIM4	globin.cse.psu.edu/ftp/dist/sim/	cDNA-genome alignment
SPC	www.ebi.ac.uk/~thanaraj/ SpliceProximalCheck.html	Discriminate false proximal splice sites
Splice_programs	www.ebi.ac.uk/~thanaraj/ gene_altSplice.html	Methods to check the involvement of splice junctions in alternative splicing
SplicedAlign and Procrustes	hto13.usc.edu/software/procrustes/#salign	Derive gene structures
TAP	http://stl.wustl.edu/~zkan/TAP/	Derive alternate gene structures
TIGR gene indices	www.tigr.org/tdb/tgi.shtml	EST clusters and contigs

4.2.1 *Finding Alternative Exons from Known Gene Structures*

To find alternative exons from known gene structures, Thanaraj (Thanaraj 1999) and Hide (Hide et al. 2001) first constructed "theoretical" mRNA parts from the gene and then compared them with the EST database. Exon skipping events that correspond to cassette exons were identified by concatenating a 50-bp tag from the 3′ terminus of the proceeding exon with a 50-bp tag from the 5′ terminus of each of succeeding exons. This creates a set of all consecutive and nonconsecutive exon-exon junctions. Each of such 100-bp exon-exon constructs is searched for similarity with the collection of EST sequences. A skipping event is reported when an EST aligns with the construct without a gap.

Exon isoform events are created by alternative 5′ and 3′ splice sites. They are detected by connecting 50-bp exon sequence tags from the 5′ and 3′ end of a given intron. These exon-exon constructs are searched for similarity against EST databases. While an EST sequence that shows uninterrupted similarity at the junction confirms the normal event, a gap in the alignment at the splice junction indicates either truncation or extension of intron sequences depending on whether the gap occurred in the EST sequence or in the exon-exon construct. The gaps are further mapped to the gene sequence and the end nucleotides are checked for splice site consensus sequences.

To detect retained introns, for every intron in the gene structure, two constructs, the exon–intron and the intron–exon constructs are built by using the 50-bp tags flanking the exon-intron junction or those flanking the intron-exon junctions. These constructs are searched for similarity with the collection of EST sequences. An EST sequence showing similarity to both the constructs belonging to an intron indicates potential candidates for intron retention events.

4.2.2 *Finding Alternative Splice Events in Genome Regions of Unknown Gene Structures*

The increasing amount of genomic sequence data makes it necessary to identify (alternative) splice sites in sequences of unknown structure. This is performed by using either consensus EST cluster sequence or self-clustering methods to map and compare individual EST sequences with genomic DNA.

Human ESTs were used for a genome-wide detection of alternative splicing in humans (Modrek et al. 2001). In this work the genome region that aligns with an EST consensus sequence is identified. Every EST from the cluster is then searched for similarity with the identified genome region and the potential splice sites are identified. The putative splice sites are then validated against splice site consensus weight matrices (generally the GT-AG type). Subsequently all the mutually exclusive splice site pairs are identified as alternative splicing events.

Two programs, procrustes-EST and transcript assembly program (TAP) have been used to identify alternative exons with a self-clustering approach.

Procrustes-EST (Gelfand et al. 1999) uses EST contigs and assemblies from TIGR Human Gene Index as the transcript information to delineate alternative intron–exon structures. First, the EST contigs that show similarity to a given genome region are selected. Then candidate splice sites are predicted on the genome region and then chains of exons with the highest similarity score with one or more of the EST contigs are identified. Such predicted exon-intron structures are merged into superstructures through the following simple procedure: Each of the triples (intron–exon–intron) from these predicted structures is considered against each other – if the right intron of a triple coincided with the left intron of a second triple, the triples are merged into a superstructure. This procedure is repeated until no triples could be added to the constructed superstructure. All such possible superstructures are constructed. A superstructure that intersected neither the annotated gene structure nor any other superstructure in a common exon is omitted as an orphan. The remaining superstructures represent the different alternative gene structures for the genome region.

TAP (Kan et al. 2001) is built to delineate gene structures using EST sequences that align in a DNA region on the genome. First, exon segments are identified through DNA-EST alignments. The gaps between exons are accepted as introns (splice junction pairs), if they have the consensus GT-AG sequences or have more than two EST sequences supporting such gaps. Such observed splice junction pairs along the length of the gene are sorted out to assemble the joint gene structures; the algorithm uses the EST-encoded connectivity and redundancy information to sort out the complex alternative splicing patterns. The connectivity information from EST sequences is decoded into four types and scored appropriately. The program scores conflicting, contiguous, transitive and gapped connections and builds a matrix to record the connectivity relationship between pairs of introns. The gene structure is assembled by tracing a path through the filled connectivity matrix from 5′ to 3′ direction in a manner that at each elongation step from an intron i to j, the connection with the maximum matrix score is chosen. Thus, at each step contiguous connection takes precedence over transitive connection, which in turn takes precedence over gapped connections. Mutually exclusive connections are thus not included in the same path. Alternative paths are generated by branching the path when multiple downstream splice pairs have the same connectivity score.

4.3 Limitations, Problems and Accuracy of Predicting Alternative Splicing Events

The expression of alternative exons in a given tissue, cell type, or developmental stage is not addressed by most computer generated databases. Attempts have been made to decipher the tissue specificity of alternative poly-A site usage (Beaudoing and Gautheret 2001). One of the major problems to delineate tissue specificity of alternative splicing events is the lack of proper classification systems for EST libraries. Currently, the anatomical site, pathology conditions, and developmental stage of EST libraries are ill-defined. However, there have been attempts to improve the situation, as NCBI has recently released its own classifiers.

Repeats, paralogous genes, and duplicated or multiple copies of the genes cause false-positive and negative exon predictions. Repeats in gene sequences as well as in intergenic regions are common in mammalian genomes (Lander et al. 2001). Furthermore, genomic regions and genes can be duplicated. Therefore, it is often difficult to unambiguously assign a transcript or EST cluster to a gene. As a result, highly similar gene products could be interpreted as the result of alternative splicing, but are generated by paralogous genes. Repeats are masked before gene-transcript alignments are carried out. Masking is done by checking every gene for the occurrence of known repeats, which requires an a priori knowledge of repeat sequences. Duplicated genes and redundant gene entries in a start-up data set are identified by checking for similarity among the entries of start-up gene set and retaining one representative entry. However, it is impossible to test whether a matching EST sequence is derived from the representative gene entry or from one of the removed gene entries. One possible solution is to identify repeats in the alignment data by observing where multiple matches to EST sequences occur. When two or more genes show matches to the same region of a transcript, these matches are considered to be repeat matches and are discarded. Similarly, a transcript is also removed if a region of it shows matches to more than one region on the gene (Clark and Thanaraj 2002). This method has the distinct advantage that no a priori knowledge of the repetitive elements is required.

Considering the quality of alignments; generally, stringent criteria such as that a high scoring segment has at least 95% of identity and the E value is $<10^{-15}$ are required to be imposed. It is our experience that these E values often remove any high scoring segment of length <30–35 bases, which is characteristic for exons generated by alternative 5′ and 3′ splice site usage. Performing local alignments with slightly relaxed criteria between the EST gap regions and the region on the gene between the partial structure could solve this problem (Clark and Thanaraj 2002).

The exact assignment of splice sites is often difficult, but is necessary for the *ab initio* gene prediction programs and to delineate gene structures from the gene-EST alignments. In gene-EST alignments, there can be more than one candidate splice site pair. Since this does not change the coding sequence, computer programs often ignore it. However, for studies related to splicing, it is essential to exactly pinpoint the splice sites and programs such as SpliceProximalCheck (SPC) have been developed for it (Thanaraj and Robinson 2000).

Most of the prediction programs rely on EST data, which have several intrinsic problems. Programs that assemble gene structures by matching and overlapping EST sequences do not discriminate whether an EST is derived from the same or from different clones. Alternative splicing events, such as mutually exclusive exons can therefore, not be identified. Alternative variants are often rare (Hide et al. 2001; Kan et al. 2001; Clark and Thanaraj 2002) and many programs that require at least two or more EST sequences before confirming an event will ignore them.

Since ESTs are derived by sequencing the ends of cDNAs, the EST-coverage is nonuniform along the length of the gene. The internal regions of the gene have low coverage; in a similar manner, the 3′ ends of the genes can be over represented. These bias the EST data sets to events in the 5′ and 3′ UTR and under represent internal exons. This bias is evident when studies that include genomic data are compared with pure EST-based studies. A genome-wide survey of alternative splicing finds 74% of alternative events in the coding region (Modrek et al. 2001), whereas datasets derived purely from ESTs find only 20% such events in the coding region (Mironov et al. 1999).

The success rate of detecting alternative splice events *in silico* has been tested by experimental validation of a limited set of the predicted alternative events (Brett et al. 2002). RT-PCR and subsequent sequencing of a limited set of the predicted alternative events indicated a success rate of 92% and a false negative rate of 2.5%.

4.4 Ad initio Prediction of Alternative Exons

Currently the efforts that derive data on alternative splice events do not "predict" the events, but instead locate them by positioning EST sequences on either the genes or mRNAs. The current gene prediction programs do not predict alternative gene structures. Though some of the available programs such as GenScan and HMMgene can predict suboptimal genes (Table 3), it has not been assessed whether such predicted suboptimal genes correspond to alternative gene structures. Predicting alternative exons from genomic data is still at an early stage. At the best, we are now able to detect alternatively spliced exons based on similarity to EST sequences. Since alternative splicing is signaled by weak signals as marked by poor consensus sequences, poor correlation among the different structural elements, and use of minor intron/exon types, prediction of them is difficult. It is also the case that alternative splicing often results in altered reading frames in downstream coding exons or to premature stop codons. Thus, while selection of optimal exons is the key in the current gene prediction programs, the emphasis on alternative gene structure prediction is the choice of suboptimal exons. Further, the EST-based methodologies locate alternative exons in a noncontextual manner along the native gene structure and do not illustrate which combinations of these exons exist in full-length isoform transcripts. Thus, efforts need to be concentrated on dynamic programming procedures that can put together suboptimal and optimal exons in a proper context.

5 Overview of Available Databases and Programs

During the last years, several databases of alternatively spliced exons were generated (Table 2). These data sets have derived from one of two main approaches: (1) examining annotated forms in databases such as EMBL/GenBank/DDBJ, Swiss-Prot and MedLine – these include ASDB, Alternative-Exon Database, and AsMamDB, or (2) examining alignments of EST/cDNA sequences with genomic DNA sequences – these include the Intronerator and AltExtron. These individual databases differ in the sizes of the generated data sets, and in the detail of the methodologies employed and hence in the quality. The first database generated was a compilation of neuron-specific alternative exons that was based on sequences published in peer reviewed journals (Stamm et al. 1994). A more general alternative splicing database (ASBD) was developed later. In the first version of ASDB, protein sequences generated by alternative splicing were identified by cluster analysis of Swiss-Prot entries (Gelfand et al. 1999). Later this database was extended with a DNA division that contains complete genes for which alternative splicing has been mentioned in the Genbank annotations (Dralyuk et al. 2000). AsMamDB was developed to systematically study alternative spliced genes in mammalian systems and contains alternative exons based on human, mouse and rat Genbank entries (Ji et al. 2001). Another specialized database is the Intronerator that compiles cDNA alignments and a catalog of alternatively spliced genes from *C. elegans* (Kent and Zahler 2000). A compilation of human genes that use GC-AG introns shows that 1 in every 20 alternative introns is a GC-AG intron (Thanaraj and Clark 2001). Finally, starting from an early compilation (Mount 1982), several databases of splice sites are available. SpliceDB is a database of mammalian splice sites (Burset et al. 2001) and mutations in splice sites that are implicated in human diseases have been compiled (Nakai and Sakamoto 1994). Splice sites used by GC-AG introns have been compiled on the AltExtron web data (Thanaraj and Clark 2001). Intronic sequences were compiled in the intron and sequence information database (ISIS; Croft et al. 1999). Finally, RNA recognition motif binding proteins (Birney et al. 1993) and all *Drosophila* splicing regulatory proteins (Mount and Salz 2000) have been compiled. The various computer programs and specialized databases that were used to predict and analyze alternative exons are compiled in Table 3.

Acknowledgement. This work was supported by the EBI (TAT) and the Deutsche Forschungsgemeinschaft (Sta399/7-1) and SFB473 C8 to SS.

References

Amendt BA, Si ZH, Stoltzfus CM (1995) Presence of exon splicing silencers within human immunodeficiency virus type 1 tat exon 2 and tat-rev exon 3: evidence for inhibition mediated by cellular factors. Mol Cell Biol 15(11):6480

Ashiya M, Grabowski PJ (1997) A neuron-specific splicing switch mediated by an array of pre-mRNA repressor sites: evidence of a regulatory role for the polypyrimidine tract binding protein and a brain-specific PTB counterpart. RNA 3(9):996–1015

Ayane M, Preuss U, Köhler G et al. (1991) A differentially expressed murine RNA encoding a protein with similarities to two types of nucleic acid binding motifs. Nucleic Acids Res 19:1273–1278

Beaudoing E, Gautheret D (2001) Identification of alternate polyadenylation sites and analysis of their tissue distribution using EST data. Genome Res 11(9):1520–1526

Bell MY, Cowper AE, Lefranc M.-P et al. (1998) Influence of intron length on alternative splicing of CD44. Mol Cell Biol 18:5930–5941

Berget SM (1995) Exon recognition in vertebrate splicing. J Biol Chem 270:2411–2414

Birney E, Kumar S, Krainer AR (1993) Analysis of the RNA-recognition motif and RS and RGG domains: conservation in metazoan pre-mRNA splicing factors. Nucleic Acids Res 21(25):5803–5816

Black DL (1991) Does steric interference between splice sites block the splicing of a short c-src neuron-specific exon in non neuronal cells? Genes Dev 5:389–402

Boggs RT, Gregor P; Idriss S et al. (1987) Regulation of sexual differentiation in *D. melanogaster* via alternative splicing of RNA from the transformer gene. Cell 50(5):739–747

Breitbart RE, Andreadis A, Nadal-Ginard B (1987) Alternative splicing: a ubiquitous mechanism for the generation of multiple protein isoforms from single genes. Annu Rev Biochem 56:467–495

Brett D, Pospisil H, Valcarcel J et al. (2002) Alternative splicing and genome complexity. Nat Genet 30(1):29–30

Bruzik JP, Maniatis T (1995) Enhancer-dependent interaction between 5′ and 3′ splice sites in trans. Proc Natl Acad Sci USA 92(15):7056–7059

Burge CB, Padgett RA, Sharp PA (1998) Evolutionary fates and origins of U12-type introns. Mol Cell 2(6):773–785

Burset M, Seledtsov IA, Solovyev VV (2000) Analysis of canonical and non-canonical splice sites in mammalian genomes. Nucleic Acids Res 28(21):4364–4375

Burset M, Seledtsov IA, Solovyev VV. (2001) SpliceDB: database of canonical and non-canonical mammalian splice sites. Nucleic Acids Res 29(1):255–259

Caceres JF, Stamm S, Helfman DM et al. (1994) Regulation of alternative splicing in vivo by overexpression of antagonistic splicing factors. Science 265(5179):1706–1709

Caputi M, Casari G, Guenzi S, Tagliabue R, Sidoli A, Melo DA, Baralle FE (1994) A novel bipartite splicing enhancer modulates the differential processing of the human fibronectin EDA exon. Nucleic Acids Res 22(6):1018–1022

Carlo T, Sterner DA, Berget SM (1996) An intron splicing enhancer containing a G-rich repeat facilitates inclusion of a vertebrate micro-exon. RNA 2(4):342–353

Chandler D, McGuffin ME, Piskur J et al. (1997) Evolutionary conservation of regulatory strategies for the sex determination factor transformer-2. Mol Cell Biol: 2908–2919

Chaudhari N, Hahn WE (1983) Genetic expression in the developing brain. Science 220:924–928

Clark F, Thanaraj TA (2002) Categorization and characterization of transcript-confirmed constitutively and alternatively spliced introns and exons from humans. Hum Mol Gent 11:1–14

Cooper TA (1998) Muscle-specific splicing of a heterologous exon mediated by a single muscle-specific splicing enhancer from the cardiac troponin T gene. Mol Cell Biol 18(8):4519–4525

Cooper TA, Mattox W (1997) The regulation of splice site selection, and its role in human disease. Am J Hum Genet 61:259–266

Coulter LR, Landree MA, Cooper TA (1997) Identification of a new class of exonic splicing enhancers by in vivo selection [published erratum appears in Mol Cell Biol 1997 June 17(6):3468]. Mol Cell Biol 17(4):2143–2150

Croft L, Schandorff S, Clark F et al. (1999) ISIS, the intron information system, reveals the high frequency of alternative splicing in the human genome. Nat Genet 24:340–341

Del Gatto F, Breathnach R (1995) Exon and intron sequences, respectively, repress and activate splicing of a fibroblast growth factor receptor 2 alternative exon. Mol Cell Biol 15(9):4825–4834

Dralyuk I, Brudno M, Gelfand MS et al. (2000) ASDB: database of alternatively spliced genes. Nucleic Acids Res 28(1):296–297

Elrick LL, Humphrey MB, Cooper TA, Berget SM (1998) A short sequence within two purine-rich enhancers determines 5′ splice site specificity. Mol Cell Biol 18(1):343–352

Epstein JA, Glaser T, Cai J et al. (1994) Two independent and interactive DNA-binding subdomains of the Pax6 paired domain are regulated by alternative splicing. Genes Dev 8(17):2022–2034

Fogel BL, McNally MT (2000) A cellular protein, hnRNP H, binds to the negative regulator of splicing element from Rous sarcoma virus. J Biol Chem 275(41):32371–32378

Fu X-D (1995) The superfamily of arginine/serine-rich splicing factors. RNA 1:663–680

Gallego ME, Gattoni R, Stevenin J et al. (1997) The SR splicing factors ASF/SF2 and SC35 have antagonistic effects on intronic enhancer-dependent splicing of the beta-tropomyosin alternative exon 6A. EMBO J 16(7):1772–1784

Gelfand MS, Dubchak I, Dralyuk I et al. (1999) ASDB: database of alternatively spliced genes. Nucleic Acids Res 27(1):301–302

Graham IR, Hamshere M, Eperon IC (1992) Alternative splicing of a human alpha-tropomyosin muscle-specific exon: identification of determining sequences. Mol Cell Biol 12(9):3872–3882

Graveley BR (2000) Sorting out the complexity of SR protein functions. RNA 6:1197–1211

Hanamura A, Caceres JF, Mayeda A et al. (1998) Regulated tissue-specific expression of antagonistic pre-mRNA splicing factors. RNA 4:430–444

Hastings ML, Krainer AR (2001) Pre-mRNA splicing in the new millennium. Curr Opin Cell Biol 13(3):302–309

Heinrichs V, Ryner LC, Baker B (1998) Regulation of sex-specific selection of fruitless 5′ splice sites by transformer and transformer-2. Mol Cell Biol 18:450–458

Helfman DM, Rici WM (1989) Branch point selection in alternative splicing of tropomyosin pre-mRNAs. Nucleic Acids Res 17:5633–5640

Hertel KJ, Maniatis T (1998) The function of multisite splicing enhancers. Mol Cell 1:449–455

Hide WA, Babenko VN, van Heusden PA et al. (2001) The contribution of exon-skipping events on chromosome 22 to protein coding diversity. Genome Res 11(11):1848–1853

Hoshijima K, Inoue K, Kiguchi I et al. (1991) Control of doublesex alternative splicing by transformer and transformer-2 in *Drosophila*. Science 1991(252):833–836

Huh GS, Hynes RO (1994) Regulation of alternative pre-mRNA splicing by a novel repeated hexanucleotide element. Genes Dev 8(13):1561–1574

Humphrey MB, Bryan J, Cooper TA, Gerget SM (1995) A 32-nucleotide exon-splicing enhancer regulates usage of competing 5′ splice sites in a differential internal exon. Mol Cell Biol 15(8):3979–3988

Jensen KB, Drege BK, Stefani G et al. (2000) Nova-1 regulates neuron-specific alternative splicing and is essential for neuronal viability. Neuron 25:359–371

Ji H, Zhou Q, Wen F et al. (2001) AsMamDB: an alternative splice database of mammals. Nucleic Acids Res 29:260–263

Jumaa H, Nielsen PJ (1997) The splicing factor SRp20 modifies splicing of its own mRNA and ASF/SF2 antagonizes this regulation. EMBO J 16:5077–5085

Kamma H, Portman DS, Dreyfuss G (1995) Cell type specific expression of hnRNP proteins. Exp Cell Res 221:187–196

Kan Z, Rouchka EC, Gish WR et al. (2001) Gene structure prediction and alternative splicing analysis using genomically aligned ESTs. Genome Res 11(5):889–900

Kawai J, Shinagawa A; Shibata K et al. (2001) Functional annotation of a full-length mouse cDNA collection. Nature 409(6821):685–690

Kawamoto S (1996) Neuron-specific alternative splicing of nonmuscle myosin II heavy chain-B pre-mRNA requires a cis-acting intron sequence. J Biol Chem 271(30):17613–17616

Kent WJ, Zahler AM (2000a) Conservation, regulation, synteny, and introns in a large-scale *C. briggsae–C. elegans* genomic alignment. Genome Res 10(8):1115–1125

Kent WJ, Zahler AM (2000b) The intronerator: exploring introns and alternative splicing in *Caenorhabditis elegans*. Nucleic Acids Res 28(1):91–93

König H, Ponta H, Herrlich P (1998) Coupling of signal transduction to alternative pre-mRNA splicing by a composite splice regulator. EMBO J 10:2904–2913

Komatsu M, Kominami E, Arahata K et al. (1999) Cloning and characterization of two neural-salient serine/arginine-rich (NSSR) proteins involved in the regulation of alternative splicing in neurons. Genes Cells 4:593–606

Kosaki A, Nelson J, Webster NJ (1998) Identification of intron and exon sequences involved in alternative slicing of insulin receptor pre-mRNA. J Biol Chem 273:10331–10337

Lander ES, Linton LM, Birren B et al. (2001) Initial sequencing and analysis of the human genome. Nature 409(6822):860–921

Levine A, Durbin R (2001) A computational scan for U12-dependent introns in the human genome sequence. Nucleic Acids Res 29(19):4006–4013

Libri D, Goux-Pelletan M, Brody E, Fiszman MY (1990) Exon as well as intron sequences are cis-regulating elements for the mutually exclusive alternative splicing of the beta tropomyosin gene. Mol Cell Biol 10(10):5036–5046

Libri D, Piseri A, Fiszman MY (1991) Tissue-specific splicing in vivo of the beta-tropomyosin gene: dependence on an RNA secondary structure. Science 252(5014):1842–1845

Lou H, Gagel RF (1999) Mechanism of tissue-specific alternative RNA processing of the calcitonin CGRP gene. Front Horm Res 25:18–33

Lou H, Gagel RF, Berget SM (1996) An intron enhancer recognized by splicing factors activates polyadenylation. Genes Dev 10(2):208–219

Lui H-X, Zhang M, Krainer AR (1998) Identification of functional exonic splicing enhancer motifs recognized by individual SR proteins. Genes Dev 12:1998–2012

Lui H-X, Chew SL, Cartegni L et al. (1999) Exonic splicing enhancer motif recognized by human SC35 under splicing conditions. Mol Cell Biol 20:1063–1071

Lynch KW, Maniatis T (1995) Synergistic interactions between two distinct elements of a regulated splicing enhancer. Genes Dev 9(3):284–293

Manley JL, Tacke R (1996) SR proteins and splicing control. Genes Dev 10:1569–1579

Mattox W, Baker BS (1991) Autoregulation of the splicing of transcripts from the transformer-2 gene of *Drosophila*. Genes Dev 5:786–796

Mattox W, Palmer MJ, Baker BS (1990) Alternative splicing of the sex determination gene transformer-2 is sex-specific in the germ line but not in the soma. Genes Dev 4:789–805

Mayeda A, Krainer AR (1992) Regulation of alternative pre-mRNA splicing by hnRNP A1 and splicing factor SF2. Cell 68(2):365–375

Mayeda A, Helfman DM, Krainer AR (1993) Modulation of exon skipping and inclusion by heterogeneous nuclear ribonucleoprotein A1 and pre-mRNA splicing factor SF2/ASF [published erratum appears in Mol Cell Biol 1993 Jul 13(7):4458]. Mol Cell Biol 13(5):2993–3001

McCarthy EM, Phillips JA 3rd (1998) Characterization of an intron splice enhancer that regulates alternative splicing of human GH pre-mRNA. Hum Mol Genet 7(9):1491–1496

McNally LM, McNally MT (1996) SR protein splicing factors interact with the Rous sarcoma virus negative regulator of splicing element. J Virol 70(2):1163–1172

McNally LM, McNally MT (1998) An RNA splicing enhancer-like sequence is a component of a splicing inhibitor element from Rous sarcoma virus. Mol Cell Biol 18(6):3103–3111

Milner RJ, Sutcliffe JG (1983) Gene expression in rat brain. Nucleic Acids Res 11:5497–5520

Min H, Turck CW, Nikolic JM, Black DL (1997) A new regulatory protein, KSRP, mediates exon inclusion through an intronic splicing enhancer. Genes Dev 11:1023–1036
Mironov AA, Fickett JW, Gelfand MS (1999) Frequent alternative splicing of human genes. Genome Res 9:1288–1293
Modafferi EF, Black DL (1997) A complex intronic splicing enhancer from the c-src pre-mRNA activates inclusion of a heterologous exon. Mol Cell Biol 17(11):6537–6545
Modrek B, Resch A, Grasso C et al. (2001) Genome-wide detection of alternative splicing in expressed sequences of human genes. Nucleic Acids Res 29(13):2850–2859
Mount SM (1982) A catalogue of splice junction sequences. Nucleic Acids Res 10(2):459–472
Mount SM, Salz HK (2000) Pre-messenger RNA processing factors in the *Drosophila* genome. J Cell Biol 150(2):F37–F44
Nagel RJ, Lancaster AM, Zahler AM (1998) Specific binding of an exonic splicing enhancer by the pre-mRNA splicing factor SRp55. RNA 4(1):11–23
Nakai K, Sakamoto H (1994) Construction of a novel database containing aberrant splicing mutations of mammalian genes. Gene 141:171–177
Nayler O, Cap C, Stamm S (1998) Human transformer-2-beta gene (SFRS10): complete nucleotide sequence, chromosomal localization, and generation of a tissue-specific isoform. Genomics 53:191–202
Neuwald AF, Liu JS, Lawrence CE (1995) Gibbs motif sampling: detection of bacterial outer membrane protein repeats. Prot Sci 4:1618–1632
Pagani F, Buratti E, Stuani C, Romano M, Zuccato E, Niksic M, Giglio L, Faraguna D, Baralle FE (2000) Splicing factors induce cystic fibrosis transmembrane regulator exon 9 skipping through a nonevolutionary conserved intronic element. J Biol Chem 275(28):21041–21047
Polydorides AD, Okano HJ, Yang YYL et al. (2000) A brain-enriched polypyrimidine tract-binding protein antagonizes the ability of Nova to regulate neuron-specific alternative splicing. Proc Natl Acad Sci USA 97:6350–6355
Reddy R (1989) Compilation of small nuclear RNA sequences. Methods Enzymol 180:521–532
Screaton GR, Caceres JF, Mayeda A et al. (1995) Identification and characterization of three members of the human SR family of pre-mRNA splicing factors. EMBO J 14(17):4336–4349
Shapiro MB, Senapathy P (1987) RNA splice junctions of different classes of eukaryotes: sequence statistics and functional implications in gene expression. Nucleic Acids Res 15(17):7155–7174
Shiga N, Takeshima Y, Sakamoto H, Inoue K, Yokota Y, Yokoyama M, Matsuo M (1997) Disruption of the splicing enhancer sequence within exon 27 of the dystrophin gene by a nonsense mutation induces partial skipping of the exon and is responsible for Becker muscular dystrophy. J Clin Invest 100(9):2204–2210
Si ZH, Rauch D, Stoltzfus CM (1998) The exon splicing silencer in human immunodeficiency virus type 1 Tat exon 3 is bipartite and acts early in spliceosome assembly. Mol Cell Biol 18(9):5404–5413
Sirand-Pugnet P, Durosay P, Brody E, Marie J (1995) An intronic (A/U)GGG repeat enhances the splicing of an alternative intron of the chicken beta-tropomyosin pre-mRNA. Nucleic Acids Res 23:3501–3507
Smith CWJ, Valcárcel J (2000) Alternative pre-mRNA splicing: the logic of combinatorial control. TIBS 25:381–388
Southby J, Gooding C, Smith CW (1999) Polypyrimidine tract binding protein functions as a repressor to regulate alternative splicing of alpha-actinin mutually exclusive exons. Mol Cell Biol 19(4):2699–2711
Staffa A, Acheson NH, Cochrane A (1997) novel exonic elements that modulate splicing of the human fibronectin EDA exon. J Biol Chem 272:33394–33401
Stamm S, Zhang MQ, Marr TG et al. (1994) A sequence compilation and comparison of exons that are alternatively spliced in neurons. Nucleic Acids Res 22(9):1515–1526
Stamm S, Casper D, Hanson V et al. (1999) Regulation of the neuron-specific exon of clathrin light chain B. Mol Brain Res 64:108–118
Stamm S, Zhu J, Nakai K et al. (2000) An alternative-exon database and its statistical analysis. DNA Cell Biol 19(12):739–756

Steitz JA (1992) Splicing takes a holiday. Science 257:888–889
Stoss O, Olbrich M, Hartmann AM et al. (2001) The STAR/GSG family protein rSLM-2 regulates the selection of alternative splice sites. J Biol Chem 276(12):8665–8673
Tacke R, Manley JL (1999) Determinants of SR protein specificity. Curr Opin Cell Biol 11:358–362
Tacke R, Chen Y, Manley JL (1997) Sequence-specific RNA binding by an SR protein requires RS domain phosphorylation: creation of an SRp40-specific splicing enhancer. Proc Natl Acad Sci USA 94(4):1148–1153
Tanaka K, Watakabe A, Shimura Y (1994) Polypurine sequences within a downstream exon function as a splicing enhancer. Mol Cell Biol 14:1347–1354
Thanaraj TA (1999) A clean data set of EST-confirmed splice sites from *Homo sapiens* and standards for clean-up procedures. Nucleic Acids Res 27(13):2627–2637
Thanaraj TA (2000) Positional characterisation of false positives from computational prediction of human splice sites. Nucleic Acids Res 28:744–754
Thanaraj TA, Clark F (2001) Human GC-AG alternative intron isoforms with weak donor sites show enhanced consensus at acceptor exon positions. Nucleic Acids Res 29(12):2581–2593
Thanaraj TA, Robinson A (2000) Prediction of exact boundaries of exons. Briefings Bioinform 1:343–356
Tian H, Kole R (1995) Selection of novel exon recognition elements from a pool of random sequences. Mol Cell Biol 15:6291–6298
Tian M, Maniatis T (1993) A splicing enhancer complex controls alternative splicing of doublesex pre-mRNA. Cell 74(1):105–114
Tsukahara T, Casciato C, Helfman DM (1994) Alternative splicing of beta-tropomyosin pre-mRNA: multiple cis-elements can contribute to the use of the 5′- and 3′-splice sites of the nonmuscle/smooth muscle exon 6. Nucleic Acids Res 22(12):2318–2325
Valentine CR (1998) The association of nonsense codons with exon skipping. Mutat Res 411(2):87–117
Weighardt F, Biamonti G, Riva S (1996) The role of heterogeneous nuclear ribonucleoproteins (hnRNP) in RNA metabolism. BioEssays 18:747–756
Wentz MP, Moore BE, Cloyd MW, Berget SM, Donehower LA (1997) A naturally arising mutation of a potential silencer of exon splicing in human immunodeficiency virus type 1 induces dominant aberrant splicing and arrests virus production. J Virol 71(11):8542–8551
Wise JA (1993) Guides to the heart of the spliceosome. Science 262:1978–1979
Wu S, Romfo CM, Nilson TW et al. (1999) Functional recognition of the 3′ splice site AG by the splicing factor U2AF. Nature 402:832–835
Zahler AM, Neugebauer KM, Lane WS et al. (1993) Distinct functions of SR proteins in alternative pre-mRNA splicing. Science 260(5105):219–222
Zandberg H, Moen TC, Baas PD (1995) Cooperation of 5′ and 3′ processing sites as well as intron and exon sequences in calcitonin exon recognition. Nucleic Acids Res 23(2):248–255
Zhang L, Liu W, Grabowski PJ (1999) Coordinate repression of a trio of neuron-specific splicing events by the splicing regulator PTB. RNA 5(1):117–130
Zhang MQ, Marr TG (1993) A weight array method for splicing signal analysis. Comput Appl Biosci 9(5):499–509
Zhang MQ, Marr TG (1994) Fission yeast gene structure and recognition. Nucleic Acids Res 22(9):1750–1759
Zheng ZM, He PJ, Baker CC (1999) Function of a bovine papillomavirus type 1 exonic splicing suppressor requires a suboptimal upstream 3′ splice site. J Virol 73(1):29–36
Zhuang Y, Weiner AM (1986) A compensatory base change in U1 snRNA suppresses a 5′ splice site mutation. Cell:827–835

Multiple Roles of the SR Protein Family in Splicing Regulation

J.R. Sanford, D. Longman, and J.F. Cáceres[1]

The serine and arginine-rich proteins (SR proteins) are a highly conserved family of essential pre-mRNA splicing factors. SR proteins have a modular domain structure consisting of RNA binding and protein–protein interaction modules. SR proteins function as molecular adapters, mediating interactions between the pre-mRNA and the assembling spliceosome. Unlike most essential splicing factors, SR proteins have acquired an inherent flexibility that allows them to function at numerous steps in spliceosome assembly and therefore, to play key roles in regulation of splice site selection. In the post-genomics era it is widely accepted that alternative splicing of pre-mRNAs may significantly expand the capacity of the genome to generate the functional complexity of the proteome. Therefore, it is essential to understand both the mechanisms of splice site selection and how *trans*-acting factors, such as the SR proteins, are regulated. Within this chapter we will discuss how the structure of SR proteins influences their roles in alternative splice site selection.

1 Introduction

Pre-mRNA splicing is an essential step in the flow of genetic information in virtually all eukaryotes. Intervening sequences, termed introns, are excised from precursor mRNAs by a large ribonucleoprotein complex, the spliceosome. The spliceosome consists of the five small nuclear ribonucleoproteins particles (snRNPs) U1, U2, U4, U5 and U6, and 50–100 non-snRNP splicing factors (Kramer 1996; Neubauer et al. 1998; Reed 2000). Each snRNP is a multi-subunit complex between an uracyl-rich small nuclear RNA (U snRNA) and several proteins, some of which are unique to the individual snRNP and others that are common to many spliceosomal snRNPs (Will and Luhrmann 2001). The SR proteins, a family of structurally and functionally related proteins, are among the most extensively characterised non-snRNP splicing factors (Fu 1995). The SR proteins function as molecular adapters, mediating interactions

[1]MRC Human Genetics Unit, Western General Hospital, Crewe Road, Edinburgh EH4 2XU, Scotland, UK

Progress in Molecular and Subcellular Biology, Vol. 31
Philippe Jeanteur (Ed.)

between the pre-mRNA and the assembling spliceosome. They are essential splicing factors and also modulate alternative splice site selection. Initial analyses of SR protein function in constitutive splicing assays in vitro suggested functional redundancies; however, this may not be the case in vivo where individual SR proteins may contribute to the alternative splicing of specific pre-mRNAs. Within this chapter we will discuss the structure-function relationship between the modular domains of SR proteins and their numerous roles in both constitutive and alternative splicing. We will also examine recent experiments designed to address these issues in genetically tractable model organisms. The hypothesis emerging from these distinct lines of study are that individual SR proteins, or combinations thereof, are required for the proper alternative splicing of pre-mRNAs that are important throughout the development of metazoans.

2 Spliceosome Assembly

Introns are classically defined by conserved sequences or splice sites, found at their 5′ and 3′ ends (Fig. 1A). Splice sites are recognised as pairs across either exons or introns, depending on which distance is shorter. For example, yeast genes have very small introns and recognition of exons seems to occur by interactions mediated across the intron itself, in a process known as intron definition (Abovich and Rosbash 1997; Romfo et al. 2000; Lim and Burge 2001). Intron definition is also the predominant mechanism in splicing of small *Drosophila* introns (Talerico and Berget 1994). In contrast, the correct identification of exons is a complex problem in vertebrate genes, which have small exons separated by large introns (Black 1995). In this case, exon recognition is facilitated by interactions between the upstream 3′ splice site and the downstream 5′ splice site facilitate, in a process known as exon definition (Berget 1995).

Spliceosome assembly is a dynamic process involving the recruitment of both snRNP and non-snRNP splicing factors to the pre-mRNA. Briefly, spliceosome assembly initiates with recognition of the 5′ splice site by the 5′ end of the U1 snRNA (Lerner et al. 1980; Black et al. 1985; Zhuang and Weiner 1986). Subsequently, the 3′ splice site is recognized by the heterodimeric splicing factor (U2AF, U2 snRNP auxiliary factor, Wu et al 1999). Interactions between U1 snRNP and U2AF are mediated by intron bridging proteins such as SF1 (BBP in yeast) or members of the SR protein family generating the E complex (Fig. 1B; Wu and Maniatis 1993; Abovich et al. 1994; Abovich and Rosbash 1997; Berglund et al. 1997). Spliceosomal A complex is formed when the U2 snRNP is recruited to the branch point sequence (BPS; Parker et al. 1987; Ruskin et al. 1988; Zamore and Green 1989). The branch point adenosine is thought to be specified as a bulged nucleotide arising from base pairing between the U2 snRNA and the BPS (Query et al. 1994). Association of the U4/U6·U5 tri-snRNP

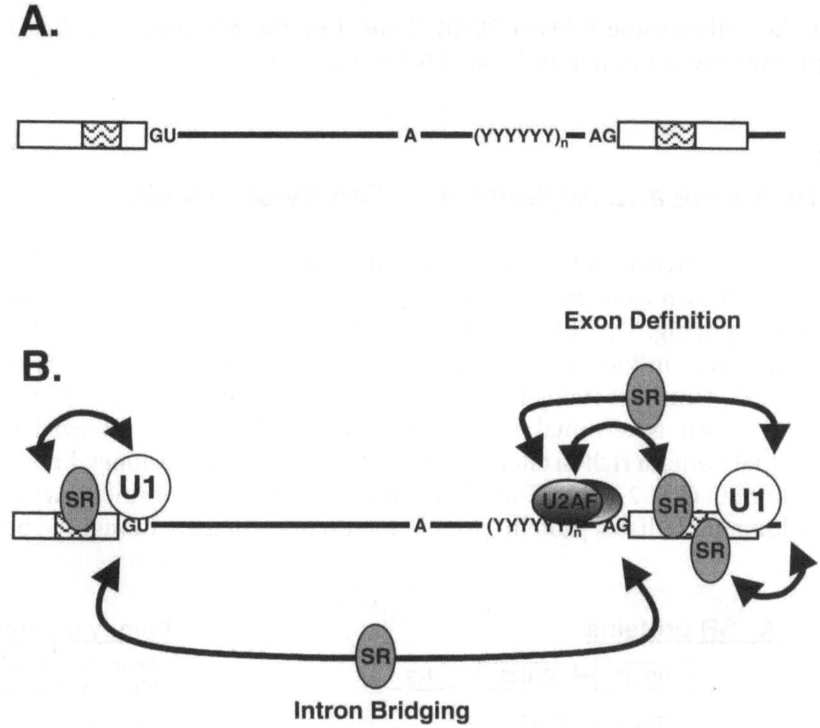

Fig. 1. SR proteins are involved at numerous steps of pre-spliceosome assembly. **A** The organisation of splicing signals within a typical pre-mRNA. Exons (*open boxes*) may contain splicing enhancer elements (*wavy boxes*) that promote recognition of the nearby 5′ or 3′ splice sites (*GU* and *AG*, respectively). Additional signals consist of the polypyrimidine tract (*Y*) that resides between the branch point adenosine (*A*) and the 3′ splice site. **B** The exon-dependent and -independent functions of SR proteins in pre-spliceosome assembly. SR proteins can promote both the recognition of 5′ and 3′ splice sites as well as communication of splice sites by exon definition or intron bridging interactions. *Arrows* indicate RS domain-mediated interactions

with the pre-spliceosome generates the pre-catalytic B complex. The mature spliceosome is formed by a complex rearrangement of RNA–RNA interactions that were established as the spliceosome assembled. As a result, both U1 and U4 snRNPs are released from the spliceosome. Once assembled, the mature spliceosome catalyses two sequential transesterification reactions (for review see Nilsen 2000). Interestingly, the recent discovery in yeast extracts of a very large particle that contains all five of the U snRNPs has challenged this dogmatic view of spliceosome assembly. The existence of this penta-snRNP particle suggests that the spliceosome may actually exist as a pre-assembled multi-snRNP particle, independently of a pre-mRNA substrate (Nilsen 2002; Stevens et al. 2002). Although protein components of the spliceosome play critical roles in recognition of splice sites and assembly of the splicing complex, it is widely hypothesised that catalysis is performed by the snRNA components of the spliceosome (Nilsen 2000). Thus, like the ribosome, the heart of the spliceosome is thought to be an RNA machine.

3
The Serine and Arginine-Rich Family of Proteins

A group of structurally and functionally related serine/arginine-rich splicing factors, known collectively as the SR proteins, have multiple functions in pre-mRNA splicing. The SR proteins are not only required for constitutive splicing, but also influence regulation of alternative splicing (Fu 1995; Graveley 2000). SR family proteins have a modular structure consisting of one or two copies of an N-terminal RNA-recognition motif (RRM) followed by a C-terminal domain rich in alternating serine and arginine residues, known as the RS domain (Fig. 2). The RRMs determine RNA binding specificity, whereas the RS domain mediates protein–protein interactions that are thought to be essen-

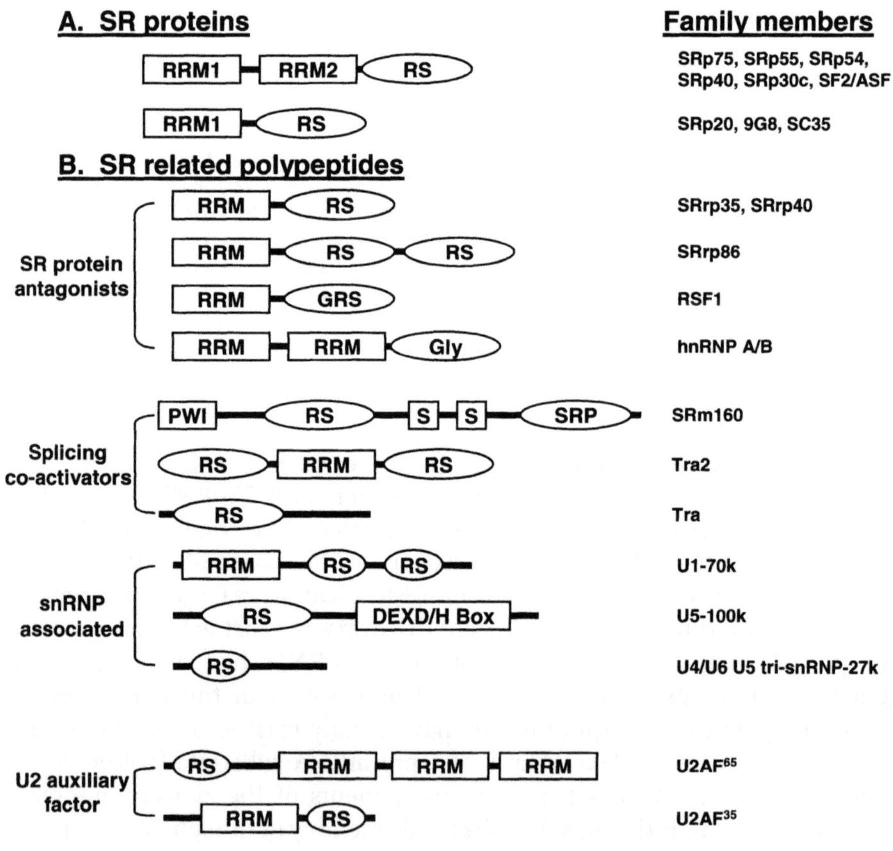

Fig. 2. A Domain structures of SR protein and B SR related polypeptides

tial for the recruitment of the splicing apparatus and for splice site pairing (Wu and Maniatis 1993; Tacke and Manley 1999). Additionally, SR family and SR-related proteins function in the recognition of exonic splicing enhancers (ESEs) leading to the activation of suboptimal adjacent 3′ splice sites (Blencowe 2000).

SF2/ASF (splicing factor 2/alternative splicing factor) was the first SR protein to be identified and cloned using two different strategies. An activity that complemented an otherwise splicing-deficient HeLa S100 cytosolic extract was purified from HeLa nuclear extracts, and termed SF2 (splicing factor 2, Krainer and Maniatis 1985; Krainer et al. 1990a). SF2 was also shown to modulate alternative splicing in vitro, in a concentration-dependent manner (Krainer et al. 1990b, 1991). Independently, a protein factor termed ASF (alternative splicing factor), which was identical to SF2, was purified from 293 cells on the basis of an alternative splicing assay (Ge and Manley 1990; Ge et al. 1991). Additional members of this family of proteins were identified using different strategies. For instance, a monoclonal antibody raised against mammalian spliceosomes allowed the identification of another human SR protein, SC35 (Fu and Maniatis 1990, 1992). Interestingly, a cDNA encoding SC35 had been initially cloned as an open reading frame (termed PR264), in the opposite strand of a *trans*-spliced *myb* exon (Vellard et al. 1992). Most other members of the SR family of proteins were identified based on their reactivity with a monoclonal antibody, mAb104, that recognises a phosphorylated epitope and reacted against active sites of RNA polymerase II transcription on lampbrush amphibian chromosomes (Roth et al. 1990). B52, a *Drosophila* protein that is a nearly identical variant of SRp55 was also identified by means of a different monoclonal antibody that recognises transcriptionally active chromatin on *Drosophila* polytene chromosomes (Champlin et al. 1991). The finding that all mAb104-immunoreactive proteins could be easily purified by a simple two-step purification involving selective precipitation in the presence of millimolar concentrations of $MgCl_2$ allowed purification of SR proteins from several different sources (Zahler et al. 1992; Sanford and Bruzik 1999; Zahler 1999). Polymerase chain reaction (PCR) based assays with degenerate primers recognising conserved features in the RNP-2 and RNP-1 submotifs of RRM1, or the conserved heptapeptide present in RRM2 allowed the identification and cloning of more members of this family of proteins (Kim et al 1992; Kim and Baker 1993; Screaton et al. 1995). The mouse homologue of SRp20, termed X16, had been initially identified as a gene that is preferentially expressed in pre-B cell lines, relative to mature B cell lines (Ayane et al. 1991). Recently, the availability of completed genomes from different organisms has led to the realisation that the SR proteins are highly conserved throughout metazoans, and in plants (Zahler et al. 1992; Lazar et al. 1995; Lopato et al. 1996a,b, 1999), with individual members of this family of proteins showing higher homology across species than to other family members within the same species.

A class of related RS domain-containing proteins termed SR protein related polypeptides (SRrp) or SR-like proteins, have a domain structure different from the SR family proteins, and may or may not contain RRMs (Fig. 2). This group of proteins, includes among others, the U1–70 K protein, both subunits of U2AF and the splicing coactivators SRm 160/300 (two SR-related nuclear matrix proteins of 160 and 300 kDa (Blencowe et al. 1999). Some of the SR-related proteins are structurally very similar to authentic SR proteins, but do not share all common functions. For instance, SRrp86 does not complement splicing-defective S100 extracts; however, it is able to inhibit individual SR proteins in S100 extracts (Barnard and Patton 2000). Likewise, two SR-protein-like factors, SRrp40 and SRrp35, antagonise authentic SR proteins and regulate alternative splicing (Cowper er al. 2001). A recent genome-wide survey of RS domain proteins has revealed a large complexity of RS domain proteins in metazoans with functions not only in pre-mRNA splicing, but also in chromatin remodelling, transcription by RNA polymerase II and cell cycle progression (Boucher et al. 2001).

4 The Many Roles of Serine and Arginine-Rich Proteins in Constitutive Splicing

Unlike most essential splicing factors, SR proteins have acquired an inherent flexibility that allows them to function at numerous steps in spliceosome assembly and, therefore, to play key roles in regulation of splice site selection. For instance, the SR proteins directly promote E complex formation by stabilising U1 snRNP bound at the 5′ splice site and U2AF bound to the polypyrimidine tract adjacent to the 3′ splice site (see Fig. 1B; Kohtz et al. 1994; Staknis and Reed 1994). SR proteins also bridge the 5′ and 3′ splice sites by promoting RS-mediated interactions with the U1snRNP-associated 70-kDa protein (U1 70 K) at the 5′ splice site and with the small subunit of the heterodimeric splicing factor, $U2AF^{35}$, at the 3′ splice site (Wu and Maniatis 1993; Kohtz et al. 1994). Thus, a network of protein–protein interactions builds a bridge between the 5′ and 3′ splice sites, effectively looping out the intron during pre-spliceosome assembly. This model is also supported by bimolecular splicing experiments demonstrating that 5′ and 3′ splice site complexes on separate transcripts can functionally interact (Bruzik and Maniatis 1995; Chiara and Reed 1995). The intron-bridging activities of SR proteins offer an attractive model for the role of SR proteins in spliceosome assembly. However, it has been difficult to test this model directly and many studies have failed to accurately separate the intron bridging activities of SR proteins from their exon-dependent activities. Thus, potential exon-dependent functions of SR proteins may have been misinterpreted as intron bridging activities. The distinction between exon-dependent and -independent functions of SR proteins has been facilitated by the observation that exonless pre-mRNAs can proceed through the first catalytic steps of the splicing reaction in nuclear extract. However, the same substrates are not processed in S100 extracts, suggesting that SR proteins play additional roles in spliceosome assembly other than the exon-dependent recruitment of essential splicing factors to the splice sites (Hertel and Maniatis 1999). One plausible hypothesis may be that SR proteins promote recruitment of the U4/U6·U5 tri-snRNP to the pre-spliceosome (Roscigno and Garcia-Blanco 1995; Fetzer et al. 1997; Teigelkamp et al. 1997). Additionally, SR proteins can also negatively regulate splicing of an intron, as shown in the adenovirus late transcript, where binding of SF2/ASF to an intronic purine-rich sequence inhibits the splicing of a 3′ splice site (Kanopka et al. 1996). Finally, the discovery that some SR and SR-like proteins remain associated with the mRNA products after the splicing reaction and that a subset of SR proteins shuttle from the nucleus to the cytoplasm suggested that SR proteins may have roles not only in nuclear pre-mRNA splicing, but also have additional functions such as mRNA transport, or be involved in cytoplasmic events (Caceres et al. 1998; Huang and Steitz 2001).

5
Actions of Serine and Arginine-Rich Proteins and hnRNP A/B Proteins in Splice Site Selection

The first evidence that the SR proteins modulate splice site selection came from experiments showing that an increased concentration of individual SR proteins resulted in the selection of intron-proximal 5′ splice sites in pre-mRNAs that contain two or more alternative 5′ splice sites. Strikingly, an excess of hnRNP A/B proteins had the opposite effect, promoting the selection of intron-distal 5′ splice sites. Thus, the antagonistic activities of SR and hnRNP A/B type proteins are important determinants of splice site selection in a concentration-dependent manner (Mayeda and Krainer 1992; Mayeda et al. 1993; Caceres et al. 1994; Yang et al. 1994; Wang and Manley 1995). These effects are position-dependent rather than sequence-dependent and have been observed with different pre-mRNA substrates both in vitro and also in vivo. Thus, the relative level and activity of members of the SR and hnRNP A/B family of proteins may represent an important determinant of alternative splicing regulation, modulating the patterns of alternative splicing in a tissue-specific, or developmentally regulated manner. Indeed, there are tissue-specific variations in the total and relative amounts of SR proteins (Zahler et al. 1993a), and in particular the molar ratio of SF2/ASF to its antagonist, hnRNPA1, varies considerably in different rat tissues (Hanamura et al. 1988). This supports the notion that changes in the ratio of these proteins can affect alternative splicing of a variety of pre-mRNAs in vivo. Mechanistically, the functional antagonism of SF2/ASF and hnRNP A1 in splice site selection is based on competitive binding to pre-mRNA (Fig. 3). Using a model pre-mRNA substrate containing two duplicated 5′ splice sites, it was shown that SF2/ASF interferes with hnRNP A1 binding

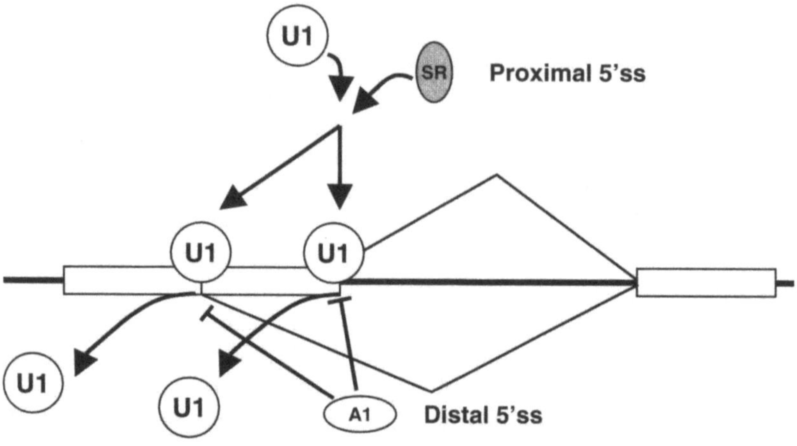

Fig. 3. Regulation of alternative 5′ splice site selection by the SR proteins and hnRNP A1. SR proteins and hnRNP A1 modulate binding of U1 snRNP to duplicated 5′ splice sites. SR proteins increase binding of U1 snRNP to both splice sites, resulting in selection of the proximal 5′ splice site. In contrast, hnRNP A1 decreases U1 snRNP binding at both splice sites, resulting in selection of the distal splice site

and enhances U1 snRNP binding at both 5′ splice sites. Thus, simultaneous occupancy of the U1 snRNP results in the selection of the 5′ proximal splice site. In contrast, hnRNP A1 binds co-operatively and indiscriminately to this pre-mRNA and interferes with U1 snRNP binding at both sites, resulting in a shift to the distal 5′ splice site (Eperon et al. 1993, 2000).

Protein activities that antagonise SR protein function in regulated splicing are not restricted to the hnRNP A/B family of proteins. In *Drosophila* an antagonist of SR protein function, termed RSF1, has been identified (Labourier et al. 1999). In addition, individual SR proteins can sometimes have opposite effects on alternative splice site selection too, as in the case of the antagonistic effects of SF2/ASF and SC35 in the regulation of β-tropomyosin (Gallego et al. 1997), and of SF2/ASF and SRp20 on the regulation of SRp20 pre-mRNA alternative splicing (Jumaa and Nielsen 1997). Whereas the antagonistic effects of SR and hnRNP A/B proteins is based on competitive binding to pre-mRNA, the mechanism by which individual SR proteins or SR-related proteins antagonise SR proteins is not fully understood. It may involve competitive binding to pre-mRNA or alternatively be due to sequestration of individual SR proteins, altering in this way the balance of SR and hnRNP proteins. For instance, RSF1, the splicing repressor identified in *Drosophila*, interacts with the RS domain of SF2/ASF and prevents stable binding of U1snRNP binding at the 5′ splice site (Labourier et al. 1999). Additionally, the SR-related protein SRrp86 interacts with a subset of SR proteins and selectively inhibits the function of individual SR proteins in constitutive splicing and also affects alternative splicing regulation by inhibiting the selection of 5′ proximal splice sites

(Barnard and Patton 2000). Two recently characterised SR-related proteins, SRrp35 and SRrp40, have a single N-terminal RRM followed by a C-terminal RS domain with a primary structure that resembles the single-RRM SR protein, SC35. These proteins inhibit authentic SR proteins in constitutive splicing and also antagonise SR proteins in alternative splicing. Increased concentrations of these proteins results in the selection of the most distal 5′ splice site in an adenovirus E1A pre-mRNA that contains three alternative 5′ splice sites competing for a unique 3′ splice site, an activity that is reminiscent of that of hnRNP A/B proteins. Thus, it is likely that the RRM, and not the RS domain, has a dominant role in this function (Cowper et al. 2001).

The general activities of SR proteins and hnRNP A/B proteins on 5′ splice site selection do not appear to require the recognition of specific target sequences. In contrast, binding to specific sites is a clear requirement when these proteins act as enhancers and silencers. Several models have been proposed to explain the function of SR proteins in enhancer-dependent splicing. Briefly, SR family proteins bound to exonic splicing enhancers (ESEs) can promote U2AF recruitment to the polypyrimidine tract and activate an adjacent 3′ splice site (Graveley et al. 2001; Romfo et al. 2001). This activity involves an RS domain-mediated interaction between an enhancer-bound SR protein and the small subunit of U2AF ($U2AF^{35}$, U2AF recruitment model; Zuo and Maniatis 1996). A role for two SR-related nuclear matrix proteins (SRm160/300) in this process has also been proposed. ESE-bound SR proteins might interact with the splicing coactivator SRm160 and establish a set of interactions which are different from those required for the recruitment of $U2AF^{65}$ to the polypyrimidine-tract (coactivator model; Blencowe 2000). The splicing coactivator SRm160 also promotes interactions with other SR proteins and snRNP particles and activates enhancer-dependent splicing. Biochemical and genetic evidence showed that SRm160-SR interactions are important for optimal splicing activity and also for proper development of the nematode *C. elegans* (Longman et al. 2001).

Interestingly, the recruitment model does not seem to be generally applicable to all enhancer-dependent splice sites. For certain pre-mRNA substrates, the presence of an ESE-bound SR protein does not act to promote recruitment of U2AF to the polypyrimidine tract, rather it helps to antagonise the negative activity of hnRNP proteins recognising exonic splicing silencer (ESS) elements (Inhibitor model, Kan and Green 1999; Graveley 2000; Hastings and Krainer 2001). The antagonistic activities of SR proteins and hnRNP A/B proteins also function in this type of regulated event, as evidenced in the HIV-1 tat exon 3, which harbours SF2/ASF and SC35-dependent ESEs and a silencer, ESS3, which is bound by hnRNP A1. Initial binding of hnRNP A1 to ESS3 causes further hnRNP A1 binding to an upstream region of this exon, which is prevented by ESE-bound SF2/ASF, but not by SC35. Thus, this splicing silencer suppressed SC35, but not SF2/ASF-dependent splicing (Zhu et al. 2001). Despite a general role for SR and hnRNP A/B proteins in a large variety of splicing decisions both in a tissue-specific manner and during development, it is most likely that tissue-specific or developmentally regulated splicing factors play an important role in the regulation of alternative splicing. The use of genetic approaches in *Drosophila* has allowed the identification of tissue or developmental stage-specific factors; however, progress in the identification of such factors in mammalian systems has been hampered by technical difficulties. One of the best characterised examples of tissue-specific alternative splicing regulators is represented by the NOVA-1 protein, a neuron-specific RNA binding protein that regulates neuron-specific alternative splicing and is essential for neuronal viability (Jensen et al. 2000; Dredge et al. 2001). Interestingly, the activity of NOVA-1 is antagonised by a tissue-specific isoform of the general splicing repressor PTB, referred to as a neurally enriched homologue of PTB (nPTB), or brain PTB (br PTB), which is closely related to PTB and is expressed in glia and neurons (Polydorides et al. 2000).

6 Serine and Arginine-Rich Modular Domains

As stated above, the SR proteins have a modular structure consisting of one or two RNA-recognition motifs (RRMs) and a C-terminal domain, rich in arginine and serine residues. In the following sections we will discuss the roles of the modular domains of SR proteins in constitutive splicing, alternative splicing and subcellular localisation.

6.1 RNA Binding Activities

The RNA recognition motif (RRM) is a conserved, modular RNA binding domain of approximately 80 amino acids in length and contains two small, highly conserved sequence elements, the RNP-1 octamer and the RNP-2 hexamer (Nagai et al. 1990; Hoffman et al. 1991). The RRM motif is present in one or several copies in many RNA-binding proteins involved in pre-mRNA and pre-rRNA processing (Kenan et al. 1991; Birney et al. 1993). A crystal structure of the N-terminal RRM of the U1 snRNP-associated protein, U1A, has been solved and shows the existence of a four-stranded antiparallel β-sheet connected by two α-helices. The highly conserved RNP-2 and RNP-1 motifs are adjacent to the central β1 and β3 strands and several solvent-exposed aromatic residues within these motifs are thought to contact bound RNA through ring-stacking interactions with single-stranded bases. A co-crystal structure of the U1A N-terminal RRM with bound RNA confirmed these stacking interactions (Nagai et al. 1995). A very similar structure has been shown for the RRM of hnRNP C1/C2 (Wittekind et al. 1992). Substitution of conserved residues within the RNP-1 and RNP-2 submotifs affects binding of the U1A and U1–70 K proteins to U1 snRNA (Scherly et al. 1989; Surowy et al. 1989). SR proteins contain one or two RRMs, however, in those SR proteins that contain two RRMs, the second RRM is atypical and not highly conserved and always contains a heptapeptide, SWQDLKD, that is a signature of this domain (Birney et al. 1993). Mutation of the RNP-2 or RNP-1 motifs of SF2/ASF inhibits binding to RNA and results in decreased activity in pre-mRNA splicing complementation assays (Caceres and Krainer 1993). As stated above, the RRMs of SR proteins have been shown to bind to RNA in a sequence-specific manner even in the absence of the RS domain (Manley and Tacke 1996). However, in the context of a full length protein, phosphorylation of the RS domain enhances sequence-specific RNA binding (Xiao and Manley 1997, 1998). Thus, there appears to be an interplay between these two domains which is likely to be important for their roles in spliceosome assembly and regulation.

SELEX protocols (selected evolution of ligands through exponential enrichment; Tuerk and Gold 1990) were employed to characterise the RNA binding specificity of individual SR proteins. This approach allows for the selection of high affinity binding sites from a pool of randomised sequences and resulted in the identification of preferred binding sites for individual SR proteins (Tacke and Manley 1995; Tacke et al. 1997; Cavaloc et al. 1999). Typically, selected sequences were purine-rich, resembling 5′ splice sites or exonic sequences, known to work as splicing enhancers. In contrast, identification of exonic splicing enhancer motifs recognised by individual SR proteins using a functional SELEX strategy gave rise to more divergent consensus sequences. This degeneracy is consistent with the fact that exonic enhancers evolved within extremely diverse protein coding sequences and are recognised by a small number of SR proteins that bind RNA with limited sequence specificity (Liu et al. 1998, 2000; Schaal and Maniatis 1999). These studies led to the conclusion that SR proteins are sequence-specific RNA binding proteins with distinct RNA binding specificities (for a review, see Tacke and Manley 1999). Individual SR proteins have been found to associate with exonic splicing enhancers (ESEs) in many different genes, leading to the activation of otherwise inefficient upstream 3′ splice sites (for a review see Blencowe 2000). As stated above, in some cases SR proteins bound to ESEs may recruit U2AF to the 3′ splice site (Zuo and Maniatis 1996; Zhu and Krainer 2000; Graveley et al. 2001; Romfo et al. 2001). This mechanism is employed in both constitutive and alternative 3′ splice site selection (Schaal and Maniatis 1999).

6.2 The RS Domain

The RS domain is a distinctive feature of the SR family of proteins, and is also present in a family of related splicing factors, the so-called SR-like proteins. It consists of simple arginine and serine repeats, occasionally interrupted by other amino acids, and its length and sequence are highly conserved for individual SR proteins (Zahler et al. 1992; Birney et al. 1993). The RS domain has been demonstrated to serve several functions. The cellular localisation of SR proteins is thought to be influenced, at least in part by the RS domain, which mediates the interaction with the SR protein nuclear import receptor, transportin-SR (Kataoka et al. 1999; Lai et al. 2000, 2001). Although the RS domain is a nuclear localisation signal, subnuclear targeting to the speckles requires at least two of the three constituent domains of SF2/ASF. In contrast, in two SR proteins that have a single RRM (SC35 and SRp20), the RS domain is both necessary and sufficient as a targeting signal to the speckles (Li and Bingham 1991; Hedley et al. 1995; Caceres et al. 1997). The RS domain is also an important determinant of nucleo-cytoplasmic shuttling; however, this domain is not sufficient to promote shuttling of an unrelated protein reporter, suggesting that additional signals are required (Caceres et al. 1998).

The RS domain of SR proteins mediates protein–protein interactions with other RS domain containing factors such as U1 70K and $U2AF^{35}$ (Wu and Maniatis 1993; Kohtz et al. 1994). However, the biochemical basis of this interaction is poorly understood. Once SR proteins are tethered to their sites of action through RNA-RRM interactions, it appears that the RS domain is responsible for mediating critical protein–protein interactions with other components of the splicing machinery (Graveley and Maniatis 1998). Since U2AF binding is thought to recruit the U2 snRNP to the branch point sequence (Ruskin et al. 1988; Valcarcel et al. 1996), it is likely that SR proteins also indirectly stabilise the spliceosomal A complex.

7 Role of the Modular Domains of Serine and Arginine-Rich Proteins in Constitutive and Alternative Splicing

7.1 Constitutive Splicing Activities

In this section we will describe how the modular domains of SR proteins contribute to splicing specificity, in both constitutive and regulated splicing. An extensive structure–function analysis of the SR protein SF2/ASF demonstrated that both the RRMs and the RS domain were essential for the constitutive splicing activity in vitro (Caceres and Krainer 1993; Zuo and Manley 1993). However, a recent study showed that the RS domain of SF2/ASF is dispensable for splicing of several substrates, including constitutive and enhancer-dependent pre-mRNAs. Thus, the RS domain is only required, in conjunction with $U2AF^{35}$, for pre-mRNA substrates that have weak splicing signals. These new findings underscore the existence of both RS domain-dependent and -independent activities of SR proteins (Zhu and Krainer 2000).

The role of the modular domains of SR proteins in constitutive splicing has been extensively studied. A functional splicing commitment assay was used to analyse the contribution of individual domains of SR proteins to pre-mRNA splicing of different substrates. In this assay, individual SR proteins are pre-incubated with different pre-mRNAs and the resulting commitment complexes are challenged by addition of splicing extracts in the presence of excess competitor RNAs. It was shown that different SR proteins were able to commit different pre-mRNAs to the splicing pathway. For example, SC35 was able to commit β-globin pre-mRNA, whereas other SR proteins had lower activity. In contrast, only SF2/ASF was able to commit HIV tat pre-mRNA to splicing, whereas other SR proteins, such as SC35, had no effect (Fu 1993). This splicing commitment assay was used with pre-mRNAs that are responsive to different SR proteins, and domain swaps between SC35 and SF2/ASF. It was found that RRM1 of SF2/ASF could function in splicing without the atypical RRM2, in a structural arrangement that is reminiscent of the structure of SC35. When assaying the activity of different domain swaps in the splicing of an IgM pre-mRNA-comprising exons C3 and C4 which depends on SC35, but not on SF2/ASF, the presence of RRM2 of SF2/ASF abolished splicing activity of the resulting chimera, demonstrating that this domain inhibits IgM splicing. Conversely, deletion of RRM2 from SF2/ASF allowed commitment of IgM pre-mRNA to splicing (Chandler et al. 1997). Similar domain swaps between SF2/ASF and SC35, or domain deletion mutants of SF2/ASF were used with a different functional assay, S100 complementation (Mayeda et al. 1999). In this assay, β-globin pre-mRNA (exons 1 and 2) is spliced indiscriminately with either SF2/ASF or SC35; human immunodeficiency virus tat pre-mRNA (exons 2 and 3) is preferentially spliced with SF2/ASF, whereas an IgM pre-mRNA (exons C3 and C4) is preferentially spliced with SC35. It was shown that the RS domains of SF2/ASF and SC35 could be exchanged without effect on substrate specificity. The RRMs of SF2/ASF were active only in the context of a two-RRM structure, and RRM2 had a dominant role in substrate specificity. In contrast, the single RRM of SC35 can function alone, but its substrate specificity can be influenced by the presence of an additional RRM. Thus, the RRMs behave as modules that, when present in different combinations, can have positive, neutral, or negative effects on splicing, depending upon the specific substrate. These experiments, using two different functional assays, clearly demonstrated that individual domains in SR proteins function as modules in constitutive splicing. Thus, substrate specificity in constitutive splicing is determined by the nature of the RRMs and the identity of the RS domain did not affect splicing specificity, suggesting that the RS domains may have redundant functions in constitutive splicing (Chandler et al. 1997; Mayeda et al. 1999).

7.2 Alternative Splicing Activities

In contrast to constitutive splicing, the roles of the modular domains of SR proteins in regulated splicing has been less extensively studied. Analysis of

SF2/ASF deletion mutant proteins in splice site switching assays in vitro established that both RRMs were necessary and sufficient to influence the selection of proximal 5′ splice sites and that the RS domain was dispensable for this activity. Since both RRMs in SF2/ASF are required for efficient binding to RNA, this suggested that RNA binding is required for alternative splicing (Caceres and Krainer 1993; Zuo and Manley 1993). However, it has not been clearly established whether general and/or sequence-specific RNA binding by SR proteins is required for their activity in alternative splicing regulation.

The activity of a series of chimeric proteins consisting of domain swaps between different SR proteins showed that splice site selection is determined by the nature of the RRMs, and that RRM2 of SF2/ASF has a dominant role and can confer specificity to a heterologous protein. SF2/ASF promotes the selection of the most proximal 5′ splice site in the adenovirus E1A pre-mRNA, which has three alternative 5′ splice sites competing for a unique 3′ splice site, giving rise to the 13S isoform. In contrast, overexpression of SRp20 led to the selection of an intermediate site, the 12S isoform (Caceres et al. 1997). Insertion of RRM2 of SF2/ASF between the RRM and the RS domain of SRp20 rendered a chimeric protein that was as active as wild-type SF2/ASF in promoting the selection of the most proximal 5′ splice site in the adenovirus E1a pre-mRNA; whereas insertion of RRM1 of SF2/ASF did not alter the activity of SRp20 (van Der Houven Van Oordt et al. 2000a). This experiment clearly demonstrated that the nature of the RRMs determines splice site selection, and in particular, that RRM2 of SF2/ASF has a dominant role. These conclusions were further confirmed by the activity of the SF2/SRp40 chimeric proteins on the alternative splicing of the fibronectin EDI exon which is dependent on the binding of SR proteins to an exonic splicing enhancer. Thus, alternative splicing specificity is determined by the RRMs and in particular, the identity of RRM2 is important. In contrast, the identity of the RS domain is not important, as the RS domains are functionally interchangeable. The contribution of the RRMs to alternative splicing specificity in vivo suggests that sequence-specific RNA binding by SR proteins is required for this activity. These results demonstrate the modularity of SR proteins in alternative splicing regulation.

Two non-mutually exclusive models could explain the dominant role of RRM2 of SF2/ASF in alternative splicing regulation. It is possible that RNA binding specificity of the chimeric proteins was altered in such a way that SF2/ASF specificity of binding was conferred to those chimeric proteins that contain RRM2 of SF2/ASF. Alternatively, RRM2 of SF2/ASF could be mediating protein–protein interactions and bringing wild-type SF2/ASF or other SR proteins into play.

It is not clear how the RRMs in the SR proteins bind to RNA. It has been shown that both RRMs of SF2/ASF bind with strong synergy to RNA, and consequently, the high affinity binding sites selected by SELEX with an SF2/ASF protein comprising both RRMs are distinct from the sites selected with the RRM1 alone (Tacke and Manley 1995; Li et al. 2000). In contrast, in other RNA binding proteins such as the U1snRNP-associated U1A protein, individual RRMs have their own RNA-binding specificity and interact independently with distinct RNA elements in pre-mRNA (Lutz and Alwine 1994). An alternative explanation for the dominant role of RRM2 of SF2/ASF in alternative splicing is that this domain may be involved in protein–protein interactions in such a way that its presence helps recruit additional splicing factors to the spliceosome and modulate splicing specificity. It is possible that RRM2 could help mediate homodimeric interactions recruiting wild-type SF2/ASF protein. This activity would explain why the sole presence of RRM2 of SF2/ASF results in EDI inclusion in the fibronectin gene, and selection of 13S 5′-splice site in the E1A adenovirus pre-mRNA, both activities that are strongly favoured by wild-type SF2/ASF protein. It should be taken into account that the alternative splicing assays, both in vivo and in vitro, are carried out in the presence of multiple endogenous SR proteins that could be recruited to the spliceosome via protein–protein interactions with the exogenously expressed proteins. A role for RRM2 of SF2/ASF in mediating protein–protein interactions is also suggested by the repressing splicing activities of this factor when bound to an intronic repressor element in the adenovirus IIIa pre-mRNA (Kanopka et al. 1996; Dauksaite and Akusjarvi 2002). The use of MS2-SF2/ASF fusion proteins showed that RRM2 is both necessary and sufficient for this repression, and in particular, the conserved heptapeptide SWQDLKD, is essential for splicing inhibition. Since this motif is not predicted to be involved in RNA binding, it would suggest that RRM2 is involved in mediating protein–protein interactions with factors that negatively affect splice site recognition (Dauksaite and Akusjarvi 2002).

Specific roles for the RS domains of individual SR proteins have been suggested by the high phylogenetic conservation of specific sequences within the RS domains (Birney et al. 1993). Individual RS domains have also been shown to have unique properties in directing subcellular localisation and influencing the ability of SR proteins to shuttle from the nucleus to the cytoplasm (Li and Bingham 1991; Hedley et al. 1995; Caceres et al. 1997, 1998).

The RS domain alone can activate splicing of an enhancer-dependent intron. In this system a heterologous MS2 RNA binding site (a high affinity-binding site for the bacteriophage MS2 coat protein) replaced the ESE from a weakly spliced substrate. This construct is not spliced in HeLa nuclear extract because endogenous SR proteins cannot recognise the MS2 binding site. Splicing activation is achieved by supplementing the reaction with an MS2 protein that had been fused to an RS domain (Graveley and Maniatis 1998). The MS2-RS domain fusion is capable of interacting with and recruiting $U2AF^{35}$ to the 3′ splice site resulting in recognition of the 3′ splice site (Graveley et al. 2001). The activity of MS2-RS fusion proteins containing different RS domains correlated directly with the number of RS dipeptides present in the RS domain, suggesting that different RS domains have unique activities (Graveley and Maniatis 1998; Graveley et al. 1998). However, the RS domains are functionally interchangeable in vivo, since chimeric proteins consisting of both RRMs of SF2/ASF fused to different RS domains are able to rescue cell viability in a chicken B cell line DT40 that has inactivated endogenous SF2/ASF (Wang et al. 1998). In contrast, different RS domains varied considerably in their ability to restore Tra2 function in *Drosophila*, suggesting that not all RS domains are functionally equivalent in vivo (Dauwalder and Mattox 1998).

8 Functional Redundancy of Serine and Arginine-Rich Proteins

The SR proteins seem to be functionally redundant in constitutive splicing as illustrated by the ability of any individual SR protein to complement an otherwise inactive cytosolic HeLa S100 extract. Does this mean that all SR proteins have been selected to play the same roles in spliceosome assembly? Not necessarily, as the primary limitation of the S100 complementation assay is that there are limiting, but detectable levels of endogenous SR proteins present in the extract (Zahler et al. 1993a). Additionally, several differences in the ability of these proteins to regulate alternative splicing, as well as the ability of individual SR proteins to commit different pre-mRNAs to the splicing pathway suggested that individual SR proteins may have unique functions in splicing regulation (Fu 1993; Caceres et al. 1994; Wang and Manley 1995; Chandler et al. 1997). More importantly, recent in vivo and genetic analyses of SR proteins have indicated that not all SR proteins are functionally redundant (see Table 1). For example, SF2/ASF is essential for cell viability in the DT40 chicken cell line, and its depletion cannot be rescued by overexpression of other SR proteins (Wang et al. 1998). Genetic disruption of SRp55/*B52* in *Drosophila* results in lethality during development, however, no general splicing defects have been found in the null background (Ring and Lis 1994; Peng and Mount 1995). It was later found that SR proteins are able to complement the loss of B52 in most tissues, except in the brain, where B52 is the predominant protein (Hoffman and Lis 2000). SRp20 is essential for mouse development (Jumaa et al. 1999), and conditional deletion of the SR protein SC35 in the thymus causes a defect in T cell maturation (Wang et al. 2001). RNA interference (RNAi) experiments with *C. elegans* SR proteins showed that, whereas the ortholog of the mammalian SF2/ASF (CeSF2/ASF) is an essential gene, functional knockouts of other SR genes resulted in no obvious phenotype, which is indicative of functional redundancy (Kawano et al. 2000; Longman et al. 2000). The RNAi protocol allows simultaneous interference of two or more gene products, thus, combinatorial RNAi is a powerful way to test for genetic interactions. Simultaneous suppression of two or more SR proteins in certain combinations caused lethality or other developmental defects. These results suggest that at least some SR proteins are functionally redundant and that the requirement for a particular SR protein may be due to specific functions in the tissue or developmental stage in which a particular SR protein is predominant.

9
Role of Transcription in Alternative Splicing

In addition, alternative splicing may not only be regulated by the relative abundance of antagonistic factors, but also by a more complex process involving the transcription machinery. Pre-mRNA splicing is co-transcriptional, and a high degree of co-ordination in both time and space exits for mRNA capping, splicing and polyadenylation, which occur in intimate association with the elongating RNA polymerase II (Bentley 1999; Hirose and Manley 2000).

It has been recently shown that changes in the promoter structure strongly affect splice site selection. Using a series of human α-globin/fibronectin (FN) minigenes that include the alternatively spliced EDI (extra domain I) exon of FN, but differ in the promoter driving their expression, it was shown that the extent of EDI exon inclusion is dependent on the promoter from which the transcript has originated. This difference is due not to promoter strength, but to promoter quality (Cramer et al. 1997). The EDI alternative exon contains an ESE, which is recognised by the SR proteins SF2/ASF and 9G8; however, the effect of these proteins in promoting exon inclusion is modulated by the promoter (Cramer et al. 1999). Two alternative, but not exclusive, models to explain the promoter control of alternative splicing have been proposed. On the one hand, different promoters may have differential ability in the recruitment of SR proteins and/or other splicing factors to the C-terminal CTD domain of RNA polymerase II. Alternatively, the processivity of the elongating RNA polymerase II itself could influence the alternative splicing. Using the fibronectin pre-mRNA substrate and different transcriptional activators, it was demonstrated that the SV40 T antigen (T-Ag) which causes a decrease in RNA polymerase II processivity, led to an increase in the fibronectin EDI exon inclusion into mature mRNA. In contrast, the viral transcriptional activator VP16, which promotes elongation by polymerase II, resulted in EDI exon skipping (Kadener er al. 2001). Thus, low RNA polymerase II processivity or internal pauses for elongation lead to exon inclusion. In contrast, a highly elongating RNA polymerase II, or the absence of internal pauses, leads to exon exclusion.

10
Alternative Splicing and Signal Transduction

SR proteins are extensively phosphorylated in vivo, and this post-translational modification has been shown to extensively affect the protein–protein

Table 1. Genetic analyses of SR proteins

Gene	Strategy	Phenotype	Comments	Reference
Chicken				
SF2/ASF	B cell line (DT40) knock-out	Cell lethal	Accumulation of incompletely processed pre-mRNA; Constitutive splicing of model introns not affected; Switch in alternative splicing pattern of model pre-mRNAs	Wang et al. (1998)
Mouse				
SRp20	Cre-loxP mediated knock-out	Lethal at blastocyst stage	Effect on splicing not tested	Jumaa et al. (1999)
SC35	Homozygous germline deletion	Essential for embryonic development	X.D. Fu (unpubl. data)	
	Cre-loxP mediated thymus-specific knock-out	Reduced thymus size, reduction of thymocytes, defect in T cell maturation	Alteration of T cell-specific regulated splicing of CD45 gene	Wang et al. (2001)
Drosophila				
B52 (SRp55)	Null mutant (homozygous)	Lethal at 1st and 2nd instar larval stages	Splicing of tested endogenous pre-mRNAs not affected.	Ring and Lis (1994)
			Splicing defects in tissues where B52 is major SR protein, e.g. *ftz* pre-mRNA in brain.	Hoffman and Lis (2000)
	Dominant mutant ($B52^{ED}$ allele)	Homozygous lethal	Enhancement of RNA processing defects, alters sex-specific splicing of *doublesex* pre-mRNA on sensitised background	Peng and Mount (1995)
C. elegans				
SF2/ASF	RNAi	Late embryonic lethal		Longman et al. (2000)
SRp20, SC35, SC35–2, SRp40, SRp75	RNAi	No phenotype; Depletion of certain combinations of SR genes led to distinct phenotypes such as embryonic lethality, defects in gonad and germ cells development	Depletion of tested SR proteins and their combinations had no effect on constitutive or alternative splicing of tested genes	Kawano et al. (2000); Longman et al. (2000)

interactions mediated by the RS domain. We will not discuss this in detail in this chapter, since it is covered in in the chapter by Soret and Tazi.

Protein kinases involved in the phosphorylation of SR protein splicing regulators could play an important role in linking alternative splicing regulation to extracellular signals. For example, serine phosphorylation of the SR protein SRp40, via a PI-3 kinase-signalling pathway, has been shown to regulate alternative splicing of PKC βII mRNA (Patel et al. 2001). A splicing silencer in CD44 alternative exon v5 is recognised by hnRNP A1 leading to its repression, which is relieved by activation of the ERK MAP-kinase pathway (Matter et al. 2000; Weg-Remers et al. 2001). It has also been recently shown that stress-induced signalling influences the ratio of SR and hnRNP proteins in the nucleus, via a mechanism that involves phosphorylation and cytoplasmic accumulation of the splicing factor hnRNP A1. The concomitant decrease in nuclear hnRNP A1 abundance is reflected in changes in alternative splicing activity. Thus, signal transduction pathways may control alternative splice site selection in vivo by regulating the subcellular localisation of antagonistic pre-mRNA splicing factors (van Der Houven Van Oordt et al. 2000). The SR proteins, together with most other splicing factors, localise in nuclear speckles, also known as splicing factor compartments (SFCs), from where they are recruited to the sites of active transcription. Thus, regulation of this recruitment process may also act to control the relative ratios of antagonistic nuclear factors in the nucleoplasm (Misteli 2000).

11 Conclusions

In the preceding sections we have summarised an extensive array of SR protein structure-function analyses. The conclusions emerging from these studies are that the modular domains of SR proteins function independently; the RNA binding domains influence splice site selection, while the RS domain mediates protein–protein interactions and functions as a regulatory domain. Clearly, the actions of SR proteins alone are not sufficient to establish patterns of alternative splicing in vivo. Characterisation of novel *trans*-acting factors that can antagonise the activities of SR proteins in alternative splice site selection is essential to elucidate the mechanisms of tissue-specific or developmentally regulated alternative splicing. One of the major challenges of future studies will be to understand how SR proteins regulate the alternative splicing of endogenous pre-mRNAs in vivo and to identify classes of pre-mRNAs that are regulated by specific SR proteins. As discussed above, the use of genetically tractable model systems coupled with modern genomic approaches should lead to dramatic increases in our knowledge of alternative splicing regulation.

Acknowledgement. Our work is supported by the Medical Research Council (MRC) (D.L. and J.F.C) and by an EMBO long-term fellowship (J.R.S.)

References

Abovich N, Rosbash M (1997) Cross-intron bridging interactions in the yeast commitment complex are conserved in mammals. Cell 89:403–412

Abovich N, Liao XC, Rosbash M (1994) The yeast MUD2 protein: an interaction with PRP11 defines a bridge between commitment complexes and U2 snRNP addition. Genes Dev 8:843–854

Ayane M, Preuss U, Kohler G, Nielsen PJ (1991) A differentially expressed murine RNA encoding a protein with similarities to two types of nucleic acid binding motifs. Nucleic Acids Res 19:1273–1278

Barnard DC, Patton JG (2000) Identification and characterization of a novel serine-arginine-rich splicing regulatory protein. Mol Cell Biol 20:3049–3057

Bentley D (1999) Coupling RNA polymerase II transcription with pre-mRNA processing. Curr Opin Cell Biol 11:347–351

Berget SM (1995) Exon recognition in vertebrate splicing. J Biol Chem 270:2411–2414

Berglund JA, Chua K, Abovich N, Reed R, Rosbash M (1997) The splicing factor BBP interacts specifically with the pre-mRNA branchpoint sequence UACUAAC. Cell 89:781–787

Birney E, Kumar S, Krainer AR (1993) Analysis of the RNA-recognition motif and RS and RGG domains: conservation in metazoan pre-mRNA splicing factors. Nucleic Acids Res 21: 5803–5816

Black DL (1995) Finding splice sites within a wilderness of RNA. RNA 1:763–771

Black DL, Chabot B, Steitz JA (1985) U2 as well as U1 small nuclear ribonucleoproteins are involved in premessenger RNA splicing. Cell 42:737–750

Blencowe BJ (2000) Exonic splicing enhancers: mechanism of action, diversity and role in human genetic diseases. Trends Biochem Sci 25:106–110

Blencowe BJ, Bowman JA, McCracken S, Rosonina E (1999) SR-related proteins and the processing of messenger RNA precursors. Biochem Cell Biol 77:277–291

Boucher L, Ouzounis CA, Enright AJ, Blencowe BJ (2001) A genome-wide survey of RS domain proteins. RNA 7:1693–1701

Bruzik JP, Maniatis T (1995) Enhancer-dependent interaction between 5′ and 3′ splice sites in trans. Proc Natl Acad Sci USA 92:7056–7059

Caceres JF, Krainer AR (1993) Functional analysis of pre-mRNA splicing factor SF2/ASF structural domains. EMBO J 12:4715–4726

Caceres JF, Stamm S, Helfman DM, Krainer AR (1994) Regulation of alternative splicing in vivo by overexpression of antagonistic splicing factors. Science 265:1706–1709

Caceres JF, Misteli T, Screaton GR, Spector DL, Krainer AR (1997) Role of the modular domains of SR proteins in subnuclear localization and alternative splicing specificity. J Cell Biol 138:225–238

Caceres JF, Screaton GR, Krainer AR (1998) A specific subset of SR proteins shuttles continuously between the nucleus and the cytoplasm. Genes Dev 12:55–66

Cavaloc Y, Bourgeois CF, Kister L, Stevenin J (1999) The splicing factors 9G8 and SRp20 transactivate splicing through different and specific enhancers. RNA 5:468–483

Champlin DT, Frasch M, Saumweber H, Lis JT (1991) Characterization of a *Drosophila* protein associated with boundaries of transcriptionally active chromatin. Genes Dev 5:1611–1621

Chandler SD, Mayeda A, Yeakley JM, Krainer AR, Fu XD (1997) RNA splicing specificity determined by the coordinated action of RNA recognition motifs in SR proteins. Proc Natl Acad Sci USA 94:3596–3601

Chiara MD, Reed R (1995) A two-step mechanism for 5′ and 3′ splice-site pairing. Nature 375: 510–513

Cowper AE, Caceres JF, Mayeda A, Screaton GR (2001) Serine-arginine (SR) protein-like factors that antagonize authentic SR proteins and regulate alternative splicing. J Biol Chem 276: 48908–48914

Cramer P, Pesce CG, Baralle FE, Kornblihtt AR (1997) Functional association between promoter structure and transcript alternative splicing. Proc Natl Acad Sci USA 94:11456–11460

Cramer P, Caceres JF, Cazalla D, Kadener S, Muro AF, Baralle FE, Kornblihtt AR (1999) Coupling of transcription with alternative splicing: RNA pol II promoters modulate SF2/ASF and 9G8 effects on an exonic splicing enhancer. Mol Cell 4:251–258

Dauksaite V, Akusjarvi G (2002) Human splicing factor ASF/SF2 encodes for a repressor domain required for its inhibitory activity on pre-mRNA splicing. J Biol Chem 277:12579-12586

Dauwalder B, Mattox W (1998) Analysis of the functional specificity of RS domains in vivo. EMBO J 17:6049–6060

Dredge BK, Polydorides AD, Darnell RB (2001) The splice of life: alternative splicing and neurological disease. Nat Rev Neurosci 2:43–50

Eperon IC, Ireland DC, Smith RA, Mayeda A, Krainer AR (1993) Pathways for selection of 5′ splice sites by U1 snRNPs and SF2/ASF. EMBO J 12:3607–3617

Eperon IC, Makarova OV, Mayeda A, Munroe SH, Caceres JF, Hayward DG, Krainer AR (2000) Selection of alternative 5′ splice sites: role of U1 snRNP and models for the antagonistic effects of SF2/ASF and hnRNP A1. Mol Cell Biol 20:8303–8318

Fetzer S, Lauber J, Will CL, Luhrmann R (1997) The [U4/U6.U5] tri-snRNP-specific 27K protein is a novel SR protein that can be phosphorylated by the snRNP-associated protein kinase. RNA 3:344–355

Fu XD (1993) Specific commitment of different pre-mRNAs to splicing by single SR proteins. Nature 365:82–85

Fu XD (1995) The superfamily of arginine/serine-rich splicing factors. RNA 1:663–680

Fu XD, Maniatis T (1990) Factor required for mammalian spliceosome assembly is localized to discrete regions in the nucleus. Nature 343:437–441

Fu XD, Maniatis T (1992) Isolation of a complementary DNA that encodes the mammalian splicing factor SC35. Science 256:535–538

Gallego ME, Gattoni R, Stevenin J, Marie J, Expert-Bezancon A (1997) The SR splicing factors ASF/SF2 and SC35 have antagonistic effects on intronic enhancer-dependent splicing of the beta-tropomyosin alternative exon 6A. EMBO J 16:1772–1784

Ge H, Manley JL (1990) A protein factor, ASF, controls cell-specific alternative splicing of SV40 early pre-mRNA in vitro. Cell 62:25–34

Ge H, Zuo P, Manley JL (1991) Primary structure of the human splicing factor ASF reveals similarities with *Drosophila* regulators. Cell 66:373–382

Graveley BR (2000) Sorting out the complexity of SR protein functions. RNA 6:1197–1211

Graveley BR, Maniatis T (1998) Arginine/serine-rich domains of SR proteins can function as activators of pre-mRNA splicing. Mol Cell 1:765–771

Graveley BR, Hertel KJ, Maniatis T (1998) A systematic analysis of the factors that determine the strength of pre-mRNA splicing enhancers. EMBO J 17:6747–6756

Graveley BR, Hertel KJ, Maniatis T (2001) The role of U2AF35 and U2AF65 in enhancer-dependent splicing. RNA 7:806–818

Hanamura A, Caceres JF, Mayeda A, Franza BR, Krainer AR (1998) Regulated tissue-specific expression of antagonistic pre-mRNA splicing factors. RNA 4:430–444

Hastings ML, Krainer AR (2001) Pre-mRNA splicing in the new millennium. Curr Opin Cell Biol 13:302–309

Hedley ML, Amrein H, Maniatis T (1995) An amino acid sequence motif sufficient for subnuclear localization of an arginine/serine-rich splicing factor. Proc Natl Acad Sci USA 92:11524–11528

Hertel KJ, Maniatis T (1999) Serine-arginine (SR)-rich splicing factors have an exon-independent function in pre-mRNA splicing. Proc Natl Acad Sci USA 96:2651–2655

Hirose Y, Manley JL (2000) RNA polymerase II and the integration of nuclear events. Genes Dev 14:1415–1429

Hoffman BE, Lis JT (2000) Pre-mRNA splicing by the essential *Drosophila* protein B52: tissue and target specificity. Mol Cell Biol 20:181–186

Hoffman DW, Query CC, Golden BL, White SW, Keene JD (1991) RNA-binding domain of the A protein component of the U1 small nuclear ribonucleoprotein analyzed by NMR spectroscopy is structurally similar to ribosomal proteins. Proc Natl Acad Sci USA 88:2495–2499

Huang Y, Steitz JA (2001) Splicing factors SRp20 and 9G8 promote the nucleocytoplasmic export of mRNA. Mol Cell 7:899–905

Jensen KB, Dredge BK, Stefani G, Zhong R, Buckanovich RJ, Okano HJ, Yang YY, Darnell RB (2000) Nova-1 regulates neuron-specific alternative splicing and is essential for neuronal viability. Neuron 25:359–371

Jumaa H, Nielsen PJ (1997) The splicing factor SRp20 modifies splicing of its own mRNA and ASF/SF2 antagonizes this regulation. EMBO J 16:5077–5085

Jumaa H, Wei G, Nielsen PJ (1999) Blastocyst formation is blocked in mouse embryos lacking the splicing factor SRp20. Curr Biol 9:899–902

Kadener S, Cramer P, Nogues G, Cazalla D, de la Mata M, Fededa JP, Werbajh SE, Srebrow A, Kornblihtt AR (2001) Antagonistic effects of T-Ag and VP16 reveal a role for RNA pol II elongation on alternative splicing. EMBO J 20:5759–5768

Kan JL, Green MR (1999) Pre-mRNA splicing of IgM exons M1 and M2 is directed by a juxtaposed splicing enhancer and inhibitor. Genes Dev 13:462–471

Kanopka A, Muhlemann O, Akusjarvi G (1996) Inhibition by SR proteins of splicing of a regulated adenovirus pre-mRNA. Nature 381:535–538

Kataoka N, Bachorik JL, Dreyfuss G (1999) Transportin-SR, a nuclear import receptor for SR proteins. J Cell Biol 145:1145–1152

Kawano T, Fujita M, Sakamoto H (2000) Unique and redundant functions of SR proteins, a conserved family of splicing factors, in *Caenorhabditis elegans* development. Mech Dev 95:67–76

Kenan DJ, Query CC, Keene JD (1991) RNA recognition: towards identifying determinants of specificity. Trends Biochem Sci 16:214–220

Kim YJ, Baker BS (1993) Isolation of RRM-type RNA-binding protein genes and the analysis of their relatedness by using a numerical approach. Mol Cell Biol 13:174–183

Kim YJ, Zuo P, Manley JL, Baker BS (1992) The *Drosophila* RNA-binding protein RBP1 is localized to transcriptionally active sites of chromosomes and shows a functional similarity to human splicing factor ASF/SF2.Genes Dev 6:2569–2579

Kohtz JD, Jamison SF, Will CL, Zuo P, Luhrmann R, Garcia-Blanco MA, Manley JL (1994) Protein–protein interactions and 5′-splice-site recognition in mammalian mRNA precursors. Nature 368:119–124

Krainer AR, Maniatis T (1985) Multiple factors including the small nuclear ribonucleoproteins U1 and U2 are necessary for pre-mRNA splicing in vitro. Cell 42:725–736

Krainer AR, Conway GC, Kozak D (1990a) Purification and characterization of pre-mRNA splicing factor SF2 from HeLa cells. Genes Dev 4:1158–1171

Krainer AR, Conway GC, Kozak D (1990b) The essential pre-mRNA splicing factor SF2 influences 5′ splice site selection by activating proximal sites. Cell 62:35–42

Krainer AR, Mayeda A, Kozak D, Binns G (1991) Functional expression of cloned human splicing factor SF2: homology to RNA-binding proteins, U1 70K, and *Drosophila* splicing regulators. Cell 66:383–394

Kramer A (1996) The structure and function of proteins involved in mammalian pre-mRNA splicing. Annu Rev Biochem 65:367–409

Labourier E, Bourbon HM, Gallouzi IE, Fostier M, Allemand E, Tazi J (1999) Antagonism between RSF1 and SR proteins for both splice-site recognition in vitro and *Drosophila* development. Genes Dev 13:740–753

Lai MC, Lin RI, Huang SY, Tsai CW, Tarn WY (2000) A human importin-beta family protein, transportin-SR2, interacts with the phosphorylated RS domain of SR proteins. J Biol Chem 275:7950–7957

Lai MC, Lin RI, Tarn WY (2001) Transportin-SR2 mediates nuclear import of phosphorylated SR proteins. Proc Natl Acad Sci USA 98:10154–10159

Lazar G, Schaal T, Maniatis T, Goodman HM (1995) Identification of a plant serine-arginine-rich protein similar to the mammalian splicing factor SF2/ASF. Proc Natl Acad Sci USA 92: 7672–7676

Lerner MR, Boyle JA, Mount SM, Wolin SL, Steitz JA (1980) Are snRNPs involved in splicing? Nature 283:220–224
Li H, Bingham PM Arginine/serine-rich domains of the su(wa) and tra RNA processing regulators target proteins to a subnuclear compartment implicated in splicing. (1991) Cell 67: 335–342
Li X, Shambaugh ME, Rottman FM, Bokar JA (2000) SR proteins Asf/SF2 and 9G8 interact to activate enhancer-dependent intron D splicing of bovine growth hormone pre-mRNA in vitro. RNA 6:1847–1858
Lim LP, Burge CB (2001) A computational analysis of sequence features involved in recognition of short introns. Proc Natl Acad Sci USA 98:11193–11198
Liu HX, Zhang M, Krainer AR (1998) Identification of functional exonic splicing enhancer motifs recognized by individual SR proteins. Genes Dev 12:1998–2012
Liu HX, Chew SL, Cartegni L, Zhang MQ, Krainer AR (2000) Exonic splicing enhancer motif recognized by human SC35 under splicing conditions. Mol Cell Biol 20:1063–1071
Longman D, Johnstone IL, Caceres JF (2000) Functional characterization of SR and SR-related genes in *Caenorhabditis elegans*. EMBO J 19:1625–1637
Longman D, McGarvey T, McCracken S, Johnstone IL, Blencowe BJ, Caceres JF (2001) Multiple interactions between SRm160 and SR family proteins in enhancer-dependent splicing and development of *C. elegans*. Curr Biol 11:1923–1933
Lopato S, Mayeda A, Krainer AR, Barta A (1996a) Pre-mRNA splicing in plants: characterization of Ser/Arg splicing factors. Proc Natl Acad Sci USA 93:3074–3079
Lopato S, Waigmann E, Barta A (1996b) Characterization of a novel arginine/serine-rich splicing factor in *Arabidopsis*. Plant Cell 8:2255–2264
Lopato S, Kalyna M, Dorner S, Kobayashi R, Krainer AR, Barta A (1999) atSRp30, one of two SF2/ASF-like proteins from *Arabidopsis thaliana*, regulates splicing of specific plant genes. Genes Dev 13:987–1001
Lutz CS, Alwine JC (1994) Direct interaction of the U1 snRNP-A protein with the upstream efficiency element of the SV40 late polyadenylation signal. Genes Dev 8:576–586
Manley JL, Tacke R (1996) SR proteins and splicing control. Genes Dev 10:1569–1579
Matter N, Marx M, Weg-Remers S, Ponta H, Herrlich P, Konig H (2000) Heterogeneous ribonucleoprotein A1 is part of an exon-specific splice-silencing complex controlled by oncogenic signaling pathways. J Biol Chem 275:35353–35360
Mayeda A, Krainer AR (1992) Regulation of alternative pre-mRNA splicing by hnRNP A1 and splicing factor SF2. Cell 68:365–375
Mayeda A, Helfman DM, Krainer AR (1993) Modulation of exon skipping and inclusion by heterogeneous nuclear ribonucleoprotein A1 and pre-mRNA splicing factor SF2/ASF. Mol Cell Biol 13:2993–3001
Mayeda A, Screaton GR, Chandler SD, Fu XD, Krainer AR (1999) Substrate specificities of SR proteins in constitutive splicing are determined by their RNA recognition motifs and composite pre-mRNA exonic elements. Mol Cell Biol 19:1853–1863
Misteli T (2000) Cell biology of transcription and pre-mRNA splicing: nuclear architecture meets nuclear function. J Cell Sci 113 (Pt 11):1841–1849
Nagai K, Oubridge C, Jessen TH, Li J, Evans PR (1990) Crystal structure of the RNA-binding domain of the U1 small nuclear ribonucleoprotein A. Nature 348:515–520
Nagai K, Oubridge C, Ito N, Jessen TH, Avis J, Evans P (1995) Crystal structure of the U1A spliceosomal protein complexed with its cognate RNA hairpin. Nucleic Acids Symp Ser 1–2
Neubauer G, King A, Rappsilber J, Calvio C, Watson M, Ajuh P, Sleeman J, Lamond A, Mann M (1998) Mass spectrometry and EST-database searching allows characterization of the multi-protein spliceosome complex. Nat Genet 20:46–50
Nilsen TW (2000) The case for an RNA enzyme. Nature 408:782–783
Nilsen TW (2002) The spliceosome: no assembly required? Mol Cell 9:8–9
Parker R, Siliciano PG, Guthrie C (1987) Recognition of the TACTAAC box during mRNA splicing in yeast involves base pairing to the U2-like snRNA. Cell 49:229–239

Patel NA, Chalfant CE, Watson JE, Wyatt JR, Dean NM, Eichler DC, Cooper DR (2001) Insulin regulates alternative splicing of protein kinase C beta II through a phosphatidylinositol 3-kinase-dependent pathway involving the nuclear serine/arginine-rich splicing factor, SRp40, in skeletal muscle cells. J Biol Chem 276:22648–22654

Peng X, Mount SM (1995) Genetic enhancement of RNA-processing defects by a dominant mutation in B52, the *Drosophila* gene for an SR protein splicing factor. Mol Cell Biol 15:6273–6282

Polydorides AD, Okano HJ, Yang YY, Stefani G, Darnell RB (2000) A brain-enriched polypyrimidine tract-binding protein antagonizes the ability of Nova to regulate neuron-specific alternative splicing. Proc Natl Acad Sci USA 97:6350–6355

Query CC, Moore MJ, Sharp PA (1994) Branch nucleophile selection in pre-mRNA splicing: evidence for the bulged duplex model. Genes Dev 8:587–597

Reed R (2000) Mechanisms of fidelity in pre-mRNA splicing. Curr Opin Cell Biol 12:340–345

Ring HZ, Lis JT (1994) The SR protein B52/SRp55 is essential for *Drosophila* development. Mol Cell Biol 14:7499–7506

Romfo CM, Alvarez CJ, van Heeckeren WJ, Webb CJ, Wise JA (2000) Evidence for splice site pairing via intron definition in *Schizosaccharomyces pombe.* Mol Cell Biol 20:7955–7970

Romfo CM, Maroney PA, Wu S, Nilsen TW (2001) 3′ splice site recognition in nematode trans-splicing involves enhancer-dependent recruitment of U2 snRNP. RNA 7:785–792

Roscigno RF, Garcia-Blanco MA (1995) SR proteins escort the U4/U6.U5 tri-snRNP to the spliceosome. RNA 1:692–706

Roth MB, Murphy C, Gall JG (1990) A monoclonal antibody that recognizes a phosphorylated epitope stains lampbrush chromosome loops and small granules in the amphibian germinal vesicle. J Cell Biol 111:2217–2223

Ruskin B, Zamore PD, Green MR (1988) A factor, U2AF, is required for U2 snRNP binding and splicing complex assembly. Cell 52:207–219

Sanford JR, Bruzik JP (1999) SR proteins are required for nematode *trans*-splicing in vitro. RNA 5:918–928

Schaal TD, Maniatis T (1999) Selection and characterization of pre-mRNA splicing enhancers: identification of novel SR protein-specific enhancer sequences. Mol Cell Biol 19:1705–1719

Scherly D, Boelens W, van Venrooij WJ, Dathan NA, Hamm J, Mattaj IW (1989) Identification of the RNA binding segment of human U1 A protein and definition of its binding site on U1 snRNA. EMBO J 8:4163–4170

Screaton GR, Caceres JF, Mayeda A, Bell MV, Plebanski M, Jackson DG, Bell JI, Krainer AR (1995) Identification and characterization of three members of the human SR family of pre-mRNA splicing factors. EMBO J 14:4336–4349

Staknis D, Reed R (1994) SR proteins promote the first specific recognition of Pre-mRNA and are present together with the U1 small nuclear ribonucleoprotein particle in a general splicing enhancer complex. Mol Cell Biol 14:7670–7682

Stevens SW, Ryan DE, Ge HY, Moore RE, Young MK, Lee TD, Abelson J (2002) Composition and functional characterization of the yeast spliceosomal penta-snRNP. Mol Cell 9:31–44

Surowy CS, van Santen VL, Scheib-Wixted SM, Spritz RA (1989) Direct, sequence-specific binding of the human U1-70K ribonucleoprotein antigen protein to loop I of U1 small nuclear RNA. Mol Cell Biol 9:4179–4186

Tacke R, Manley JL (1995) The human splicing factors ASF/SF2 and SC35 possess distinct, functionally significant RNA binding specificities. EMBO J 14:3540–3551

Tacke R, Manley JL (1999) Determinants of SR protein specificity. Curr Opin Cell Biol 11:358–362

Tacke R, Chen Y, Manley JL (1997) Sequence-specific RNA binding by an SR protein requires RS domain phosphorylation: creation of an SRp40-specific splicing enhancer. Proc Natl Acad Sci USA 94:1148–1153

Talerico M, Berget SM (1994) Intron definition in splicing of small *Drosophila* introns. Mol Cell Biol 14:3434–3445

Teigelkamp S, Mundt C, Achsel T, Will CL, Luhrmann R (1997) The human U5 snRNP-specific 100-kD protein is an RS domain-containing, putative RNA helicase with significant homology to the yeast splicing factor Prp28p.RNA 3:1313–1326

Tuerk C, Gold L (1990) Systematic evolution of ligands by exponential enrichment: RNA ligands to bacteriophage T4 DNA polymerase. Science 249:505–510

Valcarcel J, Gaur RK, Singh R, Green MR (1996) Interaction of U2AF65 RS region with pre-mRNA branch point and promotion of base pairing with U2 snRNA. Science 273:1706–1709

van Der Houven Van Oordt W, Newton K, Screaton GR, Caceres JF (2000A) Role of SR protein modular domains in alternative splicing specificity in vivo. Nucleic Acids Res 28:4822–4831

van Der Houven Van Oordt W, Diaz-Meco MT, Lozano J, Krainer AR, Moscat J, Caceres JF (2000b) The MKK(3/6)-p38-signaling cascade alters the subcellular distribution of hnRNP A1 and modulates alternative splicing regulation. J Cell Biol 149:307–316

Vellard M, Sureau A, Soret J, Martinerie C, Perbal B (1992) A potential splicing factor is encoded by the opposite strand of the *trans*-spliced c-myb exon. Proc Natl Acad Sci USA 89:2511–2515

Wang HY, Xu X, Ding JH, Bermingham JR Jr, Fu XD (2001) SC35 plays a role in T cell development and alternative splicing of CD45. Mol Cell 7:331–342

Wang J, Manley JL (1995) Overexpression of the SR proteins ASF/SF2 and SC35 influences alternative splicing in vivo in diverse ways. RNA 1:335–346

Wang J, Xiao SH, Manley JL (1998) Genetic analysis of the SR protein ASF/SF2: interchangeability of RS domains and negative control of splicing. Genes Dev 12:2222–2233

Weg-Remers S, Ponta H, Herrlich P, Konig H (2001) Regulation of alternative pre-mRNA splicing by the ERK MAP-kinase pathway. EMBO J 20:4194–4203

Will CL, Luhrmann R (2001) Spliceosomal UsnRNP biogenesis, structure and function. Curr Opin Cell Biol 13:290–301

Wittekind M, Gorlach M, Friedrichs M, Dreyfuss G, Mueller L (1992) 1H, 13C, and 15N NMR assignments and global folding pattern of the RNA-binding domain of the human hnRNP C proteins. Biochemistry 31:6254–6265

Wu JY, Maniatis T (1993) Specific interactions between proteins implicated in splice site selection and regulated alternative splicing. Cell 75:1061–1070

Wu S, Romfo CM, Nilsen TW, Green MR (1999) Functional recognition of the 3′ splice site AG by the splicing factor U2AF35. Nature 402:832–835

Xiao SH, Manley JL (1997) Phosphorylation of the ASF/SF2 RS domain affects both protein-protein and protein-RNA interactions and is necessary for splicing. Genes Dev 11:334–344

Xiao SH, Manley JL (1998) Phosphorylation-dephosphorylation differentially affects activities of splicing factor ASF/SF2. EMBO J 17:6359–6367

Yang X, Bani MR, Lu SJ, Rowan S, Ben David Y, Chabot B (1994) The A1 and A1B proteins of heterogeneous nuclear ribonucleoparticles modulate 5′ splice site selection in vivo. Proc Natl Acad Sci USA 91:6924–6928

Zahler AM (1999) Purification of SR protein splicing factors. Methods Mol Biol 118:419–432

Zahler AM, Lane WS, Stolk JA, Roth MB (1992) SR proteins: a conserved family of pre-mRNA splicing factors. Genes Dev 6:837–847

Zahler AM, Neugebauer KM, Lane WS, Roth MB (1993a) Distinct functions of SR proteins in alternative pre-mRNA splicing. Science 260:219–222

Zamore PD, Green MR (1989) Identification, purification, and biochemical characterization of U2 small nuclear ribonucleoprotein auxiliary factor. Proc Natl Acad Sci USA 86:9243–9247

Zhu J, Krainer AR (2000) Pre-mRNA splicing in the absence of an SR protein RS domain. Genes Dev 14:3166–3178

Zhu J, Mayeda A, Krainer AR (2001) Exon identity established through differential antagonism between exonic splicing silencer-bound hnRNP A1 and enhancer-bound SR proteins. Mol Cell 8:1351–1361

Zhuang Y, Weiner AM (1986) A compensatory base change in U1 snRNA suppresses a 5′ splice site mutation. Cell 46:827–835
Zuo P, Maniatis T (1996) The splicing factor U2AF35 mediates critical protein–protein interactions in constitutive and enhancer-dependent splicing. Genes Dev 10:1356–1368
Zuo P, Manley JL (1993) Functional domains of the human splicing factor ASF/SF2. EMBO J 12:4727–4737

Heterogeneous Nuclear Ribonucleoprotein Particle A/B Proteins and the Control of Alternative Splicing of the Mammalian Heterogeneous Nuclear Ribonucleoprotein Particle A1 Pre-mRNA

B. Chabot, C. LeBel, S. Hutchison, F.H. Nasim, and M.J. Simard[1]

1 Heterogeneous Nuclear Ribonucleoprotein Particle Proteins and Splicing

1.1 HnRNP Proteins

Mammalian pre-messenger RNAs (pre-mRNAs) interact with a distinct set of proteins to form heterogeneous nuclear ribonucleoprotein particles (hnRNPs). More than 25 different hnRNP proteins have been found associated with these complexes (Piñol-Roma et al. 1988; Dreyfuss et al. 1993). The core hnRNP proteins A1, A2, B1, B2, C1, and C2 are the most abundant species and are expressed in actively growing cells to levels that are comparable to core histones (Beyer et al. 1977; Kiledjian et al. 1994). Although several observations support the notion that core hnRNP proteins participate in pre-mRNA processing, the question of whether these proteins act in a sequence-independent or a sequence-specific manner is still a matter of debate (Abdul-Manan and Williams 1996; Abdul-Manan et al. 1996; McAfee et al. 1997). Early investigations have suggested that hnRNP proteins associate randomly with many different types of pre-mRNAs and ribopolymers to form particles sharing common structures and sedimentation properties (McKnight and Miller 1979; Lamb and Daneholt 1979; Pullman and Martin 1983; Thomas et al. 1983; Wilk et al. 1983). While the great abundance of core hnRNP proteins in the nucleus of growing cells (approximately 0.2 mM in a HeLa cell nucleus) can be taken as an argument in favor of generic RNA binding activity, it is unclear what fraction of this population of hnRNP proteins is free to interact with a nascent RNA sequence that exits the RNA polymerase II transcription complex. More recently, the use of antibodies recognizing individual hnRNP proteins has revealed that some hnRNP proteins can be deposited nonrandomly on nascent pre-mRNAs (Piñol-Roma et al. 1989; Matunis et al. 1993; Wurtz et al. 1996),

Département de Microbiologie et d'Infectiologie, Faculté de Médecine, Université de Sherbrooke, Sherbrooke, Québec, Canada J1H 5N4
[1] *Present address*: M.J. Simard, Program in Molecular Medicine, University of Massachusetts Medical School, Worcester, Massachusetts 01605, USA

Progress in Molecular and Subcellular Biology, Vol. 31
Philippe Jeanteur (Ed.)

suggesting that binding can be sequence-specific. Given the high abundance of hnRNP proteins and their ability to bind to many types of sequences with a range of affinities (Cobianchi et al. 1988; Conway et al. 1988; Piñol-Roma et al. 1988; LeStourgeon et al. 1990; Casas-Finet et al. 1993; Abdul-Manan and Williams 1996; Abdul-Manan et al. 1996), it has remained unclear to what extent the functions of core hnRNP proteins in nuclear RNA processing always depend on their ability to bind RNA in a sequence-specific manner.

HnRNP proteins have been implicated in many aspects of gene expression, including transcription, splicing, polyadenylation, transport, and mRNA stability. In this review, we will limit our presentation and discussion to the relationship between hnRNP proteins and splicing, and will put more emphasis on recent advances that have helped to better understand the function of the core hnRNP A/B proteins in this process.

1.2 HnRNP Proteins and Generic Splicing

Several studies have reported interactions between hnRNP proteins and components of the splicing machinery or with splicing signals on pre-mRNA. The hnRNP C protein can interact with the 5′ stem loop of the U2 snRNP (Temsamani and Pederson 1996), and can disrupt base-pairing between the U4 and U6 snRNAs (Forné et al. 1995). In pre-mRNA, hnRNP C can bind to the polypyrimidine tract of 3′ splice sites in vitro (Sebillon et al. 1995; Swanson and Dreyfuss 1988), but is apparently displaced from the pre-mRNA during spliceosome assembly (Bennett et al. 1992). Experiments designed to look at the function of the hnRNP C protein in generic splicing have also yielded mixed results: while antibodies against hnRNP C can inhibit splicing in vitro (Choi et al. 1986; Sierakowska et al. 1986), abrogating hnRNP C1/C2 expression in vivo does not affect cell viability (Williamson et al. 2000). The situation is similar with the hnRNP A1 protein: hnRNP A1 can interact with snRNPs and 3′ splice sites (Swanson and Dreyfuss 1988; Buvoli et al. 1992), but its initial interaction with pre-mRNA complexes in vitro is lost upon spliceosome assembly (Bennett et al. 1992). A mouse erythroleukemic cell line severely deficient in the expression of hnRNP A1 has been described (Ben-David et al. 1992), indicating that A1 is also not required for cell viability. Moreover, the splicing efficiency of some pre-mRNAs is not affected by depleting core hnRNP A/B proteins from mammalian nuclear extracts (Caputi et al. 1999; S. Hutchison, C. LeBel and B. Chabot, unpubl. results).

Although antibodies against hnRNP M proteins can inhibit splicing when added to a nuclear splicing extract (Gattoni et al. 1996), the role of hnRNP M in splicing has not been confirmed. In contrast, the hnRNP F protein interacts with components of the nuclear cap binding complex, and its depletion from a HeLa nuclear extract causes a splicing defect that can be partially corrected by the addition of recombinant hnRNP F (Gamberi et al. 1997). A recent study suggests that the hnRNP Q and R proteins may participate in generic splicing since their depletion from a HeLa extract decreases splicing efficiency, a situation that can be corrected by the addition of a fraction enriched in Q and R proteins (Mourelatos et al. 2001). Despite the lack of strong evidence to support a general function for hnRNP proteins in generic splicing, we have to bear in mind that in vitro splicing systems only offer a simplistic representation of the natural splicing environment. The model pre-mRNAs that have been used to derive the "rules" of splicing are often not from constitutively expressed genes. Moreover, most natural pre-mRNAs have much longer introns. Some hnRNP proteins could also be indirectly involved in splicing, for example by recycling specific splicing components (e.g., snRNPs) essential for the splicing machinery. Finally, the lack of strong evidence to support the involvement of specific core hnRNP proteins in generic splicing may also reflect redundancy in function between different members of the hnRNP family of proteins.

1.3 HnRNP Proteins and Alternative Splicing

In contrast to the lack of strong evidence to support a function for hnRNP proteins in constitutive splicing, several hnRNP proteins have been implicated in the alternative splicing of specific pre-mRNAs. In this category, the activity of hnRNP I has been particularly well documented. HnRNP I, also known as the polyrimidine-tract binding protein (PTB), can bind to UC-rich sequences that are found at the 3′ splice sites of some alternative exons, such as the human α-tropomyosin, the rat β-tropomyosin, the α-actinin, and the gamma aminobutyric acid type A receptor ($GABA_A$) pre-mRNAs (reviewed in Valcárcel and Gebauer 1997; Wagner and Garcia-Blanco 2001). In several of these cases, hnRNP I/PTB appears to antagonize the binding of the splicing factor U2AF, thereby repressing 3′ splice site use (Lin and Patton 1995; Singh et al. 1995; Ashiya and Grabowski 1997). A modulating activity for hnRNP I/PTB has also been reported for other pre-mRNAs. Such examples include the binding of hnRNP I/PTB in (1) the introns flanking the n-src alternative exon to facilitate exon skipping in nonneuronal cells (Chan and Black 1997; Chou et al. 2000), (2) the intron upstream of alternative exon 3 in the rat α-tropomyosin to favor skipping in smooth muscle tissue (Gooding et al. 1994, 1998), (3) the intron and exon sequences of the rat and human fibroblast growth factor receptor, respectively, to favor exon exclusion (Carstens et al. 2000; Le Guiner et al. 2001). A splicing variant of PTB (nPTB) expressed in neuronal cells has been suggested to facilitate the inclusion of the n-src and $GABA_A$ exons in neuronal cells (Zhang et al. 1999; Markovtsov et al. 2000). nPTB has also been proposed to interfere with the activity of the RNA binding protein NOVA implicated in the inclusion of alternative exon 3A in the glycine receptor α2 pre-mRNA (Polydorides et al. 2000). Finally, there is at least one case in which hnRNP I/PTB has been linked to an enhancer activity (Lou et al. 1999). In all these cases, the mechanisms by which hnRNP I/PTB elicits splicing control remain unknown.

Members of the family of hnRNP F/H proteins can bind RNA in a sequence-specific manner (Caputi and Zahler 2001), and have been implicated in the activity of both splicing enhancers and silencers. In the *src* pre-mRNA, the hnRNP F and H proteins are part of a complex assembling on an intronic enhancer element that promotes the neuro-specific inclusion of the N1 exon (Min et al. 1995; Chou et al. 1999). In a recent study, the binding of hnRNP H to a GGGA-containing element was shown to be essential for the activity of a SC35-bound enhancer element in the human immunodeficiency virus exon 6D (Caputi and Zahler 2002). HnRNP H is required for the activity of a silencer element located in alternative exon 7 of the rat β-tropomyosin pre-mRNA (Chen et al. 1999). It has also been implicated in the activity of repressor elements in the human immunodeficiency virus *tat* exon 2 (Jacquenet et al. 2001), and in the intron of the Rous sarcoma virus (Fogel and McNally 2000).

HnRNP G interacts with the Tra2β protein and can inhibit Tra2β-dependent splicing (Venables et al. 2000). Although hnRNP K contains a type of RNA binding motif (i.e., KH domain) that is found in a well-known splicing regulator in *Drosophila* (the PSI protein; Siebel et al. 1994, 1995), hnRNP K has mostly been described as a transcription factor in mammalian systems (Tomonaga and Levens 1995; Michelotti et al. 1996). The use of hnRNP K as a bait in a two-hybrid interaction assay has allowed the recovery of splicing proteins of the serine/arginine-rich (SR) family (9G8 and SRp20), but the biological relevance of this observation remains unknown (Shnyreva et al. 2000). Notwithstanding, some mammalian proteins containing a lysine/histidine-rich (KH) motif have been implicated in splicing regulation. The strongest evidence for a function of KH motif proteins in alternative splicing is for the KH-type splicing regulatory protein (KSRP), which associates with the n-src intron enhancer (Min et al. 1997). While the KH domain-containing protein SF1 is an essential splicing factor (Arning et al. 1996; Berglund et al. 1997), the KH domain-containing FUSE binding protein (FBP) has been recovered in association with hnRNP I/PTB (Grossman et al. 1998), but its function in splicing is unclear.

Thus, recent investigations focusing on specific examples of alternative splicing have revealed novel activities for hnRNP I, F and H proteins. These proteins bind to specific RNA sequence elements which, in some cases, interfere with the productive interaction of splicing factors. However, in most cases, the mechanisms employed by these hnRNP proteins to control splice site selection remain completely unknown.

1.4 The Function of Core hnRNP A/B Proteins in Splice Site Selection

The 34-kDa hnRNP A1 is one of the best characterized of the hnRNP core proteins. This protein was proposed to play a role in mRNA stability

(Hamilton et al. 1993, 1997), mRNA transport (Michael et al. 1995; Izaurralde et al. 1997), telomere biogenesis (LaBranche et al. 1998), and to contribute to the control of NF-κB-dependent transcription (Hay et al. 2001). Important advances have also been made concerning the role of the hnRNP A/B group of proteins, particularly the A1 protein, in splice site selection. Using model human β-globin and adenovirus E1a pre-mRNAs carrying competing 5′ splice sites, strong shifts toward distal 5′ splice sites can be elicited by the addition of purified or recombinant hnRNP A1, A2 or B1 proteins to nuclear extracts (Mayeda and Krainer 1992; Mayeda et al. 1994). In contrast to these potent effects on 5′ splice site utilization in vitro, a much smaller impact was observed in vivo (Cáceres et al. 1994), and large differences in the in vivo concentration of hnRNP A1 were only associated with a modest effect on 5′ splice site selection using the adenovirus E1a pre-mRNA as a reporter (Yang et al. 1994). The *Drosophila* hrp48 protein, which is similar to hnRNP A1, is required in collaboration with PSI and the U1 snRNP to repress splicing of the P-element pre-mRNA in somatic tissues (Hammond et al. 1997). HnRNP A1 has also been shown to affect 3′ splice site choice and exon skipping, but not all pre-mRNAs are responsive to variations in the concentration of hnRNP A1 (Mayeda et al. 1993; Jiang et al. 1998; Bai et al. 1999). The transcript specificity associated with the function of A1 could be due to the presence of high-affinity binding sites which would make some pre-mRNAs more sensitive to differences in the concentration of A1. Consistent with this view, Burd and Dreyfuss (1994) identified a high-affinity binding site for hnRNP A1 (UAGGG$^{A}/_{U}$) by selection/amplification of pools of random RNA sequences.

2
The hnRNP A1 Alternative Splicing Unit

To better understand the role and mechanism of action of hnRNP A1 in splice site selection we initially used a mouse erythroleukemic cell line severely deficient in hnRNP A1 (Ben-David et al. 1992). Although the results indicated that the A1 deficiency affected the alternative splicing of an adenovirus E1a pre-mRNA in vitro, restoring near normal levels of A1 only had a small effect in vivo (Yang et al. 1994). To find a more appropriate system that would allow us to answer questions about A1 function in splicing, we noted that feedback mechanisms are often used to control various aspects of gene expression. This is the case with the RNA binding protein Sex-lethal in *Drosophila* which has been implicated in the alternative splicing of its own pre-mRNA (Bell et al. 1988, 1991). Thus, we undertook to characterize the parameters controlling the alternative splicing of the hnRNP A1 pre-mRNA itself, which generates mRNAs encoding A1 and the A1B splice variant. We reasoned that if hnRNP A1 is indeed a modulator of splice site selection, we would likely uncover A1-dependent elements involved in controlling the fate of the hnRNP A1 alternative splicing unit. Characterizing the elements that control the alternative splicing of the hnRNP A1 pre-mRNA could therefore yield insights into the function of A1, as well as provide clues to its mechanism of action.

The mammalian hnRNP A1 gene contains 11 exons including one alternatively spliced exon (exon 7B) that can either be excluded or included to yield mRNAs encoding either the hnRNP A1 or the hnRNP A1B protein, respectively. The insertion of exon 7B into the mouse mRNA adds 53 amino acids to the glycine-rich region of the protein (52 amino acids in the case of the human A1B protein; Buvoli et al. 1990). In most cell types, the expression level of the hnRNP A1 protein is much greater than the level of the A1B isoform (Buvoli et al. 1990; Hanamura et al. 1998). To guide us in the identification of potential elements implicated in splicing control, we first compared the sequences of the mouse and human alternative splicing units. An unusually high level of conservation (between 60 and 70%) for the introns flanking the alternative exon 7B was observed, compared with approximately 25% for the intron separating constitutive exons 6 and 7 (Chabot et al. 1997). Based on this sequence comparison, highly conserved elements were grouped into ten segments (called CE1–CE10). Because these intron elements are evolutionarily conserved, we postulated that they might play a role in controlling the alternative splicing of exon 7B.

2.1 High-Affinity Binding Sites for hnRNP A/B Proteins Contribute to Exon Skipping

We first identified CE1 (150 nt) as a region in the intron upstream of exon 7B that could promote exon 7B exclusion in vivo (Fig. 1; Chabot et al. 1997). In vitro, placing this element between two competing 5′ splice sites led to a major increase in the use of the upstream 5′ splice site (Chabot et al. 1997). The insertion of CE1 upstream of the proximal 13S 5′ splice site of the adenovirus E1a gene also increased the utilization of the distal 9S 5′ splice site. Because the activity of CE1 was similar to the general effect obtained by increasing the concentration of hnRNP A1 (Mayeda and Krainer 1992), we suspected that A1 might be involved in the activity of CE1. Indeed, we have now dissected CE1 into two subregions that represent high-affinity binding sites for hnRNP A1 (Fig. 1). As demonstrated by pull down and/or gel-shift assays, the CE1a and CE1d elements are binding sites for hnRNP A1, and mutations that reduce A1 binding also abrogate the function of the CE1a and CE1d elements (Chabot et al. 1997; Hutchison et al. 2002). Immunoprecipitation assays indicate that A1 remains associated with a pre-mRNA carrying high-affinity binding sites throughout the whole splicing reaction (B. Chabot, unpubl. results). The sequestration of hnRNP A1 or its depletion from a nuclear extract eliminates the CE1a- and CE1d-dependent activity on 5′ splice site choice, while the addition of recombinant hnRNP A1 protein restores the activity associated with these elements (Blanchette and Chabot 1999; Hutchison et al. 2002).

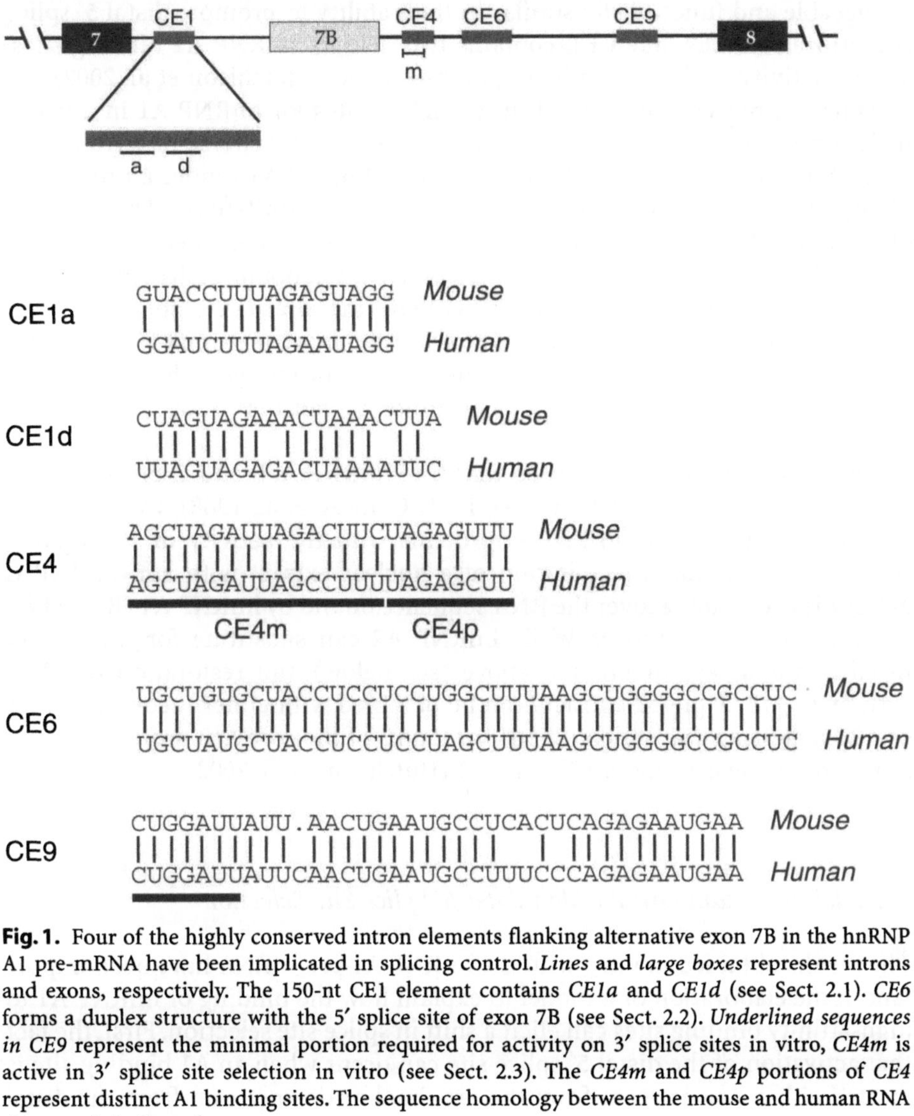

Fig. 1. Four of the highly conserved intron elements flanking alternative exon 7B in the hnRNP A1 pre-mRNA have been implicated in splicing control. *Lines* and *large boxes* represent introns and exons, respectively. The 150-nt CE1 element contains *CE1a* and *CE1d* (see Sect. 2.1). *CE6* forms a duplex structure with the 5′ splice site of exon 7B (see Sect. 2.2). *Underlined sequences in CE9* represent the minimal portion required for activity on 3′ splice sites in vitro, *CE4m* is active in 3′ splice site selection in vitro (see Sect. 2.3). The *CE4m* and *CE4p* portions of *CE4* represent distinct A1 binding sites. The sequence homology between the mouse and human RNA elements is indicated

Notably, the amount of recombinant hnRNP A1 protein required to completely shift splicing toward the distal 5′ splice site is much lower for CE1a- or CE1d-containing pre-mRNAs than for pre-mRNAs lacking these elements (Hutchison et al. 2002).

Interestingly, high-affinity hnRNP A1 binding sites are also found downstream of exon 7B (CE4; Fig. 1) and the deletion of CE4 from the A1 mini-gene, enhances the inclusion of exon 7B in vivo, as is the case with CE1 (Chabot et al. 1997; Blanchette and Chabot 1999). In vitro, CE4 and CE1 are interchangeable and functionally similar in their ability to promote distal 5′ splice site utilization. CE4, like CE1, contains two distinct hnRNP A1 binding sites, and its activity is also dependent upon hnRNP A1 (Hutchison et al. 2002).

Thus, the presence of high-affinity binding sites for hnRNP A1 in a mammalian pre-mRNA can shift the profile of 5′ splice site choice in vitro and in vivo. Moreover, high-affinity binding sites for hnRNP A1 render a pre-mRNA highly sensitive to variations in the concentration of hnRNP A1. The presence of high-affinity A1 binding sites may therefore determine whether a splicing unit will be acted upon by hnRNP A1 when the amount of free hnRNP A1 becomes limiting. Consistent with this view, the alternative splicing of the adenovirus E1a pre-mRNA is shifted toward the more proximal 5′ splice sites when a stress stimulus, such as osmotic shock, promotes a decrease in the nuclear concentration of hnRNP A/B proteins (van der Houven van Oordt et al. 2000).

Although it has been suggested that core hnRNP A, B and C proteins exist in a single complex (Lothstein et al. 1985; Conway et al. 1988), we have been unable to associate the hnRNP C proteins with the activity of A1 binding elements. An immunoprecipitation of a nuclear extract with anti-hnRNP C antibodies does not recover the RNA sequence bound by hnRNP A1 (B. Chabot, unpubl. observ.). Moreover, while hnRNP A2 can substitute for hnRNP A1 in all of the assays mentioned above (see below), full restoration of CE1d activity in A/B-depleted extracts can be obtained with either recombinant A1 or A2, suggesting that stoichiometric amounts of the core A/B proteins are not required for the activity of CE1 or CE4 (Hutchison et al. 2002).

2.1.1
A Model to Explain how A1 Modulates 5′ Splice Site Selection

The results obtained in vitro with the hnRNP A1 pre-mRNA have been important in the elaboration of a model to explain how the binding of hnRNP A1 to high-affinity binding sites can elicit a shift in splice site selection. First, the fact that activation of the distal 5′ splice site can occur when an A1 binding site is located either downstream from the proximal or downstream from the distal 5′ splice site makes it difficult to consider a model based on the direct repression of a 5′ splice site located upstream from the A1 binding site. If this were the case, placing the A1 binding element downstream from the distal 5′ splice site should decrease, not stimulate, this site. Second, while high-affinity A1 binding sites promote strong shifts in 5′ splice site utilization in vitro, this effect is not accompanied by substantial changes in the assembly of U1 snRNP-dependent complexes on the competing 5′ splice sites (Chabot et al. 1997). To explain the above observations, we have proposed the following model: the binding of A1 molecules to CE1 and CE4 would be followed by protein contacts between RNA-bound A1 proteins. This interaction would loop out the internal 5′ splice site (or the cassette exon 7B in the A1 pre-mRNA), to repress its use (Fig. 2A). The proposed change in pre-mRNA conformation, by bringing the more distant pair of splice sites (the 5′ splice site of exon 7 and the 3′ splice site of exon 8) in closer proximity, may also improve their rate of commitment. Because hnRNP A1 possibly binds with reduced affinity to other sequences in the pre-mRNA in our in vitro assays, the intermediate shift observed with substrates containing only one high-affinity A1 binding site could be attributed to less frequent A1 binding events occurring in the other half of the pre-mRNA. This situation would provide a partner A1 molecule to the one already bound to the high-affinity site. The following results and observations provide additional support for the A1-mediated looping out model.

1. A1 molecules can self-interact (Cartegni et al. 1996). We have confirmed these results and shown by yeast two-hybrid analysis that the glycine-rich domain (GRD) of A1 can interact with itself, albeit not as strongly as the complete A1 protein (C. LeBel, R.J. Wellinger and B. Chabot, unpubl. results). Moreover, we have shown that the glycine-rich domain is required for A1 molecules to interact simultaneously with two high-affinity RNA binding sites (Blanchette and Chabot 1999). Our model implies that the assembly of a stable complex between A1 molecules only occurs when A1 proteins are bound to RNA. Indeed, using a biochemical interaction assay, we have found that a short 16-nt RNA improves the interaction between A1 molecules. An important contribution of RNA "bridging" in these experiments was ruled out by performing interaction assays with a shortened version of A1 lacking the glycine-rich domain, which binds RNA as efficiently as A1, but which is not retained significantly on an A1 protein column in the presence of RNA (M. Blanchette, G. Lettre and B. Chabot, unpubl. observ.).
2. Inserting spacer elements to increase the distance between A1 binding sites and 5′ splice sites does not compromise distal 5′ splice site utilization (Nasim et al. 2002). If A1 binding sites were to interfere with the recognition of a nearby splice site or with the activity of a nearby enhancer of proximal 5′ splice site utilization, changing their relative positions in the pre-mRNA should have affected 5′ splice site selection.
3. Replacing high-affinity A1 binding sites with sequences that can form a stable RNA duplex promotes a strong shift toward the distal 5′ splice site (Nasim et al. 2002). Moreover, the insertion of spacer elements between duplex-forming sequences and 5′ splice sites does not compromise distal 5′ splice site activation. These results suggest that the mechanisms responsible for the switch in splice sites imposed by duplex-forming sequences and high-affinity A1 binding sites may be very similar.
4. Our model is also consistent with the crystal structure of UP1 bound to telomeric DNA. UP1 contains two RNA recognition motifs (RRMs) separated by a linker, but lacks the glycine-rich domain of hnRNP A1. In the co-crystal, UP1 exists as a dimer and each UP1 molecule in the dimer

A

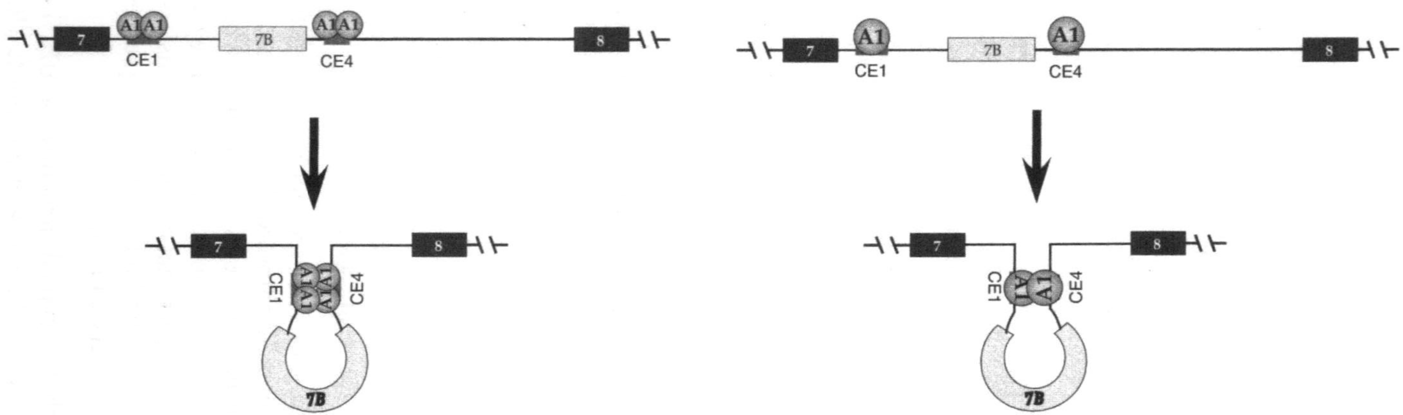

Fig. 2A–C. Models of hnRNP A1 action. **A** An interaction between hnRNP A1 molecules bound to *CE1* and *CE4* would loop out exon 7B (see Sect. 2.1.1). Although we have identified two high-affinity A1 binding sites each in CE1 and in CE4, we do not yet know whether each site is bound by a distinct A1 molecule (*left panel*) or whether a single A1 protein, through its two RRMs, can bind simultaneously to two adjacent sites (*right panel*). **B** Cross-strand interactions between RNA-bound hnRNP A1 molecules. Each of the four high-affinity A1 binding sites (*CE1a*, *CE1d*, *CE4m*, *CE4p*) is bound by the RRM1 portion of A1. The RRM2 of each A1 molecule establishes a cross-strand interaction to stabilize the formation of the complex. Although not represented here, a similar cross-strand exchange could also stabilize a complex initiated from an interaction between glycine-rich domains (GRD) involving only one A1 molecule bound to CE1 and one molecule bound to CE4. **C** HnRNP A1 and the splicing of long introns. Interactions between RNA-bound A1 molecules would loop out portions of the introns, thereby repressing splicing in these intron regions. Moreover, A1/A1 interactions within an intron would reduce the spatial distance separating a distant pair of splice sites, thereby stimulating the splicing efficiency of a long intron

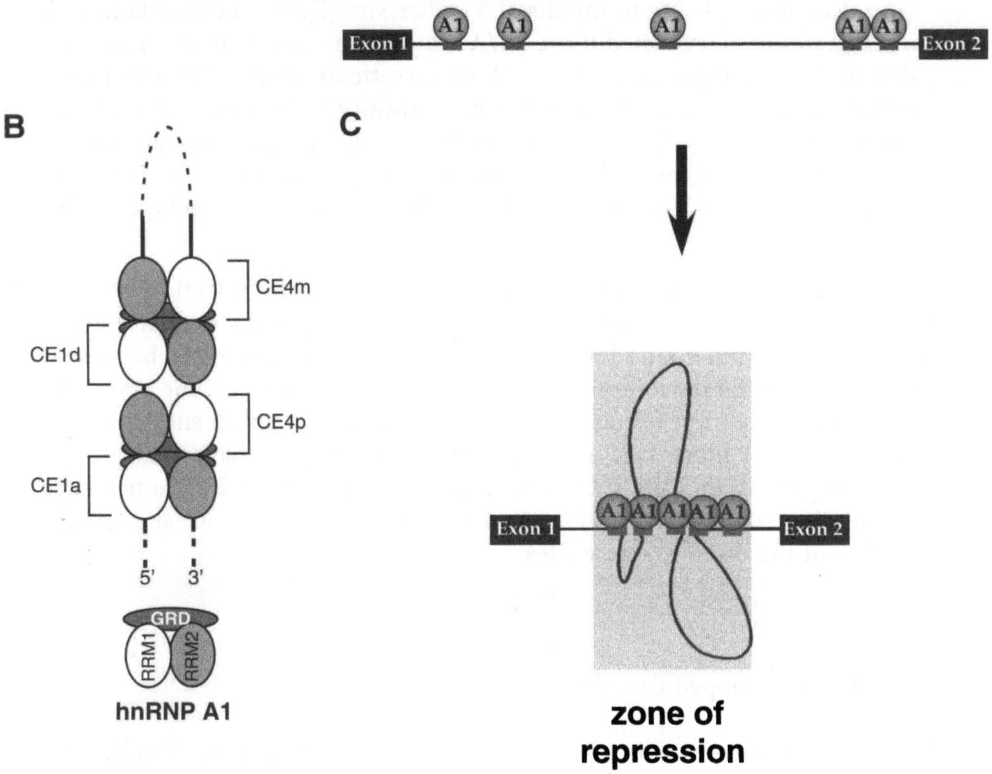

Fig. 2A–C. *Continued*

interacts with binding sites on two distinct oligonucleotides (Ding et al. 1999). We would, therefore, propose that A1 molecules bound to high-affinity sites on the pre-mRNA may interact initially through their glycine-rich domains, and that the complex may be further stabilized by "cross-strand" interactions involving both RRM domains. Because the RRM1 portion of UP1 is sufficient for specific binding (Dallaire et al. 2000), the RRM2 portion of one A1 molecule may establish the "cross-strand" interaction with sequences flanking the other RRM1-bound site (Fig. 2B). Consistent with the notion that the glycine-rich domain of A1 is not sufficient to elicit the formation of a stable complex in splicing conditions is the observation that an MS2-GRD protein cannot promote a shift in 5′ splice site choice when high-affinity A1 binding sites are replaced by MS2 binding sites (M. Blanchette and B. Chabot, unpubl. observ.).

5. The looping out model mediated by the A1/A1 interaction provides an explanation for results that have been obtained in other systems. Whereas inserting a high-affinity binding site in between two competing 5′ splice sites in an adenovirus pre-mRNA shifts splicing toward the upstream 5′ splice site, inserting an additional A1 binding site upstream of both 5′ splice

sites decreases splicing to the distal 5′ splice site (Eperon et al. 2000). Such a result would be expected if the A1/A1 interaction loops out the upstream donor site to compromise its use. Many cassette exons in other alternative splicing units are flanked by putative high-affinity A1 binding sites that may contribute to exon skipping. For example, although the theoretical distribution of a UAGRG$^{A}/_{U}$ motif is approximately one per 1000 nt, a 600-nt region containing alternative exon 17 in the rat skeletal troponin T gene contains four of these sites.

Loop formation could promote a switch in 5′ splice site selection or favor exon skipping for at least two reasons which are not mutually exclusive. First, the utilization of splice sites located in a loop may be compromised because it limits the structural flexibility required for and during spliceosome assembly. Second, loop formation would bring the external pair of splice sites in closer proximity, thereby increasing their frequency of pairing. We have made observations suggesting that the switch in splicing brought about by the presence of A1 binding sites involves both a repression of the looped out site and an activation of the distal 5′ splice sites.

2.1.1.1
Repression of a Looped Out 5′ Splice Site

The in vitro splicing of a simple pre-mRNA that contains a single 5′ splice site positioned between two high-affinity A1 binding sites is repressed (Nasim et al. 2002). The mechanism that imposes this repression is unclear. Despite a 50-fold change in 5′ splice site selection elicited by A1 binding sites, no significant difference in the assembly of U1 snRNP-dependent complexes to competing 5′ splice sites was noted in nuclear extracts (Chabot et al. 1997). Thus, although the looped out proximal 5′ splice site may be recognized efficiently at least initially, later steps of spliceosome assembly appear to be hindered.

A poorly used distal 5′ splice site can be activated simply by introducing a mutation that inactivates the proximal 5′ splice site (Nasim et al. 2002). This result indicates that repression of the proximal 5′ splice site is sufficient to stimulate the use of an alternative, distantly located 5′ splice site.

A similar situation likely occurs when an exon is placed in the loop of a hairpin in vitro and in vivo (Solnick 1985; Solnick and Lee 1987). Notably, placing a 5′ splice site in a loop formed by inverted repeats decreases the splicing efficiency of a simple one-intron pre-mRNA (Nasim et al. 2002). It is likely that a duplex structure, while probably not preventing the initial recognition of the 5′ splice site by U1 snRNP, restricts the structural flexibility that is required during spliceosome assembly. A stable interaction between A1-bound molecules may similarly hinder spliceosome formation for steric reasons.

2.1.1.2
Activation of the External Pair of Splice Sites

In vitro splicing is very inefficient when an intron greater than 1 kb is used. However, in the presence of high-affinity A1 binding sites located in the intron near the 5′ and the 3′ splice site, splicing products can now be detected (M. Cordeau, F.H. Nasim and B. Chabot, unpubl. results). Notably, the same level of stimulation is obtained when inverted repeats are inserted in the intron near the 5′ and 3′ splice sites such that most of the intron sequences are in the loop created by the RNA duplex. This result suggests that the A1-mediated change in pre-mRNA conformation stimulates pairing between distant splicing partners.

The fact that high-affinity A1 binding sites can stimulate the splicing of a pre-mRNA with a long intron raises the possibility that hnRNP A1 may play a general role in the splicing of long introns. We have examined previously the distribution of putative A1 binding elements in a subset of genes (Blanchette and Chabot 1999). Notably, we have found a higher prevalence of such elements in introns (1.2 sites per 1000 nt) compared to exons (0.35 sites per 1000 nt). Interestingly, in long introns carrying at least three putative A1 binding sites, these sites are more prevalent near 5′ and 3′ splice sites. An interaction between A1 molecules bound to intron sequences may therefore contribute to intron definition by increasing the frequency of commitment complex formation between a distant pair of splice sites (Fig. 2C). Given that the use of a looped out splice site is repressed, the gradual, A1-mediated, looping out of intron sequences as they exit from the RNA polymerase II transcription complex may also explain why intron sequences that resemble splicing signals are never used in a normal situation.

The proposed participation of A1-mediated loop formation in intron definition is reminiscent of the situation occurring in yeast where RNA duplex formation between sequences located downstream of a 5′ splice site and upstream of the branch site stimulates intron removal (Newman 1987; Libri et al. 1995; Charpentier and Rosbash 1996; Mougin et al. 1996; Howe and Ares 1997). In mammals, natural cases of duplex formation always involve nearby sequence elements (Clouet d'Orval et al. 1991; Estes et al. 1992; Blanchette and Chabot 1997; Hutton et al. 1998). Even on artificial pre-mRNAs carrying distant, but long inverted repeats, duplex formation does not occur efficiently in vivo (Solnick and Lee 1987; Eperon et al. 1988). Because mammals may have evolved to perceive duplex RNA as a sign of viral infection, proteins such as hnRNP A1 may have replaced RNA duplex formation in the promotion of intron definition.

2.1.1.3
Functional Redundancy Between Core hnRNP A/B Protein Members

The hnRNP A1 and A2 proteins are closely related in structure and the human proteins share 68% identity. For both A1 and A2, less abundant splice variants have been described (A1B and B1, respectively; Fig. 3). All the members of this family of core hnRNP proteins display a similar activity when assayed in vitro with model pre-mRNAs, although A1B is apparently less active in a 5′ splice site selection assay (Mayeda et al. 1994). Affinity chromatography assays performed with DNA or RNA molecules that contain high-affinity binding sites for hnRNP A1 also lead to the depletion of A1B, A2 and B1 proteins (Caputi et al. 1999; Hutchison et al. 2002). This co-depletion probably represents the ability of each protein to bind to these sites individually rather than reflecting the existence of a multimeric complex containing all the hnRNP A/B proteins. In support of this view, we and others have observed that a high-affinity binding site for hnRNP A1 is also bound by recombinant A1B and A2 proteins

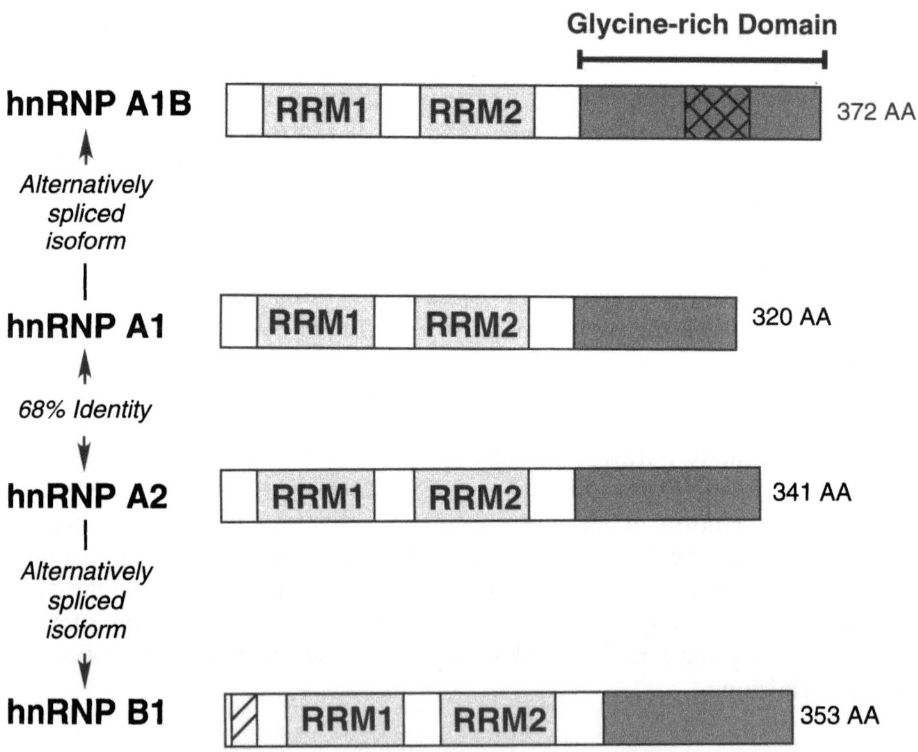

Fig. 3. The hnRNP A1 and A2 proteins and their splice variants. The position of the RNA recognition motifs (*RRM1* and *RRM2*) and the glycine-rich domain is indicated. The differences between A1 and A1B, between A2 and B1, correspond to *hatched boxes*. (Modified from Mayeda et al. 1994)

(Mayeda et al. 1994; Hutchison et al. 2002). Likewise, when the hnRNP A/B proteins are depleted from a nuclear extract by affinity chromatography, the activity of the A1 binding elements contained in CE1 and CE4 can be fully restored by the addition of recombinant hnRNP A1, A1B or A2 (Hutchison et al. 2002). Notably, in these assays, hnRNP A1B is as good an RNA binder as A1 and displays a similar 5′ splice site shifting ability. Whether the two A1 isoforms can differentially affect the alternative splicing of natural pre-mRNAs remains to be shown.

We have described previously a mouse erythroleukemic cell line (CB3) that is severely deficient in hnRNP A1 expression (Ben-David et al. 1992). In spite of a dramatic defect in A1, only modest effects were observed on 5′ splice site selection using the E1a pre-mRNA as a reporter in vivo (Yang et al. 1994). Moreover, nuclear extracts prepared from CB3 cells display no splicing defect when incubated with a model pre-mRNA carrying competing 5′ splice sites and high-affinity A1 binding sites (Hutchison et al. 2002). The biochemical depletion of hnRNP A2 and B1 proteins from a mouse CB3 nuclear extract by RNA affinity chromatography abolishes splicing to the distal 5′ splice site, and the addition of recombinant hnRNP A1, A2 or A1B protein can individually restore efficient and specific distal 5′ splice site utilization (Hutchison et al. 2002). Moreover, at least in vitro, the hnRNP A2 and B1 proteins are more active than A1 at promoting a shift in 5′ splice site usage (Mayeda et al. 1994). Thus, a redundancy of function between core hnRNP A1/A1B/A2/B1 proteins likely explains the modest impact of an A1 and A1B deficiency on 5′ splice site selection in CB3 cells.

2.1.1.4
The Looping Model and Other hnRNP Proteins

According to the looping out model of A1 action, an interaction between the glycine-rich domain (GRD) of two RNA-bound A1 proteins would initiate a change in pre-mRNA conformation that would promote a shift in splice site choice. Based on the results presented above, it is likely that each member of the group of hnRNP A/B proteins can establish similar self-interactions and, most probably, interactions between different members of that group. Because other RNA-binding proteins contain GRD domains (e.g., hnRNP H/H′/F/2H9 and hnRNP U proteins), homotypic GRD/GRD interactions could be occurring with these proteins as well. One potential case of such splicing control may occur with members of the hnRNP F/H group of proteins which recognizes GGGA-rich elements (Caputi and Zahler 2001). GGG-rich elements can modulate 5′ splice site selection and can stimulate in vitro splicing (McCullough and Berget 1997). Because these effects are reminiscent of the activity of hnRNP A1 binding sites, it is tempting to speculate that hnRNP F/H proteins bound to GGG-rich elements affect 5′ splice site selection and splicing efficiency through a mechanism that is similar to the hnRNP A/B proteins.

It is not currently known whether heterotypic interactions between the GRD of various groups of hnRNP proteins can occur (between hnRNP A1 and hnRNP F, for example). Notwithstanding, the interactions between RNA-bound GRD-containing proteins in mammalian pre-mRNAs may lead to complex changes in the structural remodeling of mammalian pre-mRNAs, with considerable impact on splice site selection.

In principle, the looping out model is applicable to any pair of proteins that can interact when bound to RNA. As mentioned earlier, binding sites for the hnRNP I/PTB have been described in many alternative splicing units, and the binding of hnRNP I/PTB in the introns flanking an alternative exon has been associated with more efficient exon skipping (reviewed in Valcárcel and Gebauer 1997; Wagner and Garcia-Blanco 2001). Because hnRNP I/PTB can self-interact, a similar model postulating an interaction between bound hnRNP I/PTB proteins has also been proposed (Chou et al. 2000). Although hnRNP I/PTB contains several RRM domains, only one has been implicated in the binding to CU-rich sequence; the other RRM domains may be involved in protein-protein interactions (Perez et al. 1997). Conceivably, RRM domains may stabilize "cross-strand" interactions, in a manner similar to what we have proposed for hnRNP A1. Consistent with this view, mutations in the upstream hnRNP I/PTB binding site affect binding of hnRNP I/PTB to the downstream site, and vice versa (Chou et al. 2000). The positioning of hnRNP I/PTB on introns on each side of the exon may place this exon in a zone of repression produced by the multimerization of hnRNP I/PTB or by placing the splice sites in a loop formed by the PTB/PTB interaction (Chou et al. 2000; Wagner and Garcia-Blanco 2001).

2.2
RNA Duplex Formation and 5′ Splice Site Selection

Possibly the strongest element involved in promoting the exclusion of exon 7B in the hnRNP A1 pre-mRNA is CE6, a 42-nt-long sequence located in the intron downstream of alternative exon 7B (Blanchette and Chabot 1997). This element inhibits the utilization of the 5′ splice site of exon 7B in vitro and its removal shifts splicing almost completely toward exon 7B inclusion in vivo. CE6 affects the selection of 5′ splice sites by forming a secondary structure that blocks recognition of the 5′ splice site of exon 7B by U1 snRNP. Although sequestering a 5′ splice site into a duplex structure was known to repress splicing (Eperon et al. 1986; Solnick and Lee 1987; Goguel et al. 1993), CE6 is the first example of a naturally occurring element capable of modulating 5′ splice site selection in vivo through RNA duplex formation. A few cases of RNA duplex formation modulating splice site decisions have now been documented in mammalian pre-mRNAs (Chebli et al. 1989; Clouet d'Orval et al. 1991; Domenjoud et al. 1991; Jiang et al. 2000). In a recent example, a mutation in tau exon 10 has been shown to disrupt the complementarity with a downstream intron region that normally prevents efficient exon inclusion (Hutton et al. 1998). Because RNA duplex formation involves *cis*-acting elements, it is likely to occur independently of the nuclear environment in which the pre-mRNA is expressed. Consequently, this mode of splicing control is unlikely to be widely used when a splice site needs to be regulated. Nevertheless, it is possible that the expression of specific RNA helicases could prevent its occurrence in some tissues. Likewise, the existence of transcriptional pause sites located in between inverted repeats could influence splicing decisions if transcription complexes with different pausing attributes are used to express the same gene in different cell types (Eperon et al. 1988; Kadener et al. 2001).

In the case of the hnRNP A1 pre-mRNA, a likely explanation for the existence of this RNA duplex is that it considerably simplifies the splicing control of alternative exon 7B. Instead of the need to coordinate events occurring at both ends of exon 7B, blocking the 5′ splice site allows control of exon 7B inclusion simply by modulating 3′ splice site usage. If the splicing decision favors exon 7/exon 7B splicing, then the exon 7B/exon 8 splice would occur later by default. As post-genomic efforts are now geared toward characterizing the diverse strategies implicated in the control of thousands of alternative splicing units, we would predict that many more examples of this type will be documented.

2.3
Elements Controlling 3′ Splice Site Selection

If the consequence of sequestering the 5′ splice site of exon 7B into a duplex structure is to relegate the control of exon 7B inclusion to the selection between the 3′ splice site of exon 7B and the 3′ splice site of exon 8, we would predict the existence of elements implicated in the control of 3′ splice site selection. We have uncovered two such elements (CE4m and CE9) in the hnRNP A1 alternative splicing unit. CE4m is located in the intron downstream of exon 7B and was identified as a repressor element based on its ability to repress the upstream 3′ splice site of exon 7B in vitro (Blanchette and Chabot 1999; Fig. 4). Because the addition of an excess of RNA carrying the CE4m sequence can stimulate splicing activity to the 3′ splice site of exon 7B in vitro, a *trans*-acting factor(s) is likely mediating the repression activity of CE4m. Although the CE4m sequence is part of the CE4 element, and thus can be bound by hnRNP A/B core proteins (M. Blanchette and B. Chabot, unpubl. results), it is unclear at the moment whether hnRNP proteins participate in the repression mediated by CE4m.

The other element involved in 3′ splice site control is contained within a 38-nt-long conserved sequence (CE9) located in the intron separating alternative exon 7B from exon 8 (Simard and Chabot 2000; Fig. 4). In vitro, the insertion of one copy of CE9 in between two 3′ splice sites is sufficient to stimulate splicing to the internal 3′ splice site of exon 7B. Although a single copy of CE9 has

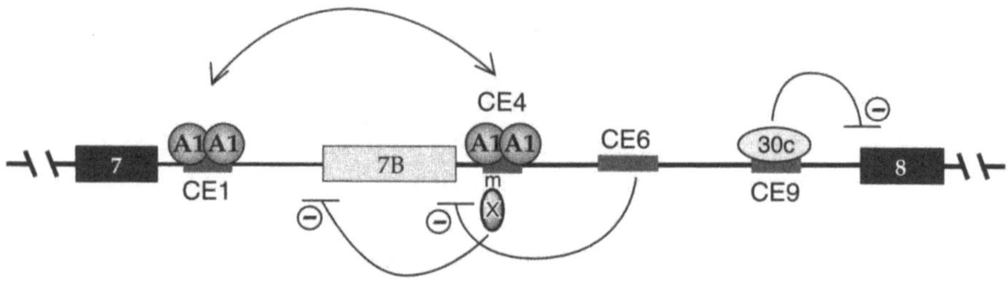

Fig. 4. The control of alternative splicing of exon 7B in the hnRNP A1 pre-mRNA. An interaction between RNA-bound hnRNP A1 molecules promotes exon 7B inclusion skipping. *CE6* forms a secondary structure with the 5′ splice site of exon 7B to block its utilization. *CE4m* represses the use of the 3′ splice site of exon 7B. Binding of SRp30c to CE9 represses the use of the 3′ splice site of exon 8

little activity when inserted in the intron of a pre-mRNA that contains only one set of splice sites, the insertion of two and three copies of CE9 elicits a sharp drop in splicing efficiency, suggesting that CE9 acts by silencing a downstream 3′ splice site. The activity of CE9 is weak in HeLa cells since deleting CE9 from an hnRNP A1 mini-gene reporter has no effect. However, consistent with its ability to repress splicing in vitro, the insertion of multiple copies of CE9 can prevent the accumulation of fully spliced mRNAs from the A1 mini-gene (Simard and Chabot 2000).

CE9 is active when inserted in heterologous pre-mRNAs, and a molar excess of a competitor RNA containing the CE9 sequence can abrogate the activity of CE9 on splicing in vitro. Thus, a *trans*-acting factor(s) is likely implicated in the activity of CE9. Using ultraviolet cross-linking and RNA affinity chromatography, we have identified a member of the SR family of proteins, namely SRp30c, as a factor that interacts with CE9 (Simard and Chabot 2002). SRp30c is most probably involved in the activity of CE9 based on the following findings. First, recombinant SRp30c specifically interacts with CE9 in gel mobility shift assays. Second, mutations in CE9 that abolish the binding of SRp30c prevent splicing inhibition. Third, the addition of recombinant SRp30c to a HeLa nuclear extract specifically represses the splicing of pre-mRNAs containing two copies of CE9 (Simard and Chabot 2002).

SR proteins have previously been implicated in the activity of another intron silencer element (3RE) which is found in the adenovirus IIIa pre-mRNA (Kanopka et al. 1996; Petersen-Mahrt et al. 1999). In this case, SRp30c and a related SR protein member, ASF/SF2, can both bind to the 3RE element and repress splicing. The situation, however, appears different for CE9 since ASF/SF2 does not strongly bind to CE9. Moreover, replacing CE9 with high-affinity binding sites for ASF/SF2 does not elicit splicing repression (Simard and Chabot 2002). CE9 is naturally located 119 nt upstream from the 3′ splice site of exon 8 (100 nt from the putative branch site), and increasing its distance from the 3′ splice site to 300 nt does not compromise the activity of CE9. It is therefore unlikely that the SRp30c/CE9 interaction prevents the recognition of the downstream 3′ splice site by sterically interfering with the binding of U2AF or U2 snRNP. Indeed, and in contrast to the 3RE element in the adenovirus IIIa pre-mRNA, CE9 does not interfere with the assembly of U2 and/or U4 snRNPs-dependent complexes (Simard and Chabot 2000).

The binding of SR proteins in an intron has also been implicated in the activity of a silencer element in the Rous sarcoma virus (RSV). As seen for CE9, the RSV intron silencer does not affect U2 snRNP-dependent complex formation (Gontarek et al. 1993). It was proposed that the binding of SR proteins to the RSV element enhances U1 snRNP binding to a nearby region of the silencer, leading to the formation of a nonfunctional spliceosome (McNally and McNally 1998; Hibbert et al. 1999; Paca et al. 2001). Whether a similar mechanism is occurring for CE9 remains to be demonstrated.

Compared to the other elements uncovered in the hnRNP A1 alternative splicing unit, CE9 appears to be a relatively weak element. Its activity is best observed in the presence of competing 3′ splice sites, or when several copies of CE9 are inserted in a pre-mRNA carrying a single pair of splice sites. Because CE9 represses the use of a downstream 3′ splice site, its natural function might be to compromise the use of the 3′ splice site of exon 8, possibly to favor exon 7/exon 7B splicing and thus, exon 7B inclusion. The fact that removing CE9 from the A1 mini-gene has no effect in HeLa cells may indicate that the activity of CE9 is offset by the activity of other intron elements that favor exon 7B skipping (i.e., CE1, CE4, CE4m and CE6). Because the concentration of SRp30c in HeLa cells may be low and the abundance of the SRp30c mRNA varies in tissues (Screaton et al. 1995), the contribution of CE9 to the control of hnRNP A1 alternative splicing may become more important in cell types expressing higher levels of SRp30c.

The identification of an element like CE9 highlights the fact that controlling the inclusion frequency of a cassette exon does not need to target solely the splice sites that flank the cassette exon itself. Modulating the use of splice sites that are used, either during exon inclusion or during exon skipping (i.e., the 5′ splice site of exon 7 or the 3′ splice site of exon 8), could be an effective strategy given that splicing decisions are the result of a competition between an alternative and a common splice site. Although most splicing control elements identified so far have been mapped in alternative exons (e.g., enhancer and silencer elements) or in flanking introns, it is likely that more examples of elements controlling the use of common splice sites will be reported in the years to come. Examples of this kind of splicing control have been suggested to occur in the neural cell adhesion molecule (NCAM) and the fibronectin pre-mRNAs (Lavigueur et al. 1993; Côté et al. 1999).

3 Conclusions

3.1 A Role for the hnRNP A1 Protein in Alternative and Constitutive Splicing

Studying the elements implicated in the control of exon 7B inclusion in the hnRNP A1 pre-mRNA has allowed us to uncover a function for the hnRNP A1 protein in the splicing of its own pre-mRNA. The proposed mechanism by which the hnRNP A1 protein controls 5′ splice site selection involves a change in the conformation of the pre-mRNA that mimics the structural change mediated by inverted repeats that loop out splice sites. The remodeling of pre-mRNA structure would compromise the use of the splice site(s) in the loop while simultaneously bringing in closer proximity distant splicing partners. Importantly, the strength of the splice sites will likely play a key role in allowing these interactions to take place. Thus, even if high-affinity A1 binding sites exist on each side of an exon, the presence of strong pairs of splice sites in the adjacent introns may kinetically improve commitment complex formation, and offset the A1-mediated exon looping. Indeed, the effects of A1 binding sites on 5′ splice site selection in vitro can be antagonized by improving the strength of the internal 5′ splice site (Nasim et al. 2002). Likewise, A1-mediated exon skipping in vitro can be prevented by a strong polypyrimidine tract at the 3′ splice site of the internal exon (Mayeda et al. 1993). Consequently, the enforcement of splice site usage by exonic enhancers may antagonize the effect of A1/A1 interactions that promote exon skipping (Fig. 5). Consistent with this

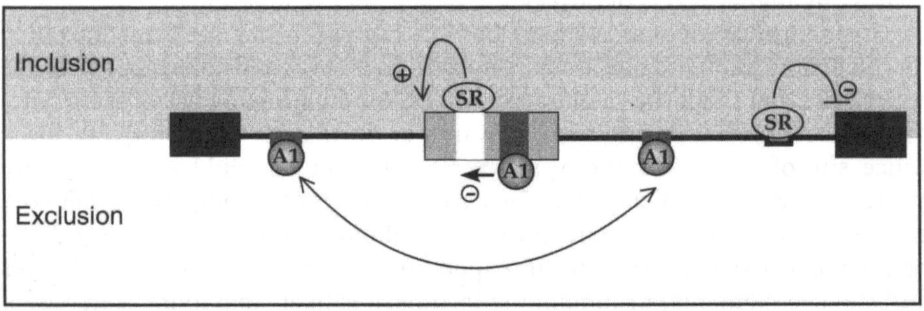

Fig. 5. Antagonism between the activity of hnRNP A1 and SR proteins. HnRNP A1 is depicted here as a repressor of exon inclusion either because it loops out the exon through an A1/A1 interaction, or because it interferes with the binding of SR proteins to a nearby exon enhancer. Depending on the identity and abundance of a given SR protein in the nucleus, exon inclusion could be stimulated by the binding of a SR protein to an exon enhancer to stimulate directly 3′ splice site use, or to an intron silencer element to reduce the use of a common exon, thereby kinetically favoring splicing to the 3′ splice site of the internal exon. These processes may occur independently of one another or in combination to enhance or antagonize the overall frequency of exon inclusion

view, only internal exons with strong splice sites or long internal exons (i.e., possibly more likely to contain enhancer elements) can offer resistance to the exon skipping activity of hnRNP A1 in vitro (Mayeda et al. 1993). The kinetics of A1 binding and A1/A1 dimer formation should also have an important impact on splicing decisions. Given recent findings indicating that the processivity of RNA polymerase II can alter the efficiency of exon inclusion (Roberts et al. 1998; Kadener et al. 2001), the more time taken before the second A1 binding site emerges from the transcription complex, the more likely a splicing complex should already have formed to commit the upstream intron as a prelude to exon inclusion.

Although the A1-mediated remodeling of pre-mRNA conformation has clear implications in alternative splicing, it is also relevant to the splicing efficiency of long introns because it brings in closer proximity distant splicing partners. Notably, intron A1 binding sites can stimulate the splicing efficiency of a pre-mRNA carrying a long intron in vitro (F.H. Nasim, M. Cordeau and B. Chabot, unpubl. observ.). Although this preliminary observation requires more exhaustive investigations, the abundance of putative high-affinity A1 binding sites and their greater prevalence near splice junctions is consistent with this view. Thus, although in vitro experiments have not yet revealed a role for hnRNP A1 in constitutive splicing, hnRNP A1 and other members of the A/B family of core hnRNP proteins may carry out an important function in the generic splicing of long constitutive introns. Splice site selection is increasingly recognized as the result of a competition between intrinsic differences in the strength of splice sites whose recognition by the splicing machinery is influenced by the antagonistic action of silencer and enhancer elements. Based on the findings described above, the ability of elements to change the conformation of the pre-mRNA and thereby affect the commitment between a pair of splice sites should also be acknowledged as being intimately associated with splicing decisions and splicing efficiency.

In addition to the implication of high-affinity A1 binding sites in 5′ splice site selection, there have been several reports showing that these sites are also involved in the negative control of 3′ splice site selection. Examples of this kind have been found in the alternative exon of fibroblast growth factor receptor 2 gene (Del Gatto-Konczak et al. 1999), in the tat and vpr exons of the human immunodeficiency virus (Caputi et al. 1999; Bilodeau et al. 2001; Tange et al. 2001), and in the V6 exon of the human CD44 gene (Konig et al. 1998; Matter et al. 2000). A role for hnRNP A1 in the activity of these elements has been demonstrated and the hnRNP A2 and B1 proteins can functionally replace A1 in these assays (Caputi et al. 1999; Bilodeau et al. 2001). In the HIV-1 tat exon 3, a recent study has shown that a high-affinity A1 binding site can antagonize the interaction of the SR protein SC35 with a nearby enhancer element (Zhu et al. 2001). In this case, it has been proposed that multimerization of A1 molecules on that exon is responsible for the displacement of SC35. It should be noted, however, that another hnRNP A1 binding site located in the upstream exon stimulates the repressing activity of the A1 site in the downstream exon.

Thus, we can envision that an interaction between A1 molecules bound in each exon could also contribute to the repression by looping out the splice sites located in the intervening portion of the pre-mRNA. In another study, the binding of hnRNP A1 in the intron near the 3′ splice junction of the tat exon can compromise U2 snRNP binding to the branch site (Tange et al. 2001). Here again, additional A1 binding sites identified in the exon downstream could loop out the 3′ splice site to compromise its use. Notwithstanding, the mechanisms by which hnRNP A1 modulates 5′ splice site selection in the hnRNP A1 pre-mRNA and compromises 3′ splice site use in other systems may be different. In favor of this interpretation, it has been noted that silencing can be promoted in vivo by the targeted binding of either a hybrid MS2-A1 protein or an MS2 protein fused to the C-terminal GRD domain of A1 (Del Gatto-Konczak et al. 1999). In contrast, the MS2-GRD protein cannot promote distal 5′ splice site selection on a model pre-mRNA carrying MS2 binding sites instead of A1 binding sites (Blanchette and Chabot, unpubl. observ.).

Thus, hnRNP A1 may employ different strategies to control splicing decisions depending on the position of high-affinity binding sites and the proximity of other *cis*-acting elements and *trans*-acting factors. Such mechanistic versatility has also been observed for the RNA binding protein ASF/SF2 which can mediate exon enhancer activity (reviewed in Tacke and Manley 1999), but can also function as a repressor in different pre-mRNAs (Kanopka et al. 1996; Jumaa and Nielsen 1997; Petersen-Mahrt et al. 1999). HnRNP A1 could therefore competitively interfere with the binding of a nearby factor, while in other situations, interactions between distantly bound hnRNP A1 molecules could change the conformation of pre-mRNAs to help carry out switches in splice site selection.

Thus, most of the functions that have been attributed to hnRNP A/B proteins are based on their ability to interact with specific sequences. However, the ability of hnRNP A1 to bind with varying affinity to many RNA molecules may indicate that A1 can associate with a broad range of sequences. Given that the UAGRG$^{U}/_{A}$ element represents a high-affinity binding site (Burd and Dreyfuss 1994; Chabot et al. 1997), it is expected that this sequence will occur frequently in introns (Blanchette and Chabot 1999). Although it is assumed that the concentration of hnRNP A/B proteins is sufficiently high to bind many, if not all, pre-mRNAs (LeStourgeon et al. 1990; Dreyfuss et al. 1993), it is more difficult to estimate how many free hnRNP A/B proteins exist in an actively transcribing nucleus. Thus, despite their high abundance, the steady-state concentration of free core hnRNP A/B proteins at the time a nascent sequence exits the RNA polymerase II transcription complex may not permit general coating and may be restricted to specific, yet abundantly distributed, sequence elements. While the binding of A1 may not always be associated with an effect on splice site choice or splicing efficiency (for example when A1 binding occurs in a splicing unit that has a small intron and/or in an intron bordered by strong splice sites), the binding of A1 in long introns flanked by strong splice sites may nevertheless stimulate intron definition. In splicing units flanked by weak splice sites, the binding of A1 in exon sequences near enhancers may contribute to exon skipping, while A1 binding sites in introns may elicit changes in the conformation of pre-mRNA that will also encourage exon skipping. Thus, the great abundance and functional redundancy of core hnRNP A/B proteins in actively growing cells may reflect their implication in the alternative splicing of many pre-mRNAs and their contribution to the removal of large introns.

3.2
Several Elements Control the Alternative Splicing of Exon 7B

Using phylogenetic comparison, we have identified several conserved intron elements, some of which were shown to have a role in modulating the alternative splicing of the hnRNP A1 pre-mRNA. As more sequences of mammalian genomes become available in the coming months and years, a worthy endeavor would be to produce comparative maps of introns that flank alternative exons. Systematic data mining of this kind will most probably lead to the identification of a limited group of sequence elements whose location and abundance contribute toward the establishment of tissue-specific or developmental-specific alternative splicing profiles.

So far, we have identified four control elements implicated in the exclusion of exon 7B from the A1 pre-mRNA. Two elements (CE1 and CE4) reside on each side of alternative exon 7B, are bound by hnRNP A1, allow the selection of exon 7 5′ splice site in vitro and favor exon 7B exclusion in vivo (Chabot et al. 1997; Blanchette and Chabot 1999; Hutchison et al. 2002). CE6 forms an RNA duplex with the 5′ splice site of exon 7B to prevent the utilization of the 5′ splice of exon 7B in vitro, and thus favors exon 7B exclusion (Blanchette and Chabot 1997). The formation of a secondary structure between CE6 and the 5′ splice site of exon 7B converts this alternative splicing unit into a simpler system where the 5′ splice site of exon 7 must select between the 3′ splice site of exon 7B and the 3′ splice site of exon 8. Because CE6-mediated duplex formation is likely to occur in a generic manner, 3′ splice site selection therefore becomes the critical decision during hnRNP A1 pre-mRNA alternative splicing. We have identified two elements that control 3′ splice site choice. First, CE4m abrogates 3′ splice site utilization of exon 7B and negatively affects the inclusion of exon 7B in vivo. Second, CE9 can repress the use of a downstream 3′ splice site, and may help shift the balance of selection toward inclusion of exon 7B in situations where the CE9-binding protein SRp30c is abundant.

Our study of the alternative splicing event occurring in the pre-mRNA of hnRNP A1 indicates that the control of a simple cassette exon can be complex. The variety of elements implicated in the alternative splicing of exon 7B suggests that the relative abundance of A1 and A1B must be tightly controlled, and possibly subjected to regulation in a tissue-specific manner. Consistent with this view, the A1/A1B mRNA and protein ratios vary in human and rat tissues (Buvoli et al. 1990; Hanamura et al. 1998). For example, the relative abundance of the A1B isoform is greater in the ovary than in the spleen (Hanamura et al. 1998). Assuming that these differences always reflect decisions in alternative splicing, controlling the production of A1B may be important because A1 and A1B may have different activities. Although we have not detected significant differences in the ability of A1 and A1B to elicit distal 5′ splice site selection on a pre-mRNA substrate carrying high-affinity A1 binding sites (C. LeBel and B. Chabot, unpubl. results), Mayeda and collaborators (1994) have reported that recombinant A1B protein binds RNA more strongly and is less active at promoting distal 5′ splice site selection on a model human β-globin pre-mRNA in vitro. In vivo approaches that will manipulate the ratio of the A1 isoforms in favor of A1B should help identify specific pre-mRNAs that are preferential targets for A1B.

Acknowledgement. M.J.S. was the recipient of a studentship from the FCAR/FRSQ. C.L. holds a studentship from the National Science and Engineering Council of Canada. B.C. is a Canada Research Chair in Functional Genomics, and is a member of the Sherbrooke RNA/RNP group supported by the Université de Sherbrooke, the FCAR and the Canadian Institutes of Health Research (CIHR). The work performed in the laboratory of B.C. was supported by a grant from the CIHR.

References

Abdul-Manan N, Williams KR (1996) hnRNP A1 binds promiscuously to oligoribonucleotides: utilization of random and homo-oligonucleotides to discriminate sequence from base-specific binding. Nucleic Acids Res 24:4063–4070

Abdul-Manan N, O'Malley SM, Williams KR (1996) Origins of binding specificity of the A1 heterogeneous nuclear ribonucleoprotein. Biochemistry 35:3545–3554

Arning S, Gruter P, Bilbe G, Krämer A (1996) Mammalian splicing factor SF1 is encoded by variant cDNAs and binds to RNA. RNA 2:794–810

Ashiya M, Grabowski PJ (1997) A neuron-specific splicing switch mediated by an array of pre-mRNA repressor sites: evidence of a regulatory role for the polypyrimidine tract binding protein and a brain-specific PTB counterpart. RNA 3:996–1015

Bai Y, Lee D, Yu T, Chasin LA (1999) Control of 3′ splice site choice in vivo by ASF/SF2 and hnRNP A1. Nucleic Acids Res 27:1126–1134

Bell LR, Maine EM, Schedl P, Cline TW (1988) Sex-lethal, a *Drosophila* sex determination switch gene, exhibits sex-specific RNA splicing and sequence similarity to RNA binding proteins. Cell 55:1037–1046

Bell LR, Horabin JI, Schedl P, Cline TW (1991) Positive autoregulation of sex-lethal by alternative splicing maintains the female determined state in *Drosophila*. Cell 65:229–239

Ben-David Y, Bani MR, Chabot B, De Koven A, Bernstein A (1992) Retroviral insertions downstream of the heterogeneous nuclear ribonucleoprotein A1 gene in erythroleukemia cells: evidence that A1 is not essential for cell growth. Mol Cell Biol 12:4449–4455

Bennett M, Pinol-Roma S, Staknis D, Dreyfuss G, Reed R (1992) Differential binding of heterogeneous nuclear ribonucleoproteins to mRNA precursors prior to spliceosome assembly in vitro. Mol Cell Biol 12:3165–3175

Berglund JA, Chua K, Abovich N, Reed R, Rosbash M (1997) The splicing factor BBP interacts specifically with the pre-mRNA branchpoint sequence UACUAAC. Cell 89:781–787

Beyer AL, Christensen ME, Walker BW, LeStourgeon WM (1977) Identification and characterization of the packaging proteins of core 40S hnRNP particles. Cell 11:127–138
Bilodeau PS, Domsic JK, Mayeda A, Krainer AR, Stoltzfus CM (2001) RNA splicing at human immunodeficiency virus type 1 3′ splice site A2 is regulated by binding of hnRNP A/B proteins to an exonic splicing silencer element. J Virol 75:8487–8497
Blanchette M, Chabot B (1997) A highly stable duplex structure sequesters the 5′ splice site region of hnRNP A1 alternative exon 7B. RNA 3:405–419
Blanchette M, Chabot B (1999) Modulation of exon skipping by high-affinity hnRNP A1-binding sites and by intron elements that repress splice site utilization. EMBO J 18:1939–1952
Burd CG, Dreyfuss G (1994) RNA binding specificity of hnRNP A1: significance of hnRNP A1 high-affinity binding sites in pre-mRNA splicing. EMBO J 13:1197–1204
Buvoli M, Cobianchi F, Bestagno MG, Mangiarotti A, Bassi MT, Biamonti G, Riva S (1990) Alternative splicing in the human gene for the core protein A1 generates another hnRNP protein. EMBO J 9:1229–1235
Buvoli M, Cobianchi F, Riva S (1992) Interaction of hnRNP A1 with snRNPs and pre-mRNAs: evidence for a possible role of A1 RNA annealing activity in the first steps of spliceosome assembly. Nucleic Acids Res 20:5017–5025
Cáceres JF, Stamm S, Helfman DM, Krainer AR (1994) Regulation of alternative splicing in vivo by overexpression of antagonistic splicing factors. Science 265:1706–1709
Caputi M, Zahler AM (2001) Determination of the RNA binding specificity of the heterogeneous nuclear ribonucleoprotein (hnRNP) H/H′/F/2H9 family. J Biol Chem 276:43850–43859
Caputi M, Zahler AM (2002) SR proteins and hnRNP H regulate the splicing of the HIV-1 tev-specific exon 6D. EMBO J 21:845–855
Caputi M, Mayeda A, Krainer AR, Zahler AM (1999) hnRNP A/B proteins are required for inhibition of HIV-1 pre-mRNA splicing. EMBO J 18:4060–4067
Carstens RP, Wagner EJ, Garcia-Blanco MA (2000) An intronic splicing silencer causes skipping of the IIIb exon of fibroblast growth factor receptor 2 through involvement of polypyrimidine tract binding protein. Mol Cell Biol 20:7388–7400
Cartegni L, Maconi M, Morandi E, Cobianchi F, Riva S, Biamonti G (1996) hnRNP A1 selectively interacts through its Gly-rich domain with different RNA-binding proteins. J Mol Biol 259:337–348
Casas-Finet JR, Smith JD Jr, Kumar A, Kim JG, Wilson SH, Karpel RL (1993) Mammalian heterogeneous ribonucleoprotein A1 and its constituent domains. Nucleic acid interaction, structural stability and self-association. J Mol Biol 229:873–889
Chabot B, Blanchette M, Lapierre I, La Branche H (1997) An intron element modulating 5′ splice site selection in the hnRNP A1 pre-mRNA interacts with hnRNP A1. Mol Cell Biol 17: 1776–1786
Chan RC, Black DL (1997) The polypyrimidine tract binding protein binds upstream of neural cell-specific c-src exon N1 to repress the splicing of the intron downstream. Mol Cell Biol 17:4667–4676
Charpentier B, Rosbash M (1996) Intramolecular structure in yeast introns aids the early steps of in vitro spliceosome assembly. RNA 2:509–522
Chebli K, Gattoni R, Schmitt P, Hildwein G, Stevenin J (1989) The 216-nucleotide intron of the E1 A pre-mRNA contains a hairpin structure that permits utilization of unusually distant branch acceptors. Mol Cell Biol 9:4852–4861
Chen CD, Kobayashi R, Helfman DM (1999) Binding of hnRNP H to an exonic splicing silencer is involved in the regulation of alternative splicing of the rat beta-tropomyosin gene. Genes Dev 13:593–606
Choi YD, Grabowski PJ, Sharp PA, Dreyfuss G (1986) Heterogeneous nuclear ribonucleoproteins: role in RNA splicing. Science 231:1534–1539
Chou MY, Rooke N, Turck CW, Black DL (1999) hnRNP H is a component of a splicing enhancer complex that activates a c-src alternative exon in neuronal cells. Mol Cell Biol 19:69–77

Chou MY, Underwood JG, Nikolic J, Luu MH, Black DL (2000) Multisite RNA binding and release of polypyrimidine tract binding protein during the regulation of c-src neural-specific splicing. Mol Cell 5:949–957

Clouet d'Orval B, d'Aubenton Carafa Y, Sirand-Pugnet P, Gallego M, Brody E, Marie J (1991) RNA secondary structure repression of a muscle-specific exon in HeLa cell nuclear extracts. Science 252:1823–1828

Cobianchi F, Karpel RL, Williams KR, Notario V, Wilson SH (1988) Mammalian heterogeneous nuclear ribonucleoprotein complex protein A1. Large-scale overproduction in *Escherichia coli* and cooperative binding to single-stranded nucleic acids. J Biol Chem 263:1063–1071

Conway G, Wooley J, Bibring T, LeStourgeon WM (1988) Ribonucleoproteins package 700 nucleotides of pre-mRNA into a repeating array of regular particles. Mol Cell Biol 8:2884–2895

Côté J, Simard MJ, Chabot B (1999) An element in the 5′ common exon of the NCAM alternative splicing unit interacts with SR proteins and modulates 5′ splice site selection. Nucleic Acids Res 27:2529–2537

Dallaire F, Dupuis S, Fiset S, Chabot B (2000) Heterogeneous nuclear ribonucleoprotein A1 and UP1 protect mammalian telomeric repeats and modulate telomere replication in vitro. J Biol Chem 275:14509–14516

Del Gatto-Konczak F, Olive M, Gesnel MC, Breathnach R (1999) hnRNP A1 recruited to an exon in vivo can function as an exon splicing silencer. Mol Cell Biol 19:251–260

Ding J, Hayashi MK, Zhang Y, Manche L, Krainer AR, Xu RM (1999) Crystal structure of the two-RRM domain of hnRNP A1 (UP1) complexed with single-stranded telomeric DNA. Genes Dev 13:1102–1115

Domenjoud L, Gallinaro H, Kister L, Meyer S, Jacob M (1991) Identification of specific exon sequence that is a major determinant in the selection between a natural and a cryptic 5′ splice site. Mol Cell Biol 11:4581–4590

Dreyfuss G, Matunis MJ, Piñol-Roma S, Burd CG (1993) hnRNP proteins and the biogenesis of mRNA. Annu Rev Biochem 62:289–321

Eperon IC, Makarova OV, Mayeda A, Munroe SH, Cáceres JF, Hayward DG, Krainer AR (2000) Selection of alternative 5′ splice sites: role of U1 snRNP and models for the antagonistic effects of SF2/ASF and hnRNP A1. Mol Cell Biol 20:8303–8318

Eperon LP, Estibeiro JP, Eperon IC (1986) The role of nucleotide sequences in splice site selection in eukaryotic pre-messenger RNA. Nature 324:280–282

Eperon LP, Graham IR, Griffiths AD, Eperon IC (1988) Effects of RNA secondary structure on alternative splicing of pre-mRNA: is folding limited to a region behind the transcribing RNA polymerase? Cell 54:393–401

Estes PA, Cooke NE, Liebhaber SA (1992) A native RNA secondary structure controls alternative splice-site selection and generates two human growth hormone isoforms. J Biol Chem 267:14902–14908

Fogel BL, McNally MT (2000) A cellular protein, hnRNP H, binds to the negative regulator of splicing element from Rous sarcoma virus. J Biol Chem 275:32371–32378

Forné T, Rossi F, Labourier E, Antoine E, Cathala G, Brunel C, Tazi J (1995) Disruption of base-paired U4.U6 small nuclear RNAs induced by mammalian heterogeneous nuclear ribonucleoprotein C protein. J Biol Chem 270:16476–16481

Gamberi C, Izaurralde E, Beisel C, Mattaj IW (1997) Interaction between the human nuclear cap-binding protein complex and hnRNP F. Mol Cell Biol 17:2587–2597

Gattoni R, Mahe D, Mahl P, Fischer N, Mattei MG, Stevenin J, Fuchs JP (1996) The human hnRNP-M proteins: structure and relation with early heat shock-induced splicing arrest and chromosome mapping. Nucleic Acids Res 24:2535–2542

Goguel V, Wang Y, Rosbash M (1993) Short artificial hairpins sequester splicing signals and inhibit yeast pre-mRNA splicing. Mol Cell Biol 13:6841–6848

Gontarek RR, McNally MT, Beemon K (1993) Mutation of an RSV intronic element abolishes both U11/U12 snRNP binding and negative regulation of splicing. Genes Dev 7:1926–1936

Gooding C, Roberts GC, Moreau G, Nadal-Ginard B, Smith CW (1994) Smooth muscle-specific switching of alpha-tropomyosin mutually exclusive exon selection by specific inhibition of the strong default exon. EMBO J 13:3861–3872

Gooding C, Roberts GC, Smith CW (1998) Role of an inhibitory pyrimidine element and polypyrimidine tract binding protein in repression of a regulated alpha-tropomyosin exon. RNA 4:85–100

Grossman JS, Meyer MI, Wang YC, Mulligan GJ, Kobayashi R, Helfman DM (1998) The use of antibodies to the polypyrimidine tract binding protein (PTB) to analyze the protein components that assemble on alternatively spliced pre-mRNAs that use distant branch points. RNA 4:613–625

Hamilton BJ, Nagy E, Malter JS, Arrick BA, Rigby WF (1993) Association of heterogeneous nuclear ribonucleoprotein A1 and C proteins with reiterated AUUUA sequences. J Biol Chem 268:8881–8887

Hamilton BJ, Burns CM, Nichols RC, Rigby WFC (1997) Modulation of AUUUA response element binding by heterogeneous nuclear ribonucleoprotein A1 in human T lymphocytes. The roles of cytoplasmic location, transcription, phosphorylation. J Biol Chem 272:28732–28741

Hammond LE, Rudner DZ, Kanaar R, Rio DC (1997) Mutations in the hrp48 gene, which encodes a *Drosophila* heterogeneous nuclear ribonucleoprotein particle protein, cause lethality and developmental defects and affect P-element third-intron splicing in vivo. Mol Cell Biol 17:7260–7267

Hanamura A, Cáceres JF, Mayeda A, Franza BR Jr, Krainer AR (1998) Regulated tissue-specific expression of antagonistic pre-mRNA splicing factors. RNA 4:430–444

Hay DC, Kemp GD, Dargemont C, Hay RT (2001) Interaction between hnRNPA1 and IkappaBalpha is required for maximal activation of NF-kappaB-dependent transcription. Mol Cell Biol 21:3482–3490

Hibbert CS, Gontarek RR, Beemon KL (1999) The role of overlapping U1 and U11 5′ splice site sequences in a negative regulator of splicing. RNA 5:333–343

Howe KJ, Ares M Jr (1997) Intron self-complementarity enforces exon inclusion in a yeast pre-mRNA. Proc Natl Acad Sci 94:12467–12472

Hutchison S, LeBel C, Blanchette M, Chabot B (2002) Distinct sets of multiple A1 binding sites control 5′ splice site selection in the hnRNP A1 pre-mRNA. J Biol Chem 277: (in press)

Hutton M, Lendon CL, Rizzu P, Baker M, Froelich S, Houlden H, Pickering-Brown S, Chakraverty S, Isaacs A, Grover A et al. (1998) Association of missense and 5′-splice-site mutations in tau with the inherited dementia FTDP-17. Nature 393:702–705

Izaurralde E, Jarmolowski A, Beisel C, Mattaj IW, Dreyfuss G, Fischer U (1997) A role for the M9 transport signal of hnRNP A1 in mRNA nuclear export. J Cell Biol 137:27–35

Jacquenet S, Mereau A, Bilodeau PS, Damier L, Stoltzfus CM, Branlant C (2001) A second exon splicing silencer within human immunodeficiency virus type 1 tat exon 2 represses splicing of Tat mRNA and binds protein hnRNP H. J Biol Chem 276:40464–40475

Jiang Z, Cộté J, Kwon JM, Goate AM, Wu JY (2000) Aberrant splicing of tau pre-mRNA caused by intronic mutations associated with the inherited dementia frontotemporal dementia with parkinsonism linked to chromosome 17. Mol Cell Biol 20:4036–4048

Jiang ZH, Zhang WJ, Rao Y, Wu JY (1998) Regulation of Ich-1 pre-mRNA alternative splicing and apoptosis by mammalian splicing factors. Proc Natl Acad Sci 95:9155–9160

Jumaa H, Nielsen PJ (1997) The splicing factor SRp20 modifies splicing of its own mRNA and ASF/SF2 antagonizes this regulation. EMBO J 16:5077–5085

Kadener S, Cramer P, Nogues G, Cazalla D, de la Mata M, Fededa JP, Werbajh SE, Srebrow A, Kornblihtt AR (2001) Antagonistic effects of T-Ag and VP16 reveal a role for RNA pol II elongation on alternative splicing. EMBO J 20:5759–5768

Kanopka A, Mühlemann O, Akusjarvi G (1996) Inhibition by SR proteins of splicing of a regulated adenovirus pre-mRNA. Nature 381:535–538

Kiledjian M, Burd CG, Gorlach M, Portman DS, Dreyfuss G (1994) Structure and function of hnRNP proteins. In: Mattaj IW, Nagai K (eds) RNA protein interactions. Oxford University Press, Oxford, pp 127–149

Konig H, Ponta, H, Herrlich P (1998) Coupling of signal transduction to alternative pre-mRNA splicing by a composite splice regulator. EMBO J 17:2904–2913

LaBranche H, Dupuis S, Ben-David Y, Bani MR, Wellinger RJ, Chabot B (1998) Telomere elongation by hnRNP A1 and a derivative that interacts with telomeric repeats and telomerase. Nat Genet 19:199–202

Lamb MM, Daneholt B (1979) Characterization of active transcription units in Balbiani rings of *Chironomus tentans*. Cell 17:835–848

Lavigueur A, La Branche H, Kornblihtt AR, Chabot B (1993) A splicing enhancer in the human fibronectin alternate ED1 exon interacts with SR proteins and stimulates U2 snRNP binding. Genes Dev 7:2405–2417

Le Guiner C, Lejeune F, Galiana D, Kister L, Breathnach R, Stevenin J, Del Gatto-Konczak F (2001) TIA-1 and TIAR activate splicing of alternative exons with weak 5′ splice sites followed by a U-rich stretch on their own pre-mRNAs. J Biol Chem 276:40638–40646

LeStourgeon WM, Barnett SF, Northington SJ (1990) Tetramers of the core proteins of 40 S nuclear ribonucleoprotein particles assemble to package nascent transcripts into a repeating array of regular particles. In: Strauss PR, Wilson SH (eds) The eukaryotic nucleus: molecular biochemistry and macromolecular assemblies. The Telford Press, Caldwell, NJ, pp 477–502

Libri D, Stutz F, McCarthy T, Rosbash M(1995) RNA structural patterns and splicing: molecular basis for an RNA-based enhancer. RNA 1:425–436

Lin CH, Patton JG (1995) Regulation of alternative 3′ splice site selection by constitutive splicing factors. RNA 1:234–245

Lothstein L, Arenstorf HP, Chung SY, Walker BW, Wooley JC, LeStourgeon WM (1985) General organization of protein in HeLa 40 S nuclear ribonucleoprotein particles. J Cell Biol 100: 1570–1581

Lou H, Helfman DM, Gagel RF, Berget SM (1999) Polypyrimidine tract-binding protein positively regulates inclusion of an alternative 3′-terminal exon. Mol Cell Biol 19:78–85

Markovtsov V, Nikolic JM, Goldman JA, Turck CW, Chou MY, Black DL (2000) Cooperative assembly of an hnRNP complex induced by a tissue-specific homolog of polypyrimidine tract binding protein. Mol Cell Biol 20:7463–7479

Matter N, Marx M, Weg-Remers S, Ponta H, Herrlich P, Konig H (2000) Heterogeneous ribonucleoprotein A1 is part of an exon-specific splice-silencing complex controlled by oncogenic signaling pathways. J Biol Chem 275:35353–35360

Matunis EL, Matunis MJ, Dreyfuss G (1993) Association of individual hnRNP proteins and snRNPs with nascent transcripts. J Cell Biol 121:219–228

Mayeda A, Krainer AR (1992) Regulation of alternative pre-mRNA splicing by hnRNP A1 and splicing factor SF2. Cell 68:365–375

Mayeda A, Helfman DM, Krainer AR (1993) Modulation of exon skipping and inclusion by heterogeneous nuclear ribonucleoprotein A1 and pre-mRNA splicing factor SF2/ASF [published erratum appears in Mol Cell Biol 1993 Jul; 13(7):4458]. Mol Cell Biol 13:2993–3001

Mayeda A, Munroe SH, Cáceres JF, Krainer AR (1994) Function of conserved domains of hnRNP A1 and other hnRNP A/B proteins. EMBO J 13:5483–5495

McAfee JG, Huang M, Soltaninassab S, Rech JE, Iyengar S, LeStourgeon WN (1997) The packaging of pre-mRNA. In: Krainer AR (ed) Eukaryotic mRNA processing. Oxford University Press, Oxford pp 68–102

McCullough AJ, Berget SM (1997) G triplets located throughout a class of small vertebrate introns enforce intron borders and regulate splice site selection. Mol Cell Biol 17:4562–4571

McKnight SL, Miller OL Jr (1979) Post-replicative nonribosomal transcription units in *D. melanogaster* embryos. Cell 17:551–563

McNally LM, McNally MT (1998) An RNA splicing enhancer-like sequence is a component of a splicing inhibitor element from Rous sarcoma virus. Mol Cell Biol 18:3103–3111

Michael WM, Choi M, Dreyfuss G (1995) A nuclear export signal in hnRNP A1: a signal-mediated, temperature-dependent nuclear protein export pathway. Cell 83:415–422

Michelotti EF, Michelotti GA, Aronsohn AI, Levens D (1996) Heterogeneous nuclear ribonucleoprotein K is a transcription factor. Mol Cell Biol 16:2350–2360

Min H, Chan RC, Black DL (1995) The generally expressed hnRNP F is involved in a neural-specific pre-mRNA splicing event. Genes Dev 9:2659–2671

Min H, Turck CW, Nikolic JM, Black DL (1997) A new regulatory protein, KSRP, mediates exon inclusion through an intronic splicing enhancer. Genes Dev 11:1023–1036

Mougin A, Gregoire A, Banroques J, Segault V, Fournier R, Brule F, Chevrier-Miller M, Branlant C (1996) Secondary structure of the yeast *Saccharomyces cerevisiae* pre-U3A snoRNA and its implication for splicing efficiency. RNA 2:1079–1093

Mourelatos Z, Abel L, Yong J, Kataoka N, Dreyfuss G (2001) SMN interacts with a novel family of hnRNP and spliceosomal proteins. EMBO J 20:5443–5452

Nasim FH, Hutchison S, Cordeau M, Chabot B (2002) High-affinity hnRNP A1 binding sites and duplex-forming inverted repeats have similar effects on 5′ splice site selection in support of a common looping out and repression mechanism. RNA 8: (in press)

Newman A (1987) Specific accessory sequences in *Saccharomyces cerevisiae* introns control assembly of pre-mRNAs into spliceosomes. EMBO J 6:3833–3839

Paca RE, Hibbert CS, O'Sullivan CT, Beemon KL (2001) Retroviral splicing suppressor requires three nonconsensus uridines in a 5′ splice site-like sequence. J Virol 75:7763–7768

Perez I, McAfee JG, Patton JG (1997) Multiple RRMs contribute to RNA binding specificity and affinity for polypyrimidine tract binding protein. Biochemistry 36:11881–11890

Petersen-Mahrt SK, Estmer C, Ohrmalm C, Matthews DA, Russell WC, Aküsjarvi G (1999) The splicing factor-associated protein, p32, regulates RNA splicing by inhibiting ASF/SF2 RNA binding and phosphorylation. EMBO J 18:1014–1024

Piñol-Roma S, Choi YD, Matunis MJ, Dreyfuss G (1988) Immunopurification of heterogeneous nuclear ribonucleoprotein particles reveals an assortment of RNA-binding proteins [published erratum appears in Genes Dev 1988 Apr; 2(4):490]. Genes Dev 2:215–227

Piñol-Roma S, Swanson MS, Gall JG, Dreyfuss G (1989) A novel heterogeneous nuclear RNP protein with a unique distribution on nascent transcripts. J Cell Biol 109:2575–2587

Polydorides AD, Okano HJ, Yang YY, Stefani G, Darnell RB (2000) A brain-enriched polypyrimidine tract-binding protein antagonizes the ability of Nova to regulate neuron-specific alternative splicing. Proc Natl Acad Sci 97:6350–6355

Pullman JM, Martin TE (1983) Reconstitution of nucleoprotein complexes with mammalian heterogeneous nuclear ribonucleoprotein (hnRNP) core proteins. J Cell Biol 97:99–111

Roberts GC, Gooding C, Mak HY, Proudfoot NJ, Smith CW (1998) Co-transcriptional commitment to alternative splice site selection. Nucleic Acids Res 26:5568–5572

Screaton GR, Cáceres JF, Mayeda A, Bell MV, Plebanski M, Jackson DG, Bell JI, Krainer AR (1995) Identification and characterization of three members of the human SR family of pre-mRNA splicing factors. EMBO J 14:4336–4349

Sebillon P, Beldjord C, Kaplan JC, Brody E, Marie J (1995) A T to G mutation in the polypyrimidine tract of the second intron of the human beta-globin gene reduces in vitro splicing efficiency: evidence for an increased hnRNP C interaction. Nucleic Acids Res 23:3419–3425

Shnyreva M, Schullery DS, Suzuki H, Higaki Y, Bomsztyk K(2000) Interaction of two multifunctional proteins. Heterogeneous nuclear ribonucleoprotein K and Y-box-binding protein. J Biol Chem 275:15498–15503

Siebel CW, Kanaar R, Rio DC (1994) Regulation of tissue-specific P-element pre-mRNA splicing requires the RNA-binding protein PSI. Genes Dev 8:1713–1725

Siebel CW, Admon A, Rio DC (1995) Soma-specific expression and cloning of PSI, a negative regulator of P element pre-mRNA splicing. Genes Dev 9:269–283

Sierakowska H, Szer W, Furdon PJ, Kole R (1986) Antibodies to hnRNP core proteins inhibit in vitro splicing of human beta-globin pre-mRNA. Nucleic Acids Res 14:5241–5254

Simard MJ, Chabot B (2000) Control of hnRNP A1 alternative splicing: an intron element represses use of the common 3′ splice site. Mol Cell Biol 20:7353–7362

Simard MJ, Chabot B (2002) SRp30c is a repressor of 3′ splice site utilization. Mol Cell Biol 22:4001–4010

Singh R, Valcárel J, Green MR (1995) Distinct binding specificities and functions of higher eukaryotic polypyrimidine tract-binding proteins. Science 268:1173–1176

Solnick D (1985) Alternative splicing caused by RNA secondary structure. Cell 43:667–676
Solnick D, Lees SI (1987) Amount of RNA secondary structure required to induce an alternative splice. Mol Cell Biol 7:3194–3198
Swanson MS, Dreyfuss G (1988) RNA binding specificity of hnRNP proteins: a subset bind to the 3′ end of introns. EMBO J 7:3519–3529
Tacke R, Manley JL (1999) Determinants of SR protein specificity, Curr Opin Cell Biol 11:358–362
Tange TO, Damgaard CK, Guth S, Valcárcel J, Kjems J (2001) The hnRNP A1 protein regulates HIV-1 tat splicing via a novel intron silencer element. EMBO J 20:5748–5758
Temsamani J, Pederson T (1996) The C-group heterogeneous nuclear ribonucleoprotein proteins bind to the 5′ stem-loop of the U2 small nuclear ribonucleoprotein particle. J Biol Chem 271:24922–24926
Thomas JO, Glowacka SK, Szer W (1983) Structure of complexes between a major protein of heterogeneous nuclear ribonucleoprotein particles and polyribonucleotides, J Mol Biol 171:439–55
Tomonaga T, Levens D(1995) Heterogeneous nuclear ribonucleoprotein K is a DNA-binding transactivator. J Biol Chem 270:4875–4881
Valcárcel J, Gebauer F (1997) Post-transcriptional regulation: the dawn of PTB. Curr Biol 7:R705–R708
van der Houven van Oordt W, Diaz-Meco MT, Lozano J, Krainer AR, Moscat J, Cáceres JF. (2000) The MKK(3/6)-p38-signaling cascade alters the subcellular distribution of hnRNP A1 and modulates alternative splicing regulation. J Cell Biol 149:307–316
Venables JP, Elliott DJ, Makarova OV, Makarov EM, Cooke HJ, Eperon IC (2000) RBMY, a probable human spermatogenesis factor, other hnRNP G proteins interact with Tra2beta and affect splicing. Hum Mol Genet 9:685–694
Wagner EJ, Garcia-Blanco MA (2001) Polypyrimidine tract binding protein antagonizes exon definition. Mol Cell Biol 21:3281–3288
Wilk HE, Angeli G, Schafer KP (1983) In vitro reconstitution of 35 S ribonucleoprotein complexes. Biochemistry 22:4592–4600
Williamson DJ, Banik-Maiti S, DeGregori J, Ruley HE (2000) hnRNP C is required for post-implantation mouse development but is dispensable for cell viability. Mol Cell Biol 20:4094–4105
Wurtz T, Kiseleva E, Nacheva G, Alzhanova-Ericcson A, Rosen A, Daneholt B (1996) Identification of two RNA-binding proteins in Balbiani ring premessenger ribonucleoprotein granules and presence of these proteins in specific subsets of heterogeneous nuclear ribonucleoprotein particles. Mol Cell Biol 16:1425–1435
Yang X, Bani MR, Lu SJ, Rowan S, Ben-David Y, Chabot B (1994) The A1 and $A1^B$ proteins of heterogeneous nuclear ribonucleoparticles modulate 5′ splice site selection in vivo. Proc Natl Acad Sci 91:6924–6928
Zhang L, Liu W, Grabowski PJ (1999) Coordinate repression of a trio of neuron-specific splicing events by the splicing regulator PTB. RNA 5:117–130
Zhu J, Mayeda A, Krainer AR (2001) Exon identity established through differential antagonism between exonic splicing silencer-bound hnRNP A1 and enhancer-bound SR proteins. Mol Cell 8:1351–1361

Phosphorylation-Dependent Control of the Pre-mRNA Splicing Machinery

J. Soret and J. Tazi[1]

1 Introduction

In this chapter, experiments will be reviewed which lead from the initial hypothesis of the role of protein phosphorylation in pre-mRNA splicing to the elucidation of mechanism(s) by which this modification alters the function of splicing factors. Phosphorylation has been implicated in the assembly of the spliceosome (Tazi et al. 1992; Mermoud et al. 1994a), regulation of splice site selection (Du et al. 1998; Xiao and Manley 1998; Prasad et al. 1999; Stojdl and Bell 1999; Pilch et al. 2001) and subcellular localization of splicing factors (Misteli and Spector 1996; Misteli et al. 1998). Several splicing factors are known to be phosphorylated in vivo, including a number of the serine/arginine-rich (SR) protein family (Fu 1995; Graveley 2000), protein components of the small nuclear ribonucleoproteins (snRNPs; Tazi et al. 1993; Woppmann et al. 1993; Fetzer et al. 1997) and proteins associated with the spliceosome (SAPs; Gozani et al. 1994; Chiara et al. 1996; Reed 1996). Phosphorylation and dephosphorylation of these factors might act as a modulator of protein-protein (Xiao and Manley 1997) and protein-RNA interactions at multiple stages during the biogenesis of mRNAs (Tazi et al. 1997). A major advance towards understanding the regulation of pre-mRNA splicing by protein phosphorylation has been to isolate and characterize potential protein kinases and phosphatases that specifically target components of the splicing apparatus. We will review the molecular components of the splicing machinery which are affected by phosphorylation and consider the mechanisms by which specific kinases and phosphatases seem to contribute to a dynamic organization of pre-mRNA splicing factory, not only during transcription, but also in response to physiological stimuli.

[1] Institut de Génétique Moléculaire, UMR5535 du CNRS, IFR 24, 1919 Route de Mende, 34293 Montpellier Cedex 5, France

Progress in Molecular and Subcellular Biology, Vol. 31
Philippe Jeanteur (Ed.)

2 Spliceosome Assembly and Pre-mRNA Splicing

2.1 Dynamic Assembly of the Spliceosome

Since the discovery of messenger RNA precursor (pre-mRNA) splicing in 1977 by Berget et al. (Berget et al. 1977), it has been postulated that the accurate excision of intron sequences must result from a complex ribonucleoprotein structure. Much progress has been made in understanding the steps leading to the formation of this structure and elucidating the splicing reaction mechanism. Rapid progress has been due to in vitro studies of the splicing reaction using mammalian or yeast cell free extracts which can perform the removal of an intron from an exogenously added simple substrate which consists of a single intron flanked by two exons (Grabowski et al. 1984; Krainer et al. 1984; Padgett et al. 1984; Ruskin et al. 1984). Although it has been appreciated that ATP hydrolysis is required for intron excision in this system, the phosphates at the splice junctions were not exchanged from exogenous ATP. Indeed, characterization of the splicing intermediates and products has established that splicing is a two step process leading to release of the intron in the form of a lariat by two transesterification reactions. In the first reaction, cleavage of a phosphodiester bond at the 5′ splice site involves the nucleophilic attack of the phosphodiester linkage by 2′ oxygene of an adenosine located a few nucleotides from the 3′ splice site. This results in the formation of two intermediates, an exon 1 with a 3′-hydroxyl and an intron-exon 2 lariat structure formed concurrently by a 2′,5′ phosphodiester bond. In the second reaction, cleavage at the 3′ splice site and ligation between the two exons by a 3′,5′ phosphodiester bond lead to the release of the lariat intron.

The four abundant small nuclear ribonucleoproteins U1, U2, U5 and U4/U6 (snRNPs) are the major components of the spliceosome, the ribonucleoprotein structure where the intron excision takes place (Steitz et al. 1987; Guthrie and Patterson 1988; Lührmann et al. 1990). They contain one (U1, U2 and U5) or two (U4/U6) snRNAs and two classes of proteins, a first class common to all snRNPs and designated either B, B′, D, D′, E, F and G or "core snRNP proteins", and a second class being unique to a particular snRNP (Lührmann 1988; Lührmann et al. 1990). The precision and efficiency of the splicing reaction are the result of a dynamic series of interactions between the snRNPs, non snRNP proteins and the pre-mRNA. By assessing the kinetics of appearance of splicing complexes by native gel electrophoresis (Grabowski et al. 1985; Konarska and Sharp 1986, 1987; Pikienlny and Rosbash 1986; Cheng and Abelson 1987; Legrain et al. 1988), glycerol gradient sedimentation (Brody and Abelson 1985; Frendewey and Keller 1985; Bindereif and Green 1986), gel filtration (Reed et al. 1988; Michaud and Reed 1991) or affinity chromatography (Grabowski and Sharp 1986; Ruby and Abelson 1988), it was possible to determine discrete steps in the spliceosome assembly. The earliest detectable mammalian pre-spliceosome complex (E) is formed in an ATP-independent manner (Michaud and Reed 1991). U1 snRNP is the only snRNP present in this complex, together with the non-snRNP splicing factor U2AF (U2 snRNP auxiliary factor), splicing factor 1 (SF1; also mammalian branch point binding protein [mBBP]; hereafter SF1/mBBP) and several spliceosome-associated proteins (SAPs), including serine/arginine-rich (SR) protein family members (Bennett et al. 1992). The assembly of this complex is a major control point for the initial recognition and pairing of splice sites (Séraphin and Rosbash 1989; Michaud and Reed 1993) and is therefore thought to be an important step in the regulation of alternative splicing (Michaud and Reed 1993; Reed 1996). The next step involves the stable binding of U2 snRNP to the pre-mRNA sequences around the branch point adenosine which leads to an A complex formation (Frendewey et al. 1987; Konarska and Sharp 1987; Jamison et al. 1992; Query et al. 1996). The multimeric components SF3a (composed of SAPs 61, 62 and 114) and SF3b (composed of SAPs 49, 130, 145 and 155) are involved in this binding (Krämer and Utans 1991; Behrens et al. 1993b; Brosi et al. 1993; Legrain and Chapon 1993; Hogges and Beggs 1994; Gozani et al. 1996; Krämer 1996; Reed 1996). Site-specific labelling of the pre-mRNA combined with UV-cross-linking showed that all subunits of SF3a and SF3b, with the exception of SAP 130, bind to sequences surrounding the branch site adenosine (Gozani et al. 1996) and are, thereby, likely contribute to the stable binding of U2 snRNP (Hogges and Beggs 1994; Reed 1996). However, in the A complex, U1 snRNP, SF1/mBBP and U2AF appear to be weakly bound (Konarska and Sharp 1987; Michaud and Reed 1991, 1993) and have not been consistently detected in complexes resolved by native gel electrophoresis. Subsequently, a tri-snRNP particle U4/U5/U6, rather than the individual snRNPs, joins the assembled complex and gives rise to the B1 complex (Cheng and Abelson 1987; Konarska and Sharp 1987). A conformational change probably involving disruption of base-paired U4 and U6 snRNAs appears to be a critical step in the initiation of splicing reactions by the assembled B2 complex (Cheng and Abelson 1987; Lamond et al. 1988).

2.2 Dissecting the Requirement for ATP in Pre-mRNA Splicing

The dynamic assembly of the spliceosome as well as the cloning of required gene products suggested that the associated proteins may play two important roles in this process: The first one is structural and involves several RNA binding proteins that recognize specific sequences and/or structures in the snRNAs and the pre-mRNA (Boggs et al. 1987; Amrein et al. 1988; Goralski et al. 1989; Ge et al. 1991; Krainer et al. 1991; Patton et al. 1991, 1993; Shannon and Guthrie 1991; Fu and Maniatis 1992; Mayeda and Krainer 1992; Zahler et al. 1992; Zamore et al. 1992; Brosi et al. 1993; Screaton et al. 1995; Krämer 1996). The second one is enzymatic and may direct the ordered association and dissociation of splicing components during the splicing process. There is evidence that protein kinases and protein phosphatases are involved in the pre-mRNA splicing process (Mermoud et al. 1992, 1994a,b; Tazi et al. 1992, 1993; Woppmann et al. 1993; Gui et al. 1994a,b; Colwill et al. 1996a; Rossi et al. 1996a). These enzymes likely influence protein-protein interactions at early stages of the spliceosome assembly pathway.

Apparent evidence for the involvement of protein phosphorylation in the control of pre-mRNA splicing came from the early finding that inhibiting protein dephosphorylation, on serine or threonine residues, blocks the catalytic steps of splicing, but not spliceosome assembly (Mermoud et al. 1992, 1994a,b; Tazi et al. 1992). It was initially shown that replacement of ATP in normal splicing extracts by ATP(γS) leads to the inhibition of the second step of splicing (Tazi et al. 1992), implying that the thiophosphorylation of spliceosomal proteins was detrimental for splicing. Furthermore, thiophosphorylation of U1 snRNP-specific 70K (U1–70K) protein inhibits pre-mRNA splicing at a pre-catalytic step (Tazi et al. 1993). Since phosphorothioate from thiophosphorylated groups of splicing factors failed to be released by phosphatases, it was proposed that a cycle of phosphorylation-dephosphorylation controls the activity of splicing factors (Tazi et al. 1993). This idea was further supported by the finding that highly specific inhibitors (okadaic acid, tautomycin and microcystin-LR) of protein phosphatases PP1 and PP2A differentially inhibit the two catalytic steps (Mermoud et al. 1992). Inhibition of PP2A predominantly inhibits the second step, whereas inhibition of PP1 and PP2A affects both steps. Splicing can be reactivated in extracts treated with phosphatase inhibitors by adding exogenous catalytic subunits of PP1 or PP2A indicating that both activities are relevant to splicing (Mermoud et al. 1992, 1994b). A form of PP1 was proposed to be a candidate for the phosphatase involved in U1–70K-specific dephosphorylation (Mermoud et al. 1992; Cardinali et al. 1994).

In addition, it has been shown that stimulating Ser/Thr-specific protein dephosphorylation selectively inhibits an early step during mammalian spliceosome assembly. HeLa nuclear extract failed to form earlier splicing complexes (E complex) when treated with purified PP1 phosphatase (Mermoud et al. 1994b). A crucial determinant of the inhibition appears to be the dephosphorylation of SR proteins (see below), as purified phosphorylated SR proteins restored splicing activity to the inactivated extract (Mermoud et al. 1994a). It is not clear, however, whether other factors participating in the formation of E complex are not affected by this dephosphorylation. Besides SR proteins, U1–70K, both subunits of U2AF and SF1/mBBP are also candidate phosphoproteins whose activities are required at early steps during spliceosome assembly. Since all these factors are expected to drive the connection between the 5′ and the 3′ splice sites due to their ability to interact with each other, it is possible that these bridging functions may be affected by changes in phosphorylation mediated by PP1.

Recent observations have also demonstrated that signal transduction pathways which involve a cascade of phosphorylation/dephosphorylation events, modulate alternative splice site selection. For example, insulin and phorbol ester treatment of myocytes induces a switch at the mRNA and protein levels between two alternatively spliced isoforms of protein kinase C from a βI to a βII isoform (Ono et al. 1987; Chalfant et al. 1995; Patel et al. 2001). The regulation of alternative splicing of CD44 antigen has also been demonstrated to be regulated by both the synergistic action of protein kinase C (PKC) and calcium and mitogen-activated protein (MAP)-kinase pathways (Konig et al. 1998; Weg-Remers et al. 2001). The reader is referred to Akker et al. (2001) for a more exhaustive catalogue of these hormonally regulated alternative splicing events.

3 Phosphorylation of Serine/Arginine-Rich Protein Splicing Factors

3.1 Regulation of Splicing

As discussed by others in this volume, the SR proteins constitute a family of splicing factors that are highly conserved in Metazoa, and have been studied extensively both in vitro and in vivo (Fu 1995; Graveley 2000). SR proteins are required for constitutive pre-mRNA splicing and also regulate alternative splice site selection in a concentration-dependent manner. Functionally, many of the SR proteins are able to bind several classes of specific RNA motifs known as exonic splicing elements (ESEs), which play a key role in the activation of weak upstream introns (Watakabe et al. 1993; Tacke and Manley 1999). An important approach in the identification and characterization of SR proteins made use of available monoclonal antibodies. Several SR protein species, defined as SRp20, SRp30a-SRp30b-SRp30c, SRp40, SRp55 and SRp75, from their apparent molecular weight in SDS gels, were easily isolated and characterized because they were recognized by mAb104 (Zahler et al. 1992) which stains lateral loops on amphibian lampbrush (Roth and Gall 1987; Roth et al. 1990) and puffs on *Drosophila* polytene chromosomes (Roth et al. 1991). The mAb104 epitope includes a phosphate indicating that these proteins are phosphorylated in vivo (Roth et al. 1990, 1991; Zahler et al. 1992). This phosphorylation concerns serine residues located within the arginine/serine-rich (RS) domain.

The RS domain is responsible for specific protein-protein interactions between RS domain-containing proteins (Wu and Maniatis 1993; Amrein et al. 1994; Kohtz et al. 1994). Such physical interactions indeed promote the binding of U1 snRNP to the 5′ splice site (see below) and constitute a bridge between 5′ and 3′ splice sites during splice site selection (Wu and Maniatis 1993) at the earliest stages of the spliceosome assembly (Reed 1996). Therefore, specific phosphorylation of serine residues located within the RS domain may be one of the key determinants regulating splicing events, because it modulates homophylic and heterophylic RS domain interactions. Indeed, phosphorylation was shown to have a positive effect on the interaction of the SR protein ASF/SF2 (SRp30a) with U1 snRNP-specific protein 70K, while its interaction with the small subunit of U2AF ($U2AF^{35}$) was not affected. In contrast, phosphorylated SF2/ASF was shown to bind itself, SRp40 and the SR related protein hTra2 (see below) poorly compared to mock phosphorylated ASF/SF2 (Xiao and Manley 1997, 1998), implying that phosphorylation of SF2/ASF decreases its binding to other SR proteins.

Not only protein–protein interactions but also RNA–protein interactions are affected by phosphorylation of the RS domain. The unphosphorylated RS domain of SF2/ASF can interact with RNA non-specifically, whereas phosphorylation prevents such interaction (Xiao and Manley 1997). This observation is also consistent with the finding that the unphosphorylated form of SRp40 protein bound the RNA avidly, but could not function in a SELEX experiment to identify a specific high affinity binding site, like the phosphorylated version of the same protein does (Tacke et al. 1997). Furthermore, p32, a protein which was initially co-purified with SF2/ASF (Krainer et al. 1991), can inhibit SR protein phosphorylation by specific kinases and impedes ASF/SF2 function as both a splicing enhancer and splicing repressor protein by preventing stable ASF/SF2 interaction with RNA (Petersen-Mahrt et al. 1999).

These observations bring up the intriguing possibility that phosphorylation of SR proteins may play a role as a regulatory switch, whereby initial interactions with constitutive splicing factors like U1 snRNP, U2AF and specific RNA sequences are enhanced to allow specific selection of an authentic splice site, while dephosphorylation of SR proteins weakens these interactions, allowing subsequent steps of the spliceosome assembly to occur. Two findings support this scenario. First, phosphatase treatment of nuclear extracts could block splicing at an early step in spliceosome assembly and inhibition could be reversed by addition of a mixture of SR proteins (Mermoud et al. 1992, 1994a,b). Second, dephosphorylation, specifically of SF2/ASF and U1 snRNP-specific 70K has been shown to be required for splicing, although at a step subsequent to spliceosome assembly (Tazi et al. 1993; Cao et al. 1997). Furthermore, both hyper- and hypophosphorylation of SR proteins have been shown to inhibit splicing (Prasad et al. 1999), demonstrating that the level of SR protein phosphorylation is critical for splicing. Although the structural basis for the differential responses to SR proteins phosphorylation is presently unknown, changes in SR protein phosphorylation play a major role in the activation of pre-mRNA splicing during early development (Sanford and Bruzik 1999) and genetic inactivation of genes coding for SR protein-specific kinases led to a lethal phenotype (see below). Thus, the control of SR protein phosphorylation is likely to be a means by which splicing is activated co-ordinately with the onset of gene expression during development.

3.2 Regulation of Cellular Localization

Many aspects of cellular localization of SR proteins are also determined, at least in part, by the phosphorylation status of the RS domain. Recently, new members of importin β (impβ) or transportin (TRN) family, termed transportin-SR1 (TRN-SR1), transportin-SR2 (TRN-SR2) and *Drosophila* transportin-SR (dTRN-SR), have been shown to interact with human and *Drosophila* SR proteins (Kataoka et al. 1999; Lai et al. 2000, 2001). Remarkably, TRN-SR2 and dTRN-SR have a strong preference for phosphorylated RS domains (Allemand et al. submitted) and TRN-SR2 mediates nuclear import of phosphorylated, but not unphosphorylated SR proteins (Lai et al. 2000, 2001).

Phosphorylation of serine residues at the RS domain releases SR proteins from storage/assembly loci (nuclear speckles) and recruits them to the sites of active transcription (Misteli et al. 1998). In contrast, both inhibition of RNA polymerase II and excess of PP1 activity caused increased dephosphorylation of SR proteins in nuclear speckles. Furthermore, overexpression of either SR protein kinase (SRPK) or Clk/Sty, a prototypical kinase with dual specificity, capable of phosphorylating tyrosines as well as serines and threonines (Colwill et al. 1996b), causes cytoplasmic accumulation of SR proteins (Gui et al. 1994a; Colwill et al. 1996a). The *Drosophila* homologue of human SF2/ASF (dASF) which lacks phosphorylation sites for *Drosophila* SRPK (dSRPK) at the RS domain, was unable to shuttle between the nucleus and the cytoplasm while imported to the nucleus (Allemand et al. 2001). Thus; the differential effects of phosphorylation on SR protein interactions have the potential to affect both splicing activity and subcellular trafficking.

4 Serine/Arginine-Rich Protein Kinases

4.1 SRPK Protein Family

The search for kinases involved in the regulation of SR proteins has resulted in the identification of several proteins which have little more in common than the ability to phosphorylate the RS domain of SR proteins. The SRPK1 was purified and cloned on the basis of its capacity to phosphorylate SC35 or other SR proteins in vitro and mediate splicing factor redistribution during cell cycle (Gui et al. 1994a,b; Wang HY et al. 1998). This kinase appears to be the prototype of a larger family of diffentially expressed kinases in mammals (*human SRPK1*, U09564; *human SRPK2*, U88666A; *mouse SRPK1*, AJ224115; *mouse SRPK2*, B006036; Gui et al. 1994a,b; Kuroyanagi et al. 1998; Wang HY et al. 1998, 1999; Papoutsopoulou et al. 1999) that has homologues in nematode (*Caenorhabditis elegans SPK-1*, AF241656; Kuroyanagi et al. 2000), in fruit fly (*Drosophila SRPK1*, AF01149; Allemand et al. 2001; Mount and Salz 2000), in plants (*Arabidopsis thaliana SRPK1*, AJ292978; SRPK2, AJ292979; SRPK3, AJ292980; SRPK4, AJ292981) and in yeast (*Saccharomyces cerevisiae Sky 1*, S55098; *Schizosaccaromyces pombe Dsk1*, D13447; Siebel et al. 1999; Tang et al. 2000). Recent characterization of novel SRPKs and genomic inspection of predicted SRPK candidates revealed that several isoforms expressed from *SRPK1* and *SRPK2* genes were generated by alternative splicing (Kuroyanagi et al. 1998; Papoutsopoulou et al. 1999; Nikolakaki et al. 2001).

In vitro, SRPK1 exhibits a strong preference for Ser-Arg sites (Colwill et al. 1996b) and catalyses a sequential phosphorylation reaction of the consecutive RS dipeptide repeats. These characteristics are consistent with recent modeling of a substrate peptide bound to the three-dimensional structure of a fully active truncated Sky1p (Nolen et al. 2001). Both inactive and active SRPKs can bind SF2/ASF in glutathione S-transferase (GST)-SF2/ASF pull down experiments, implying that SR proteins interact directly with the enzyme. This interaction is abolished following SF2/ASF phosphorylation (Wang HY et al. 1998; Yeakley et al. 1999).

SRPK1 was supposed to regulate splicing because addition of the purified enzyme to in vitro reactions has been shown to inhibit splicing (Gui et al. 1994b). However, there is no evidence that this enzyme directly influences splicing by phosphorylating a specific target protein(s) and changing its activity in splicing. Indeed, immunolocalization experiments showed that mammalian SRPKs are primarily localized in the cytoplasm, despite having putative nuclear localization signals (Wang HY et al. 1998; Koizumi et al. 1999). Addition of purified SRPK1 to permeabilized cells or overexpression in transfected cells results in an apparent disassembly of the nuclear speckles and redistribution of SR proteins into cytoplasm (Gui et al. 1994a; Colwill et al. 1996a), suggesting that SRPKs-mediated phosphorylation strongly influences the cellular distribution of SR proteins. However, the SR protein SF2/ASF accumulated in the cytoplasm, even in cells expressing the catalytically inactive kinase SRPK2^{K108R}, implying that interaction with the kinase rather than phosphorylation was responsible for cytoplasmic accumulation (Koizumi et al. 1999). Furthermore, *Drosophila* SF2/ASF (dSF2/ASF) which lacks eight repeating RS dipeptides at its RS domain was not phosphorylated at all in vitro by *Drosophila* or human SRPK1, while its RS domain was as efficient as the RS domain of hSF2/ASF to trigger GFP fusions to the nucleus (Allemand et al. 2001). This result makes it unlikely that SRPK1 is involved in the subcellular localization of dSF2/ASF. Interestingly, dSF2/ASF, unlike its human homologue, does not shuttle between the nucleus and the cytoplasm, suggesting that part of the shuttling properties could be mediated by the RS repeats at the RS domain and that these cellular events might be subjected to regulation by phosphorylation from SRPK1. In keeping with this suggestion, deletion of the RS repeats from SF2/ASF abolished both shuttling and phosphorylation by SRPK1 (Allemand et al. 2001). Even more interestingly, Npl3, a yeast protein which is a major substrate for Sky1p and SRPK1 (Siebel et al. 1999) is also a shuttling protein whose shuttling activity is regulated by Sky1p (Gilbert et al. 2001). Sky1p-mediated phosphorylation was also shown to promote the interaction between Npl3 and its nuclear import receptor Mtr10 (Senger et al. 1998). It is, therefore, possible that SRPKs regulate shuttling properties of some SR proteins (Siebel et al. 1999). This possibility is compatible with the recent discovery that SR proteins SRp20 and 9G8, like Npl3, promote the nucleocytoplasmic export of mRNAs (Huang and Steitz 2001).

The absence of authentic SR protein-encoding genes in the yeast genome and the absence of alternative splicing are both consistent with the notion that SRPKs, originally thought to be enzymes only involved in pre-mRNA splicing, play a broader role in cellular regulation. Further support for this hypothesis arose from the finding that *SKY1* disruption in yeast cells resulted in a dramatic reduction of polyamines uptake, implying that this enzyme plays a major role in polyamine transport (Erez and Kahana 2001). Moreover, Dsk1, the functional fission yeast homologue of SRPK1, was shown to associate with microtubules and the spindle pole body (Takeuchi and Yanagida 1993; Nabeshima et al. 1995), suggesting that this protein could be involved in the regulation of chromosome segregation at the metaphase/anaphase transition. In keeping with this suggestion, the *DSK1* gene was shown to be a suppressor of cold-sensitive dis1 mutants that are defective in sister chromatids separation at restrictive temperature. Implication of SRPKs in the regulation of the cell-cycle was also inferred from initial characterization of SRPK1 activity which was found to be three- to fivefold higher in extracts prepared from metaphase compared to interphase cells (Gui et al. 1994a). High levels of SRPK1 expression were also observed in human and mouse testis where the enzyme was thought to be responsible for the phosphorylation of protamines (Papoutsopoulou et al. 1999). The increasing number of different substrates that SRPKs may phosphorylate and/or interact with, together with their differential distribution in various tissues of multicellular organisms, indicate that these enzymes might have a more versatile rather than specialized cellular function such as phosphorylation of SR proteins.

4.2
Clk/Sty Kinase Family

The Clk or Sty kinase, the prototype of this family, is a cdc2-like kinase that was originally identified in an anti-phosphotyrosine antibody screen of mouse expression libraries (Ben-David et al. 1991). Clk/Sty has the unusual property of phosphorylating both serine/threonine and tyrosine residues and has been shown to autophosphorylate on all three hydroxy-amino acids (Ben-David et al. 1991; Howell et al. 1991; Lee et al. 1996). The C-terminal catalytic kinase domain of Clk/Sty carries the conserved EHLAMMERILG motif common to all members of this family which were, hence classified as LAMMER kinases (Yun et al. 1994; Nayler et al. 1997). This family has been conserved through evolution and members were found in diverse organisms including yeast (*Saccharomyces cerevisae, KNS1*; Padmanabha et al. 1991), plant (*Arabidopsis thaliana, AFC1; AFC2; AFC3*; Bender and Fink 1994), fruit fly (*Drosophila, Doa*; Yun et al. 1994) and mammals (human hClk1, hClk2, hClk3 and hClk4; mouse mClk/Sty; Ben-David et al. 1991; Johnson and Smith 1991; Hanes et al. 1994; Nayler et al. 1998; Schultz et al. 2001). Several isoforms resulting from alternative splicing have also been identified, indicating that expression of the LAMMER family may be regulated at the post-transcriptional level (Hanes et al. 1994; Duncan et al. 1995; Nayler et al. 1997). Interestingly, both in mammals and in *Drosophila*, alternative splicing generates two transcripts encoding a catalytically active kinase and a truncated protein lacking the kinase domain (Duncan et al. 1995, 1997). The ratio of these splice products appears to be regulated during *Drosophila* development and in a tissue- and cell-specific manner in mammals (Yun et al. 1994, 2000; Nayler et al. 1997; Menegay et al. 1999). Homozygous loss-of-function *doa* mutants die at an early larval stage, indicating that the single gene encoding a LAMMER kinase in *Drosophila* is critical for development (Rabinow and Birchler 1989; Rabinow et al. 1993; Yun et al. 1994, 2000).

In a two-hybrid system, Clk/Sty was shown to interact with SR proteins and with RNA binding proteins heterogeneous nuclear ribonucleoprotein G and ribonucleoprotein S1 (Colwill et al. 1996a). This interaction is thought to be mediated by a stretch of arginine-serine-rich residues at the amino-terminal region of the Clk/Sty protein which resembles the RS domain of SR proteins (Colwill et al. 1996a). Clk was also shown to interact directly with hPrp4, a kinase which contains an N-terminal RS domain and co-localizes with SR proteins in speckles (Kojima et al. 2001). Both SR proteins and hPrp4 were phosphorylated at their RS domain by Clk in vitro. Supporting the functional significance of these phosphorylations, transient overexpression of Clk/Sty has been shown to influence alternative splicing of a cotransfected reporter transcript (Prasad et al. 1999). In vitro, complementation of HeLa nuclear extracts with purified wild type and inactive form of Clk/Sty was used to demonstrate that the activity of SR proteins, but not other essential splicing factors was affected by this enzyme. Both hyper- and hypophosphorylation inhibit SR protein splicing activity, repressing constitutive splicing and switching alternative splice site selection (Prasad et al. 1999).

The genetic approach in *Drosophila* has also been used to test whether Doa-mediated phosphorylation plays a significant role in the regulation of alternative splicing (Du et al. 1998). Thus, mutations inhibiting *Doa* function reduce the phosphorylation of Rbp1, an SR protein that binds *Drosophila doublesex* splicing enhancer (dsxRE; Du et al. 1998), a *cis*-acting regulatory sequence required for *doublesex* (*dsx*) pre-mRNA splicing regulation in the cascade of splicing events leading to female sexual differentiation. In addition to SR proteins, the regulation of *dsx* pre-mRNA splicing requires Tra (transformer) and Tra2 whose subnuclear distribution is affected by Doa mutations. Given that specific interaction of each of these splicing factors with RNA is highly dependent on the presence of the other proteins (Du et al. 1998), the finding that Doa mutations cause sexual transformations in adult flies and cause changes in *doublesex* (*dsx*) pre-mRNA splicing suggested that reduced phosphorylation of RBP1 and possibly other SR proteins impairs splicing enhancer functions (Du et al. 1998). However, failure of Doa mutations to affect the sex-specific splicing of *frutless* (fru) pre-mRNA was rather surprising, given that fru and dsx splicing otherwise appears to be similarly regulated (Ryner et al. 1996; Heinrichs et al. 1998). Therefore, it is still possible that Doa has some selectivity for some SR proteins and/or other substrates than SR proteins and the effect of reduced Rbp1 phosphorylation on enhancer function remains to be tested directly.

While both Clk and SRPKs were able to phosphorylate the RS domain of SF2/ASF in vitro on sites phosphorylated in vivo, SRPK displayed greater specific activity toward this substrate than Clk. In vitro, Clk phosphorylated more substrates, and more sites within SF2/ASF RS domain (Colwill et al. 1996b), indicating that Clk may play a broader regulatory role than the phosphorylation of SR proteins. This conclusion was also inferred from previous genetic evidence showing that Doa is required for normal differentiation of a variety of developmental functions which are not necessarily limited by impaired pre-mRNA splicing (Yun et al. 1994). As differences in the patterns of expression of both SR proteins and Clk kinases were observed between cell types and tissues (Zahler et al. 1993; Nayler et al. 1997), it may be speculated that some aspects of development can be modulated by targeted phosphorylation of specific SR proteins. The relevance of the level of SR protein phosphorylation in mediating alternative or constitutive splicing needs more studies at both the biochemical and cellular levels.

Clk kinases appear to localize exclusively in the nucleus and autophosphorylation of Clk2 at Ser141, which is highly conserved among all Clk proteins, has been shown to influence its subnuclear localization (Nayler et al. 1998). When overexpressed, a catalytically active form of Clk/Sty causes a redistribution of SR proteins in the nucleoplasm of transformed cells (Colwill et al. 1996a) and may even result in the accumulation of SF2/ASF in the cytoplasm. A catalytically inactive form, however, co-localized with SR proteins, but did not induce their redistribution (Colwill et al. 1996a). Due to mislocalization of splicing factors following overexpression of Clk in transfected cells, it is still possible that the observed effects on splicing could be indirectly caused by depletion of specific SR proteins from the nucleus. Because SR proteins are known to affect splice site selection in a concentration-dependent manner, such phosphorylation-dependent localization of particular SR proteins among different nuclear pools may be a mechanism to regulate alternative splicing.

This underlines the need for a better understanding of the function and regulation of SR proteins by specific kinases in cells. Indeed, the potential of LAMMER kinases to be regulated through signal transduction cascades will provide an important and general avenue for physiological and developmental control of RNA splicing patterns.

4.3 DNA Topoisomerase I

Purification of an activity that phosphorylates individual purified SR proteins from a nuclear extract enriched with snRNPs resulted in the identification of DNA topoisomerase I as a kinase of SR proteins, raising the intriguing possibility that this enzyme would have protein kinase activity in addition to its already well known DNA relaxing activity (Rossi et al. 1996a). Thus, we now refer to this enzyme as topo I/kinase. Like SRPK1, topo I/kinase exclusively phosphorylates serine residues within the RS domain, but the number of serines phosphorylated by the two enzymes are different. SRPK1 adds nine phosphates per molecule of SF2/ASF (Gui et al. 1994b), whereas topo I/kinase transfers, at the most, three phosphates (Rossi et al. 1996a). Kinetic analysis also revealed differences in the Km and Vm values of the two enzymes which clearly established that topo I/kinase is distinct from other SR protein kinases (Gui et al. 1994b). A more striking feature of this kinase is the absence of obvious sequence motifs homologous to known protein kinases (e.g. ATP binding site). It is worth noting, however, that DNA topoisomerase I binds ATP with a kDa of 50 nM consistent with that of known protein kinases (Rossi et al. 1996a). Photoaffinity labeling with 8-azidoadenosine-5′-triphosphate (α-^{32}P) combined with mutational analysis were used to identify peptides responsible for ATP binding of purified DNA topoisomerase I expressed in the baculovirus system. This domain mapped to the 10-kDa carboxyl-terminal region of the protein. Additionally, the amino-terminal domain is required for the interaction with ASF/SF2 protein, given that deletion of the N-terminal 174 amino acids of the enzyme destroys SF2/ASF binding and kinase activity, but not ATP binding (Labourier et al. 1998). Altogether, these results support the conclusion that at least two distinct domains of DNA topoisomerase I are necessary for kinase activity; one at the carboxy-terminal region that corresponds to the ATP binding site and a second at the amino-terminal region that binds ASF/SF2. The fact that mammalian and yeast topoisomerase I possess different N-terminal domains, gives a plausible explanation for the lack of kinase activity associated with the yeast enzyme.

Human DNA topoisomerase I is a nuclear phosphoprotein composed of 765 amino acids with a predicted molecular mass of 91 kDa (D'Arpa et al. 1988). It shares 40–50% identical amino acid sequences with other eukaryotic DNA topoisomerases I and its sequence can be divided into four domains based on regions of extensive homology (Gupta et al. 1995; Stewart et al. 1996). The carboxy-terminal domain, which contains the active tyrosine at position 723 and includes the ATP-binding domain (Madden and Champoux 1992; Stewart et al. 1996), is the most conserved and spans from residues 697–765 (Stewart et al. 1996). This domain is preceded by a linker region of positively charged residues that are not conserved (Stewart et al. 1996). Residues from positions 198–651 form the conserved core domain which is resistant to proteolysis with subtilysin (Stewart et al. 1996) and chymotrypsin. The amino-terminal domain is highly charged and contains four putative nuclear localization signals which are conserved in this apparently divergent region of DNA topoisomerase I (Alsner et al. 1992). Accumulated evidence demonstrated that the amino-terminal domain is dispensable for topoisomerase I activity in vitro. Deletion of this domain, by spontaneous proteolysis (Liu and Miller 1981) or by induced mutagenesis of the wild-type protein (Stewart et al. 1996), did not impede the relaxing activity of the enzyme. This domain, however, has an essential function for nuclear localization of DNA topoisomerase I (Alsner et al. 1992). More strikingly, the same sequence spanning amino acids 139–175 is sufficient for interaction with ASF/SF2 (Labourier et al. 1998) and nuclear localization. Since the sequences in this region deviate from NLS consensus sequences, it is possible that binding of DNA topoisomerase I to SR proteins or other nuclear factors, may be the mechanism by which topoisomerase I enters the nucleus.

Relaxation-deficient mutants were obtained by site-directed mutagenesis of highly conserved amino acids at the carboxy-terminal domain of DNA topoisomerase I (Jensen and Svejstrup 1996), demonstrating that this latter is involved in cleavage-ligation reactions catalyzed by DNA topoisomerase I. However, none of these mutations affected the DNA binding properties of DNA topoisomerase I (Jensen and Svejstrup 1996). A series of filter binding assays indicated that the protein binds preferentially to superhelical rather than to relaxed DNA (Madden et al. 1995), a function exhibited by the conserved core domain (amino acids 175–659; Madden et al. 1995; Stewart et al. 1996). Although our understanding of the domains involved in topo I/kinase is still rudimentary, it is clear that important residues for DNA-catalyzed reactions are not involved in the kinase reaction. In particular, mutation of tyrosine 723 at the active site was not detrimental for topo I/kinase activity, despite the binding of ATP near to this site (Rossi et al. 1998), indicating that this tyrosine is not involved in the transfer of the gamma phosphate from ATP to ASF/SF2. More work will be necessary to provide a rigorous definition of the sequences required for the kinase activity of DNA topoisomerase I.

It is not yet clear why impaired capacity to express DNA topoisomerase I is lethal to the *Drosophila* and mouse embryo (Lee et al. 1993; Morham et al. 1996). A substantial part of our understanding of the intracellular function of DNA topoisomerase I in multicellular organisms has been gleaned, either through cytological observations (Fleischmann et al. 1984; Muller et al. 1985; Gilmour et al. 1986; Buckwalter et al. 1996) or, from inhibitor studies (D'Arpa and Liu 1989; Gupta et al. 1995; Pommier 1996). DNA topoisomerase I may play a major role in transcription as the enzyme associates with puffs in polytene chromosomes, which are regions with active transcription (Fleischmann et al. 1984; Gilmour et al. 1986). Studies of c-fos (Stewart et al. 1990), HSP70 (Kroeger and Rowe 1989) and ribosomal genes (Garg et al. 1987; Zhang et al. 1988; Cavalli et al. 1996) transcription showed that DNA topoisomerase I binding and cleavage sites are scattered along regions of DNA which are actively transcribed, but are not evident when the same genes are silent. Additionally, since DNA topoisomerase I interacts with the transcription machinery, including RNA polymerase I (Rose et al. 1988), nucleolin (Bharti et al. 1996), p53 (Solary et al. 1996) and TATA transcription factors (Kretzschmar et al. 1993; Merino et al. 1993), DNA topoisomerase I may act by relaxing the positive supercoils generated upstream of a transcribing RNA polymerase as well as the negative supercoils generated downstream (Liu and Wang 1987; Wang and Lynch 1993). Several independent observations, however, cast doubt on this simple model. First, Madden and collaborators provided evidence that human DNA topoisomerase I preferentially binds to superhelical DNA, suggesting that DNA conformation itself may target the enzyme to those regions of the genome where its activity is required (Madden et al. 1995). Second, there is no evidence that DNA topoisomerase I directly drives the elongation of transcription through its relaxing activity. In fact, the DNA relaxation activity was shown to be dispensable for both repression and activation of transcription in reconstituted transcription reactions (Merino et al. 1993).

During DNA replication, the movement of the large polymerase complex is presumed to lead to the formation of local domains of torsional stress ahead of the DNA replication fork. As specific inhibitors of type I topoisomerase induce DNA breakage at replication, it is believed that the enzyme escorts the DNA polymerase. In favor of this hypothesis are the findings that the inhibitory drugs also interfere with different stages of simian virus 40 DNA replication (Gilmour et al. 1986), and that DNA topoisomerase I is preferentially associated with normal SV40 replicative intermediates and nonreplicating SV40 DNAs which are deficient in histones (Champoux 1988, 1992; Tsao et al. 1993). Using yeast strains with mutations in the DNA topoisomerase I and DNA topoisomerase II genes, Brill and his colleagues reported that DNA and ribosomal RNA synthesis are dramatically inhibited at the restrictive temperature, suggesting that DNA replication and at least ribosomal RNA synthesis require an active topoisomerase (Brill et al. 1987). The simplest interpretation of these data is that one or the other topoisomerase can function as a swivel to allow replication fork movement. It appears then that DNA topoisomerase I plays a role in DNA replication that can be mediated by DNA topoisomerase II. Why the latter cannot substitute for the former in higher eukaryotic systems remains an unanswered question.

The discovery that DNA topoisomerase I has an intrinsic protein kinase activity will certainly allow novel analyses to be performed (Rossi et al. 1996a). Topo I/kinase may be required to achieve specific phosphorylation of proteins that associate either directly or indirectly with the transcription machinery.

We now know that topo I/kinase phosphorylates SR proteins that are associated with actively transcribed regions of *Drosophila* polytene chromosomes (Roth et al. 1991). As SR proteins are involved in the splice site choice of alternatively regulated genes in *Drosophila* (Kraus and Lis 1994; Ring and Lis 1994; Allemand et al. 2001), variations in the phosphorylation state of these factors may be instrumental in the correct *Drosophila* development. The finding that reversible phosphorylation can modulate the activity of splicing factors during the course of the reaction (Mermoud et al. 1992, 1994a,b; Tazi et al. 1992) is consistent with this possibility. In addition, SR proteins, extracted from HeLa cells treated with DNA topoisomerase I inhibitors, have a different pattern of phosphorylation as compared to those extracted from untreated cells (Rossi et al. 1996a). In vitro, the drug blocks topo I/ kinase activity showing that failure to appropriately phosphorylate SR proteins in treated cells is likely due to the inhibition of this kinase. The incomplete inhibition of SR protein phosphorylation suggests that other protein kinase(s) not affected by the drug are also active in vivo, but have a specificity different from that of topo I/kinase. Along this line, the phosphorylation mediated by topo I/kinase depends on structural features of the RS domain rather than on specific sequences (Labourier et al. 1998).

DNA topoisomerase I is also the nuclear target for a number of anticancer agents derived from the plant alkaloid camptothecin (Pommier et al. 1998) and for indolocarbazole derivatives (Bailly et al. 1999a,b) such as the antibiotic rebeccamycin, the antitumor agent NB-506 which is undergoing phase II clinical trials (Arakawa et al. 1995; Yoshinari et al. 1995; Akinaga et al. 2000; Ohkubo et al. 2000; Urasaki et al. 2001). While NB-506 bears a structural analogy with the specific protein kinase C (PKC) inhibitor staurosporine (Anizon et al. 1998), it has no significant effect on PKC (Anizon et al. 1998; Pommier et al. 1998; Bailly et al 1999b; Labourier et al. 1999a), but inhibits both relaxation (Arakawa et al. 1995; Bailly et al. 1997; Fukasawa et al. 1998; Bailly et al. 1999a,b; Labourier et al. 1999a) and kinase activities of topo I/kinase (Pilch et al. 2001). Recently, NB-506 has been shown to block, in vitro, spliceosome assembly and splicing through inhibition of SR protein phosphorylation. NB-506 also leads to specific inhibition of SR protein phosphorylation in cultured cells, suggesting that the drug could modulate gene expression by changing the splicing pattern of protein encoding genes. Consistent with this suggestion, NB-506 induces dramatic changes of the mRNAs distribution in P388 leukemia-sensitive cells, but not in P388CPT5 cells which are resistant to the drug. NB-506 treatment also leads to pronounced changes in the alternative splicing pattern of several genes, among which *Bcl-X* and *CD 44* gene products are known to affect apoptosis and tumor progression (Pilch et al. 2001). Such alterations of regulated splicing may well be a key step in accounting for the remarkable antineoplastic activities exhibited by the drug both in cell culture systems and in xenograft models of human tumor (Arakawa et al. 1995; Pommier et al. 1998). Recently, cell lines highly resistant to camptothecin were described to have hardly any detectable DNA topoisomerase I as analyzed by Western blotting (Pourquier et al. 2000). Analyses of splicing patterns of pre-mRNAs and phosphorylation status of SR proteins in these cells lines will prove to be decisive for the elucidation of the topo I/kinase role in splicing. This challenging and important issue is being actively pursued in our laboratory.

5 Phosphorylation of SR-Like Proteins

In addition to the set of phosphorylated SR proteins recognized by mAb104, several RS containing proteins have been recently identified through screening of EST databases using known SR protein sequences. These include SRrp86, an atypical SR-related protein that contain a single N-terminal RRM and two C-terminal RS-rich regions (Barnard and Patton 2000). This protein was shown to repress activation of splicing in extract complemented with SR proteins and it was also able to antagonize the effect of SR proteins on the use of competing 5′ splice sites (Barnard and Patton 2000). The RS domain of SRrp86 mediates specific interaction with a subset of SR proteins, suggesting that its effects on splicing were due to the modulation of activity of a subset of the authentic SR proteins. While initial characterization of SRrp86 indicated that the protein was phosphorylated at the RS domain, no data are now available concerning the role of this phosphorylation in splicing and/or cellular localization.

SRrp40 and SRrp35 are proteins highly homologous to the authentic SR proteins, but share similar properties with SRrp86. These proteins are proposed to antagonize authentic SR proteins in the modulation of alternative 5′ splice sites choice. In complementation assays, purified SRrp40 protein was shown to repress splicing in a dose-dependent manner (Cowper et al. 2001). Like SR proteins, SRrp40 run with higher apparent molecular weight on SDS-PAGE due to phosphorylation of the RS domain (Cowper et al. 2001). Given that SRrp40 is a shuttling protein and that shuttling activity of SR proteins is determined by the level of the RS domain phosphorylation, it may be anticipated that the phosphorylation of the RS domain of SRrp40 regulates its cellular distribution.

Several functions of SR proteins can also be antagonized by RSF1, a splicing repressor isolated from *Drosophila* which also exhibits a modular organization with an N-terminal RRM-type RNA binding domain and a C-terminal part enriched in glycine (G), arginine (R) and serine (S) residues (GRS domain; Labourier et al. 1999b). RSF1 can bind SR proteins and acts at early stages of splicing to prevent U1 snRNP binding to the 5′ splice site. The RRM domain of RSF1 mediates specific recognition of ESE sequences, whereas the GRS domain is responsible for specific protein-protein interactions, which are instrumental for the splicing repression as well as for the regulation of its subcellular localization (Labourier et al. 1999b,c). Phosphorylation of the GRS domain of RSF1 enhances its interaction with the same transportin involved in *Drosophila* SR proteins, dTRN-SR, implying that this modification may contribute to its spatial and temporal regulation (Allemand et al., submitted).

The U2AF factor, a stable component of the E complex involved in the selection of 3′ splice sites (Ruskin et al. 1988), can also be considered as an SR like protein. U2AF purified from HeLa cells comprises two subunits, 35K and 65K (Zamore and Green 1991). $U2AF^{65}$ binds the pre-mRNA in a sequence-dependent manner, with significant preference for pyrimidine stretches at the 3′ splice site (Nelson and Green 1989; Zamore and Green 1991; Singh et al. 1995), whereas $U2AF^{35}$ interacts with the 3′ splice site dinucleotide AG (Merendino et al. 1999; Wu et al. 1999; Zorio and Blumenthal 1999) and is essential for regulated splicing (Wu and Maniatis 1993; Graveley et al. 2001). The deduced amino acid sequence of $U2AF^{65}$ contains three carboxy-terminal RRMs and an amino-terminal RS motif (Zamore et al. 1992). However, only the RS motif characterizes the amino acid sequence of $U2AF^{35}$ (Zhang et al. 1992). Despite the presence of these functionally important regions in both $U2AF^{65}$ and SR proteins, in contrast to SR proteins, $U2AF^{65}$ has no effect on the selection of the 5′ splice site and cannot restore the splicing activity of post-nuclear S100 extracts. It is, however, an essential splicing factor because splicing extracts depleted of U2AF were unable to perform the splicing reaction unless they were complemented with U2AF (Zamore and Green 1989, 1991; Zamore et al. 1992; Zuo and Maniatis 1996). Moreover, the genes encoding the *Drosophila* homologue of U2AF, which complements U2AF activity in HeLa cell extract (Kanaar et al. 1993), is essential for viability (Kanaar et al. 1993; Rudner et al. 1996, 1998a,b). The RS domain of either 35 or 65 subunits is sufficient for viability (Rudner et al. 1998a), splicing activity and also to mediate nuclear import of the U2AF to the nucleus as a heterodimer (Gama-Carvalho et al. 2001). However, only $U2AF^{35}$ was shown to mediate specific protein-protein interaction with SR proteins (Wu and Maniatis 1993; Zuo and Maniatis 1996). While there is evidence that U2AF is phosphorylated by specific kinases (Zamore and Green 1989), the role of this phosphorylation awaits further clarification. The same hold true for other regulators of pre-mRNA splicing with RS-like domains, for instance, the *Drosophila* tra and tra2 (Du et al. 1998) and SRm160/300 (Blencowe et al. 1998).

6 Phosphorylation of hnRNP Proteins

6.1 HnRNP A1

Several reports showed that the ratio of the SR protein SF2/ASF to heterogeneous ribonucleoprotein A/B proteins specifies the selection of alternative splice sites. The hnRNP A1 was first shown to antagonize SF2/ASF on 5′ splice site selection. Indeed, hnRNP A1 leads to the activation of distal 5′splice sites (Mayeda and Krainer 1992; Mayeda et al. 1993, 1994; Sun et al. 1993; Caceres et al. 1994; Yang et al. 1994), whereas SF2/ASF and several members of the SR protein family promote use of proximal 5′ splice sites. The ratio of SF2/ASF to hnRNP A1 also regulates exon skipping and exon inclusion of natural and model pre-mRNAs. While an excess of SF2/ASF effectively prevents inappropriate exon skipping, hnRNP A1 rather promotes alternative exon skipping (Mayeda and Krainer 1992; Mayeda et al. 1993; Caceres et al. 1994; Yang et al. 1994; Shen et al. 1995). These antagonistic activities of SR proteins and hnRNP A1 were also demonstrated in transfected cells, suggesting that these two families of RNA binding proteins play prominent roles in the regulation of alternative splicing of a wide variety of transcripts in vivo.

The mechanism by which hnRNP A1 alters pre-mRNA splicing is not fully understood (Eperon et al. 2000; Zhu et al. 2001), mainly because it is bound to many, if not all, pre-mRNAs and not exclusively to pre-mRNAs subjected to alternative splicing (Swanson and Dreyfuss 1988b; Buvoli et al. 1990, 1992; Burd and Dreyfuss 1994). It is possible that high affinity binding of hnRNP A1 to specific sequences that may be present on only certain pre-mRNAs could have an effect on the kinetics of binding by hnRNP A1, relative to other splicing factors, and thereby determine the precise outcome of splice site selection for each substrate. Alternatively, given that hnRNP A1, like other RNA binding proteins, promotes rapid annealing of complementary strands of both RNA and DNA (Kumar et al. 1986; Kumar and Wilson 1990; Pontius and Berg 1990, 1992; Munroe and Dong 1992), it is conceivable that this activity also contributes to RNA-RNA interactions occurring during spliceosome formation. Both RNA binding and strand annealing activities of hnRNP A1 are regulated by phosphorylation (Cobianchi et al. 1993; Idriss et al. 1994). Metabolic labeling revealed that hnRNP A1 is phosphorylated, but has a high dephosphorylation rate (Cobianchi et al. 1993). The ability of recombinant hnRNP A1 to act as substrate for a number of purified Ser/thr protein kinases showed the protein to be phosphorylated exclusively on serine residues located at both N-terminal (Ser95) and C-terminal portions (Ser192 and Ser199). Phosphorylation of Ser199 by PKA resulted in the suppression of inter-strand annealing activity of hnRNP A1 (Idriss et al. 1994), without any detectable effect on its nucleic acid binding capacity. However, PKC phosphorylation, which was more abundant at site Ser192, was associated with an inhibition of RNA annealing activity and a decrease in affinity of hnRNP A1 for single-stranded polynucleotides (Idriss et al. 1994). Interestingly, protein phosphatase PP2A, but not PP1, dephosphorylated hnRNP A1 and reversed the inhibitory effect of PKC phosphorylation on the strand annealing activity (Idriss et al. 1994). Given that both RNA binding and re-annealing activity of hnRNP A1 might have a modulatory effect in pre-mRNA splicing, for example by facilitating and/or destabilizing RNA-RNA base pairing, it is tempting to propose that a cascade of phosphorylation and dephosphorylation mediated by PKC and PP2A could be responsible for sequential binding of splicing factors. The findings that hnRNP

A1 negatively regulates the binding of U1 snRNP at the 5′ splice sites (Eperon et al. 2000) and that U6 snRNP replaces U1 snRNP for binding the same sites (Kuhn et al. 1999; Murray and Jarrel 1999; Staley and Guthrie 1999; see below), make it a likely candidate whose dephosphorylation is important for subsequent catalysis by the spliceosome (Murray and Jarrel 1999; Eperon et al. 2000).

Phosphorylation of hnRNP A1 could also be a way of regulating its subcellular distribution. While hnRNP A1 has a general nucleoplasmic localization excluded from nucleoli, one of the most intriguing properties of this protein is that it shuttles between the nucleus and cytoplasm in a transcription-dependent manner (Pinol-Roma and Dreyfuss 1991, 1992; Ghetti et al. 1992). Recently, it has been shown that exposure of cells to stress stimuli such as osmotic shock or UVC irradiation induced a marked cytoplasmic accumulation of hnRNP A1, concomitant with an increase in its phosphorylation (van der Houven van Ordt et al. 2000). While the kinase responsible for the large increase in phosphorylation of hnRNP A1 is presently unknown, it could be that this hyperphosphorylation is part of the signal transduction pathway activated by osmotic shock. Indeed, the subcellular localization of hnRNP A1 is modulated by the classical stress-response kinase cascades involving $MKK_{3/6}$ and p38, but not by mitogenic activators such as PDGF or EGF (van der Houven van Ordt et al. 2000). However, it is not clear at the moment whether the increased concentration of hnRNP A1 in the cytoplasm is a result of increased export, reduced import or both. Nevertheless, these findings open novel issues to understand how signal-transduction pathways interface with the splicing machinery.

6.2
SF1/BBP, Member of the hnRNP-K Family

SF1/BBP is a 75-kDa protein that specifically recognizes the intron branch point sequence (BPS) in the pre-mRNA transcripts and allows stable binding of U2AF (Krämer and Utans 1991; Abovich and Rosbash 1997; Berglund et al. 1998; Guth and Valcarcel 2000). Mammalian SF1 contains an hnRNP K homology (KH) domain, a motif found in a variety of proteins associated with hnRNPs (Nagai 1996), a zinc knuckle and a C-terminal part rich in proline residues (Arning et al. 1996). Different isoforms of SF1 that differ in the length of the proline-rich region are generated by alternative splicing and are expressed in various mammalian cell types (Arning et al. 1996). This domain is thought to mediate protein-protein interactions with SH2 containing proteins (Taylor and Shalloway 1994) and thereby connect this splicing factor to signaling pathways. The KH motif of both yeast and mammalian SF1 is sufficient for specific recognition of the BPS, whereas the zinc knuckle supplies nonspecific general binding (Berglund et al. 1998; Rain et al. 1998; Peled-Zehavi et al. 2001).

SF1 is a substrate of cGMP-dependent protein kinase-I (PGK-I) both in vitro and in vivo. This phosphorylation concerns Ser 20, which inhibits the SF1-U2AF^{65} interaction leading to a block of pre-spliceosome assembly. Mutational analysis of amino acids 20–22 (Ser-Arg-Trp) further confirmed that this region of SF1 is necessary for the interaction with U2AF^{65} (X. Wang et al. 1999). While this region overlaps with a consensus phosphorylation site for cAMP-dependent kinase, SF1 was not phosphorylated in vitro by this kinase, making it unlikely that SF1 is a target substrate in vivo. The phosphorylation of Ser 20 by PKG did not affect the RNA binding activity of SF1 (X. Wang et al. 1999). This fits well with known abilities of SF1 and U2AF^{65} to bind independently of each other to the branch site and the polypyrimidine tract, respectively (Zamore et al. 1992; Abovich and Rosbash 1997; Berglund et al. 1997). Upon interaction, the proteins cooperate to form a ternary complex with the pre-mRNA, and the affinity of SF1 and U2AF^{65} for the RNA is increased several fold (Berglund et al. 1998). By preventing interaction between SF1 and U2AF, PKG-I indirectly weakens the binding of these factors to pre-mRNA.

Given that Ser 20 is immediately adjacent to a putative nuclear localization signal (NLS) in SF1, it is possible that PKG phosphorylation also regulates the intracellular localization of SF1. Consistent with this possibility, PKG-I is primarily found in the cytoplasm and translocates to the nucleus upon stimulation of cells with cGMP which exposes a cryptic NLS in the ATP-binding domain of the protein (Gudi et al. 1997). Again, the parallel with other proteins involved at initial stages of the spliceosome assembly is intriguing in that phosphorylation of splicing factors regulates both their localization and activity.

6.3
hnRNP C

The C-group of hnRNP proteins are abundant nuclear proteins of 42–44 kDa that bind to nascent pre-mRNA and are thought to participate in an early step of the pre-mRNA splicing pathway (Pederson 1974; Beyer et al. 1977; Karn et al. 1977; Choi et al. 1986; Mayrand et al. 1986). The hnRNP-C is known to bind avidly to poly(U) sequences (Swanson and Dreyfuss 1988a,b; Wilusz et al. 1988; Wilusz and Shenk 1990), including uridylate stretches occurring at the 3′ end of mammalian U6 snRNA (Forné et al. 1995). Given that hnRNP-C induces disruption of base-paired U4/U6 snRNAs of the U4/U6 snRNP particle, it is proposed to be involved in a step leading to dissociation of U4 snRNP during the spliceosome assembly (Forné et al. 1995). Surprisingly, a discrete region of U6 snRNA was demonstrated to modulate the phosphorylation cycle of hnRNP-C (Mayrand et al. 1996). Cleavage of nucleotides 78–95, but not other regions of U6 snRNA, nor other snRNAs, resulted in an inhibition of the dephosphorylation step of the hnRNP C protein phosphorylation cycle. This inhibition was as pronounced as that seen with the serine/threonine protein phosphatase inhibitor okadaic acid.

Unlike other hnRNP proteins, hnRNP-C is not a shuttling protein and thus has an exclusive nuclear localization (Weighardt et al. 1995; Nakielny and Dreyfuss 1996). It is also known that hnRNP C is phosphorylated in vivo by a casein kinase II-type activity (Fung et al. 1997). This phosphorylation weakens its interaction with RNA and dephosphorylation of hnRNP was shown to be required for binding target sequences (Mayrand et al. 1993, 1996). In addition to modulating the binding of hnRNP-C to pre-mRNA, the phosphorylation/dephosphorylation may also influence its homotypic associations (Barnett et al. 1989) or its binding to other spliceosome or nuclear proteins as well as import receptors (Georgatos et al. 1988; Meier and Blobel 1992).

7 Phosphorylation of snRNP Components

7.1 The U1 snRNP-Specific Protein 70K

A role for U1 snRNP in the splicing process was proposed soon after the discovery of RNA splicing in higher eukaryotes. The sequence complementarity of the 5′ end of U1 snRNA with the 5′ splice site consensus led to the suggestion that base pairing might serve to align the two ends of an intron during the splicing process (Lerner et al. 1980; Rogers and Wall 1980). The demonstration that this base pairing is indeed critical for splicing was provided by experiments in which defective splicing of precursors with mutated 5′ splice sites was rescued by compensatory mutations which restored complementarity with U1 snRNA (Zhuang and Weiner 1986). Interestingly, these experiments also revealed that not all the compensatory mutations restored splicing with the same efficiency, indicating that mechanisms other than base pairing, possibly involving proteins or other snRNPs, are directing the selection of the 5′ splice site. In vitro studies proved that U1 snRNP-specific proteins (C and 70K) are indeed required for efficient recognition of the 5′ splice site by U1 snRNP (Heinrichs et al. 1990; Jamison et al. 1995; Rossi et al. 1996b).

Remarkably, the U1–70K protein shares structural features with SR proteins, containing an N-terminal RRM and C-terminal RS domain (Theissen et al. 1986; Query et al. 1989; Krainer et al. 1991). This similarity suggested that U1–70K might interact directly with SR proteins in the selection of the 5′ splice site, a hypothesis which is supported by the finding that highly purified U1 snRNP and ASF/SF2 protein cooperate to form a ternary complex with pre-mRNA (Kohtz et al. 1994). These U1 snRNP-ASF/SF2 complexes can distinguish between a functional 5′ splice site and an inactive site where the GU dinucleotide is mutated to AU (Jamison et al. 1995). Furthermore, the RRM and RS domains of ASF/SF2 are required, indicating that specific sequences of the pre-mRNA as well as interaction of ASF/SF2 with U1 snRNP protein(s) mediate selection of the functional 5′ splice site (Kohtz et al. 1994). Analysis of U1 snRNPs in which specific proteins were deleted provided evidence that the U1–70K-specific protein is necessary for cooperative binding (Jamison et al. 1995). Because ASF/SF2 and U1–70K RS domains can interact together (Wu and Maniatis 1993; Kohtz et al. 1994), it is thought that complex formation results from direct interactions between these two factors. As stated above, this interaction is positively regulated by phosphorylation and the U1–70K protein has clearly been shown to exist as a phosphoprotein in vivo (Woppman et al. 1993). SF2/ASF and the U1–70K can be phosphorylated in vitro by a U1 snRNP-associated kinase whose identity is presently unknown (Tazi et al. 1993; Woppman et al. 1993).

Cross-linking experiments in mammalian (Sawa and Shimura 1992; Wyatt et al. 1992; Sontheimer and Steitz 1993) and in yeast (Sawa and Abelson 1992) extracts showed that the association of U1 snRNP with the 5′ splice site diminishes during the formation of the mature spliceosome, excluding the possibility that U1 snRNA itself participates in the catalysis of splicing reactions. Moreover, disruption of base pairing between the 5′ splice site and the 5′ end of U1 snRNA is required after the initial step of spliceosome assembly has been performed (Konforti et al. 1993). Genetic tests provide support for models, whereby a conserved domain in the U6 snRNA interacts with the intron positions that base pair with U1 snRNP early in spliceosome assembly. In this model, U6 snRNA is responsible for positioning the 5′ splice site in the catalytic center after being selected by U1 snRNP (Kandels-Lewis and Séraphin 1993; Lesser and Guthrie 1993). It is, therefore, possible that weakening the interaction between SF2/ASF and U1–70K by dephosphorylation has the potential to favor this exchange of snRNPs and hence, to allow the spliceosome transition required for catalysis. This possibility is consistent with the observations that thiophosphorylated U1–70K (Tazi et al. 1993) and SF2/ASF (Cao et al. 1997; Xiao and Manley 1998) inhibit catalysis, but not spliceosome assembly.

U1 snRNP appears to be the factor that recruits the 5′ splice site pre-mRNA sequences to the catalysis by the spliceosome. The alternative splicing activity of SF2/ASF has also been shown to correlate with its ability to promote multiple occupancy of alternative 5′ splice site by U1 snRNP (Eperon et al. 1993, 2000; Kohtz et al. 1994; Jamison et al. 1995). However, deletion of the RS domain did not alter this property of SF2/ASF, suggesting that the two RRMs of SF2/ASF determine alternative splicing (Caceres and Krainer 1993; Zuo and Manley 1993; Wang and Manley 1995; Caceres et al. 1997). Moreover, the RS domain has also been shown to be dispensable for in vitro splicing of several, but not all, pre-mRNAs (Zhu and Krainer 2000). Thus, the role of RS domain-mediated interaction with U1–70K in regulated and constitutive splicing could be more complex than initially suggested. Additional factors may, indeed, participate in the selection of 5′ splice sites, some of these regulatory proteins, i.e., hnRNP A1, TIA1 and RSF1, also being phosphoproteins.

7.2
The U2 snRNP 155K

The binding of U2 snRNP triggers the formation of pre-spliceosome complex A, where U2 snRNA forms an essential duplex with the branch site. As mentioned above, $U2AF^{65}$, SF1 and SR proteins facilitate this base pairing, allowing the positioning of U2 snRNP. In addition, a set of U2 snRNP-associated proteins plays a critical role in tethering the U2 snRNP to the pre-mRNA. These proteins include the heteromeric splicing factors SF3a and SF3b which interact with the 3′ and 5′ portions of U2 snRNA, respectively. SF3a consists of three subunits of 60, 66 and 120 kDa, whereas SF3b comprises four polypeptides of 50, 130, 145 and 155 kDa (Brosi et al. 1991, 1993; Bennett et al. 1992; Behrens et al. 1993a,b; Krämer et al. 1999).The U2 snRNP could be guided to the branch site by interaction of one or more of its polypeptides with $U2AF^{65}$. Such protein-protein interactions have been uncovered both in yeast and mammals (Bennett and Reed 1993; Legrain and Chapon 1993; Legrain et al. 1993; Chiara et al. 1994; Krämer et al. 1995; Rain et al. 1996; Fromont-Racine et al. 1997). For instance, the 155-kDa subunit of SF3b has recently been shown to interact directly with $U2AF^{65}$ (Gozani et al. 1998) and probably this interaction is involved in the recruitment of U2 snRNP to the BPS (Gozani et al. 1998). This U2 snRNP component also contacts pre-mRNA on both sides of BPS and is therefore positioned near or at the spliceosome catalytic center (C. Wang et al. 1998). Significantly, the 155-kDa protein is phosphorylated concomitant with or just after the first catalytic step one and is a target for cyclin E/cyclin-dependent kinase 2 (Seghezzi et al. 1998). However, phosphorylated isoforms of the 155-kDa protein were only detected in purified spliceosome and only at a specific time in the splicing pathway, implying that this phosphorylation is tightly regulated. This fit nicely with early indications, described above, showing that phosphorylation/dephosphorylation events are critical for catalysis by the spliceosome.

Interestingly, the C-terminal two thirds of the 155-kDa protein are similar to the regulatory subunit A of the phosphatase PP2A (C. Wang et al. 1998), suggesting that interaction with this phosphatase could trigger a rapid turnover of its phosphorylation status. Lack of phosphatase activity associated with the purified 155-kDa protein and absence of interaction with the PP2A domain make it unlikely that PP2A is involved in dephosphorylating the 155-kDa protein. Nevertheless, a phosphatase activity required for the 155-kDa dephosphorylation must exist and would be expected to play an important role in splicing.

7.3
The U4/U5/U6 Associated Proteins

The mature spliceosome requires the association of the 25S [U4/U6.U5] tri-snRNP complex (Bindereif and Green 1987; Cheng and Abelson 1987; Konarska and Sharp 1987; Black and Pinto 1989) with pre-assembled complex A containing U2 snRNP. Given that the U6 and U5 snRNAs make weak contacts with the pre-mRNA sequences at the splice junctions, the recruitment of the tri-snRNP complex into the spliceosome is thought to be mediated by both protein-protein and protein-RNA interactions. Consistent with this idea the 5′ splice site and, later in the reaction, the 3′ splice site is contacted by the U5-snRNP-specific 220-kDa protein, the mammalian homologue of Prp8 (Wyatt et al. 1992; MacMillan et al. 1994; Teigelkamp et al. 1995a,b; Umen and Guthrie 1995a,b,c; Siatecka et al. 1999). This interaction is likely to stabilize the binding of U5 snRNA at the splice junctions. It has also been shown that SR proteins escort the tri-snRNP particle to the pre-spliceosome, suggesting that protein-protein interactions involving SR proteins and target proteins of the tri-snRNP complex potentially mediate the integration of the tri-snRNP complex into the spliceosome (Roscigno and Garcia-Blanco 1995). Consistent with this suggestion, four proteins 27, 65, 100 and 110 kDa which are part of the tri-snRNP complex are SR-like proteins with N-terminal RS domains (Fetzer et al. 1997; Makarova et al. 2001). Although all these proteins are phosphorylated in vitro by unidentified purified snRNP-associated kinase and SR protein kinase Clk/Sty, evidence that they are phosphorylated in vivo is lacking. Further studies are required to determine whether the phosphorylation of the RS domain of these factors is important to modulate their nuclear import and/or splicing activities.

8
Concluding Remarks

The studies reported in this chapter indicate that phosphorylation-dephosphorylation of spliceosomal components constitutes a key mechanism required for accurate splicing of pre-mRNAs. Indeed, the phosphorylation level of members of the SR protein family not only affects their interactions with other splicing factors and with specific sequences of the pre-mRNA, but also controls their subcellular localization. In the case of alternative splicing, which has to be tightly regulated in a tissue-specific or developmental way, phosphorylation of splicing factors also turns out to represent a powerful means to control the selection/activation of alternative splice sites in the early steps of spliceosome assembly. Given that SR proteins regulate alternative splicing events in a dose-dependent way, phosphorylation likely contributes, subsequently to transcriptional and post-transcriptional control of SR protein expression, to modulate the levels of SR proteins involved in splicing. In the light of these observations, it is tempting to speculate that altering the phosphorylating activity of SR protein-specific kinases could lead to a reversal of dysregulated splicing events associated with a growing number of diseases. In support of this, Hartmann et al. (2001) reported a change in human Tau exon 10 splicing by phosphorylation of splicing factors. Moreover, we have recently reported that specific inhibition of the DNA topoisomerase I kinase activity by an indolocarbazole derivative can affect gene expression by modulating pre-mRNA splicing (Pilch et al. 2001). Development of high throughput screening systems allowing identification of drugs interfering specifically either with each of the SR protein kinases characterized so far, or with the signal transduction pathway(s) activating them could then open up a promising avenue of investigation towards new therapies of human diseases.

Acknowledgement. The work was supported by grants from the Association pour la Recherche sur le Cancer, GEFLUC and CNRS.

References

Abovich N, Rosbash M (1997) Cross-intron bridging interactions in the yeast commitment complex are conserved in mammals. Cell 89:403–412

Akinaga S, Sugiyama K, Akiyama T (2000) UCN-01 (7-hydroxystaurosporine) and other indolocarbazole compounds: a new generation of anti-cancer agents for the new century? Anticancer Drug Des 15:43–52

Akker SA, Smith PJ, Chew SL (2001) Nuclear post-transcriptional control of gene expression. J Mol Endocrinol 27:123–131

Allemand E, Gattoni R, Bourbon HM, Stevenin J, Caceres JF, Soret J, Tazi J (2001) Distinctive features of *Drosophila* alternative splicing factor RS domain: implication for specific phosphorylation, shuttling, and splicing activation. Mol Cell Biol 21:1345–1359

Alsner J, Svejstrup JQ, Kjeldsen E, Sorensen BS, Westergaard O (1992) Identification of an N-terminal domain of eukaryotic DNA topoisomerase I dispensable for catalytic activity but essential for in vivo function. J Biol Chem 267:12408–12411

Amrein H, Gorman M, Nöthinger R (1988) The sex-determining gene tra-2 of *Drosophila* encodes a putative RNA binding domain. Cell 55:1025–1035

Amrein H, Hedley ML, Maniatis T (1994) The role of specific protein-RNA and protein-protein interactions in positive and negative control of pre-mRNA splicing by Transformer 2. Cell 76:735–746

Anizon F, Moreau P, Sancelme M, Voldoire A, Prudhomme M, Ollier M, Severe D, Riou JF, Bailly C, Fabbro D, Meyer T, Aubertin AM (1998) Syntheses, biochemical and biological evaluation of staurosporine analogues from the microbial metabolite rebeccamycin. Bioorg Med Chem 6:1597–1604

Arakawa H, Iguchi T, Morita M, Yoshinari T, Kojiri K, Suda H, Okura A, Nishimura S (1995) Novel indolocarbazole compound 6-N-formylamino-12,13-dihydro-1,11-dihydroxy-13-(beta-D-glucopyranosyl)-5H-indolo[2,3-a]pyrrolo-[3,4-c]carbazole-5,7(6H)-dione (NB-506): its potent antitumor activities in mice. Cancer Res 55:1316–1320

Arning S, Gruter P, Bilbe G, Krämer A (1996) Mammalian splicing factor SF1 is encoded by variant cDNAs and binds to RNA. RNA 2:794–810

Bailly C, Riou JF, Colson P, Houssier C, Rodrigues-Pereira E, Prudhomme M (1997) DNA cleavage by topoisomerase I in the presence of indolocarbazole derivatives of rebeccamycin. Biochemistry 36:3917–3929

Bailly C, Carrasco C, Hamy F, Vezin H, Prudhomme M, Saleem A, Rubin E (1999a) The camptothecin-resistant topoisomerase I mutant F361S is cross-resistant to antitumor rebeccamycin derivatives. A model for topoisomerase I inhibition by indolocarbazoles. Biochemistry 38:8605–8611

Bailly C, Dassonneville L, Colson P, Houssier C, Fukasawa K, Nishimura S, Yoshinari T (1999b) Intercalation into DNA is not required for inhibition of topoisomerase I by indolocarbazole antitumor agents. Cancer Res 59:2853–2860

Barnard DC, Patton JG (2000) Identification and characterization of a novel serine-arginine-rich splicing regulatory protein. Mol Cell Biol 20:3049–3057

Barnett SF, Friedman DL, LeStourgeon WM (1989) The C protein of HeLa 40S nuclear ribonucleoprotein particles exist as anisotropic tetramers of (C1)3 C2. Mol Cell Biol 9:492–498

Behrens SE, Galisson F, Legrain P, Luhrmann R (1993a) Evidence that the 60-kDa protein of 17S U2 small nuclear ribonucleoprotein is immunologically and functionally related to the yeast PRP9 splicing factor and is required for the efficient formation of prespliceosomes. Proc Natl Acad Sci USA 90:8229–8233

Behrens SE, Tyc K, Kastner B, Reichelt J, Lührmann R (1993b) Small nuclear ribonucleoprotein (RNP) U2 contains numerous additional proteins and has a bipartite RNP structure under splicing conditions. Mol Cell Biol 13:307–319

Ben-David Y, Letwin K, Tannock L, Bernstein A, Pawson T (1991) A mammalian protein kinase with potential for serine/threonine and tyrosine phosphorylation is related to cell cycle regulators. EMBO J 10:317–325

Bender J, Fink GR (1994) AFC1, a LAMMER kinase from *Arabidopsis thaliana*, activates STE12-dependent processes in yeast. Proc Natl Acad Sci USA 91:12105–12109

Bennett M, Reed R (1993) Correspondence between a mammalian spliceosome component and an essential yeast splicing factor. Science 262:105–108

Bennett M, Michaud S, Kingston J, Reed R (1992) Protein components specifically associated with prespliceosome and spliceosome complexes. Genes Dev 6:1986–2000

Berget SM, Moore C, Sharp PA (1977) Spliced segments at the 5′ terminus of adenovirus 2 late mRNA. Proc Natl Acad Sci USA 74:3171–3175

Berglund JA, Chua K, Abovich N, Reed R, Rosbash M (1997) The splicing factor BBP interacts specifically with the pre-mRNA branchpoint sequence UACUAAC. Cell 89:781–787

Berglund JA, Abovich N, Rosbash M (1998) A cooperative interaction between U2AF65 and mBBP/SF1 facilitates branchpoint region recognition. Genes Dev 12:858–867

Beyer AL, Christensen ME, Walker BW, LeStourgeon WM (1977) Identification and characterization of the packaging proteins of core 40S hnRNP particles. Cell 11:127–138

Bharti AK, Olson MOJ, Kufe DW, Rubin EH (1996) Identification of a Nucleolin binding site in human topoisomerase I. J Biol Chem 271:1993–1996

Bindereif A, Green MR (1986) Ribonucleoprotein complex formation during pre-mRNA splicing in vitro. Mol Cell Biol 6:2582–2592

Bindereif A, Green MR (1987) An ordered pathway of snRNP binding during mammalian pre-mRNA splicing complex assembly. EMBO J 6:2415–2424

Black DL, Pinto AL (1989) U5 small nuclear ribonucleoprotein: RNA structure analysis and ATP-dependent interaction with U4/U6. Mol Cell Biol 9:3350–3359

Blencowe BJ, Issner R, Nickerson JA, Sharp PA (1998) A coactivator of pre-mRNA splicing. Genes Dev 12:996–1009

Boggs RT, Gregor P, Idriss S, Belote JM, McKeown M (1987) Regulation of sexual differentiation in *D. melanogaster* via alternative splicing of RNA from the transformer gene. Cell 50:739–747

Brill SJ, DiNardo S, Voelkel-Meiman K, Sternglanz R (1987) Need for DNA topoisomerase activity as a swivel for DNA replication and for transcription of ribosomal DNA. Nature 326: 414–416

Brody E, Abelson J (1985) The "spliceosome": yeast pre-messenger RNA associates with a 40S complex in a splicing-dependent reaction. Science 228:963–967

Brosi R, Hauri HP, Krämer A (1991) Separation of splicing factor SF3 into two components and purification of SF3a activity. J Biol Chem 268:17640–17646

Brosi R, Gröning K, Behrens SE, Lührmann R, Krämer A (1993) Interaction of mammalian splicing factor SF3a with U2 snRNP and relation of its 60-kD subunit to yeast PRP9. Science 262:102–105

Buckwalter CA, Lin AH, Tanizawa A, Pommier YG, Cheng YC, Kaufmann SH (1996) RNA synthesis inhibitors alter the subnuclear distribution of DNA topoisomerase I. Cancer Res 56: 1674–1681

Burd CG, Dreyfuss G (1994) RNA binding specificity of hnRNP A1: significance of hnRNP A1 high-affinity binding sites in pre-mRNA splicing. EMBO J 13:1197–1204

Buvoli M, Cobianchi F, Biamonti G, Riva S (1990) Recombinant hnRNP protein A1 and its N-terminal domain show preferential affinity for oligodeoxynucleotides homologous to intron/exon acceptor sites. Nucleic Acids Res 18:6595–6600

Buvoli M, Cobianchi F, Riva S (1992) Interaction of hnRNP A1 with snRNPs and pre-mRNAs: evidence for a possible role of A1 RNA annealing activity in the first steps of spliceosome assembly. Nucleic Acids Res 20:5017–5025

Caceres JF, Krainer AR (1993) Functional analysis of pre-mRNA splicing factor SF2/ASF structural domains. EMBO J 12:4715–4726

Caceres JF, Stamm S, Helfman DM, Krainer AR (1994) Regulation of alternative splicing in vivo by overexpression of antagonistic splicing factors. Science 265:1706–1709

Caceres JF, Misteli T, Screaton GR, Spector DL, Krainer AR (1997) Role of the modular domains of SR proteins in subnuclear localization and alternative splicing specificity. J Cell Biol 138:225–238

Cao W, Jamison SF, Garcia-Blanco MA (1997) Both phosphorylation and dephosphorylation of ASF/SF2 are required for pre-mRNA splicing in vitro. RNA 3:1456–1467

Cardinali B, Cohen PTW, Lamond AI (1994) Protein phosphatase 1 can modulate alternative 5′ splice site selection in a HeLa splicing extract. FEBS Lett 352:276–280

Cavalli G, Bachmann D, Thoma F (1996) Inactivation of topoisomerases affects transcription-dependent chromatin transitions in rDNA but not in a gene transcribed by RNA polymerase II. EMBO J 15:590–597

Chalfant CE, Mischak H, Watson JE, Winkler BC, Goodnight J, Farese RV, Cooper DR (1995) Regulation of alternative splicing of protein kinase C beta by insulin. J Biol Chem 270:13326–13332

Champoux JJ (1988) Topoisomerase I is preferentially associated with isolated replicating simian virus 40 molecules after treatment of infected cells with camptothecin. J Virol 62:3675–3683

Champoux JJ (1992) Topoisomerase I is preferentially associated with normal SV40 replicative intermediates, but is associated with both replicating and nonreplicating SV40 DNAs which are deficient in histones. Nucleic Acids Res 20:3347–3352

Cheng SC, Abelson J (1987) Spliceosome assembly in yeast. Genes Dev 1:1014–1027

Chiara MD, Champion-Arnaud P, Buvoli M, Nadal-Ginard B, Reed R (1994) Specific protein-protein interactions between the essential mammalian spliceosome-associated proteins SAP 61 and SAP 114. Proc Natl Acad Sci USA 91:6403–6407

Chiara MD, Gozani O, Bennett M, Champion-Arnaud P, Palandjian L, Reed R (1996) Identification of proteins that interact with exon sequences, splice sites, and the branchpoint sequence during each stage of spliceosome assembly. Mol Cell Biol 16:3317–3326

Choi YD, Grabowski PJ, Sharp PA, Dreyfuss G (1986) Heterogeneous nuclear ribonucleoproteins: role in RNA splicing. Science 231:1534–1539

Cobianchi F, Calvio C, Stoppini M, Buvoli M, Riva S (1993) Phosphorylation of human hnRNP protein A1 abrogates in vitro strand annealing activity. Nucleic Acids Res 21:949–955

Colwill K, Pawson T, Andrews B, Prasad J, Manley J, Bell JC, Duncan PI (1996a) The Clk/Sty protein kinase phosphorylates SR splicing factors and regulates their intranuclear distribution. EMBO J 15:265–275

Colwill K, Feng LL, Yeakley JM, Gish GD, Caceres JF, Pawson T, Fu XD (1996b) SRPK1 and Clk/Sty protein kinases show distinct substrate specificities for serine/arginine-rich splicing factors. J Biol Chem 271:24569–24575

Cowper AE, Caceres JF, Mayeda A, Screaton GR (2001) Serine-arginine (SR) protein-like factors that antagonize authentic SR proteins and regulate alternative splicing. J Biol Chem 276: 48908–48914

D'Arpa P, Machlin PS, Ratrie H, Rothfield NF, Cleveland DW, Earnshaw WC (1988) cDNA cloning of human DNA topoisomerase I: catalytic activity of a 67.7-kDa carboxyl-terminal fragment. Proc Natl Acad Sci USA 85:2543–2547

D'Arpa P, Liu LF (1989) Topoisomerase-targeting antitumor drugs. Biochim Biophys Acta 989: 163–177

Du C, McGuffin ME, Dauwalder B, Rabinow L, Mattox W (1998) Protein phosphorylation plays an essential role in the regulation of alternative splicing and sex determination in *Drosophila*. Mol Cell 2:741–750

Duncan PI, Howell BW, Marius RM, Drmanic S, Douville EM, Bell JC (1995) Alternative splicing of STY, a nuclear dual specificity kinase. J Biol Chem 270:21524–21531

Duncan PI, Stojdl DF, Marius RM, Bell JC (1997) In vivo regulation of alternative pre-mRNA splicing by the Clk1 protein kinase. Mol Cell Biol 17:5996–6001

Eperon IC, Ireland DC, Smith RA, Mayeda A, Krainer AR (1993) Pathways for selection of 5′ splice sites by U1 snRNPs and SF2/ASF. EMBO J 12:3607–3617

Eperon IC, Makarova OV, Mayeda A, Munroe SH, Caceres JF, Hayward DG, Krainer AR (2000) Selection of alternative 5′ splice sites: role of U1 snRNP and models for the antagonistic effects of SF2/ASF and hnRNP A1. Mol Cell Biol 20:8303–8318

Erez O, Kahana C (2001) Screening for modulators of spermine tolerance identifies Sky1, the SR protein kinase of *Saccharomyces cerevisiae*, as a regulator of polyamine transport and ion homeostasis. Mol Cell Biol 21:175–184

Fetzer S, Lauber J, Will CL, Luhrmann R (1997) The [U4/U6.U5] tri-snRNP-specific 27K protein is a novel SR protein that can be phosphorylated by the snRNP-associated protein kinase. RNA 3:344–355

Fleischmann G, Pflugfelder G, Steiner EK, Javaherian K, Howard GC, Wang JC, Elgin SCR (1984) *Drosophila* DNA topoisomerase I is associated with transcriptionally active regions of the genome. Proc Natl Acad Sci USA 81:6958–6962

Forné T, Rossi F, Labourier E, Antoine E, Cathala G, Brunel C, Tazi J (1995) Disruption of base-paired U4.U6 small nuclear RNAs induced by mammalian heterogeneous nuclear ribonucleoprotein C protein. J Biol Chem 270:16476–16481

Frendewey D, Keller W (1985) Stepwise assembly of a pre-mRNA splicing complex requires U-snRNPs and specific intron sequences. Cell 42:355–367

Fromont-Racine M, Rain JC, Legrain P (1997) Toward a functional analysis of the yeast genome through exhaustive two-hybrid screens. Nat Genet 16:277–282

Fu XD (1995) The superfamily of arginine/serine-rich splicing factors. RNA 1:663–680

Fu XD, Maniatis T (1992) Isolation of a complementary DNA that encodes the mammalian splicing factor SC35. Science 256:535–538

Fukasawa K, Komatani H, Hara Y, Suda H, Okura A, Nishimura S, Yoshinari T (1998) Sequence-selective DNA cleavage by a topoisomerase I poison, NB-506. Int J Cancer 75:145–150

Fung PA, Labrecque R, Pederson T (1997) RNA-dependent phosphorylation of a nuclear RNA binding protein. Proc Natl Acad Sci USA 94:1064–1068

Gama-Carvalho M, Carvalho MP, Kehlenbach A, Valcarcel J, Carmo-Fonseca M (2001) Nucleo-cytoplasmic shuttling of heterodimeric splicing factor U2AF. J Biol Chem 276:13104–13112

Garg LC, DiAngelo S, Jacob ST (1987) Role of DNA topoisomerase I in the transcription of supercoiled rRNA gene. Proc Natl Acad Sci USA 84:3185–3188

Ge H, Zuo P, Manley JL (1991) Primary structure of the human splicing factor ASF reveals similarities with *Drosophila* regulators. Cell 66:373–382

Georgatos SD, Stournaras C, Blobel G (1988) Heterotypic and homotypic associations between the nuclear lamins: site-specificity and control by phosphorylation. Proc Natl Acad Sci USA 85:4325–4329

Ghetti A, Pinol-Roma S, Michael WM, Morandi C, Dreyfuss G (1992) hnRNP I, the polypyrimidine tract-biding protein: distinct nuclear localization and association with hnRNAs. Nucleic Acids Res 20:3671–3678

Gilbert W, Siebel CW, Guthrie C (2001) Phosphorylation by Sky1p promotes Npl3p shuttling and mRNA dissociation. RNA 7:302–313

Gilmour DS, Pflugfelder G, Wang JC, Lis JT (1986) Topoisomerase I interacts with transcribed regions in *Drosophila* cells. Cell 44:401–407

Goralski TJ, Edström JE, Baker BS (1989) The sex determination locus transformer-2 of *Drosophila* encodes a polypeptide with similarity to RNA binding protein. Cell 56:1011–1018

Gozani O, Patton JG, Reed R (1994) A novel set of spliceosome-associated proteins and the essential splicing factor PSF bind stably to pre-mRNA prior to catalytic step II of the splicing reaction. EMBO J 13:3356–3367

Gozani O, Feld R, Reed R (1996) Evidence that sequence-independent binding of highly conserved U2 snRNP proteins upstream of the branch site is required for assembly of spliceosomal complex A. Genes Dev 10:233–243

Gozani O, Potashkin J, Reed R (1998) A potential role for U2AF-SAP 155 interactions in recruiting U2 snRNP to the branch site. Mol Cell Biol 18:4752–4760

Grabowski PJ, Sharp PA (1986) Affinity chromatography of splicing complexes: U2, U5, and U4+U6 small nuclear ribonucleoprotein particles in the spliceosome. Science 233:1294–1299

Grabowski PJ, Padgett RA, Sharp PA (1984) Messenger RNA splicing in vitro: an excised intervening sequence and a potential intermediate. Cell 37:415–427

Grabowski PJ, Seiler SR, Sharp PA (1985) A multicomponent complex is involved in the splicing of messenger RNA precursors. Cell 42:345–353

Graveley BR (2000) Sorting out the complexity of SR protein functions. RNA 6:1197–1211

Graveley BR, Hertel KJ, Maniatis T (2001) The role of U2AF35 and U2AF65 in enhancer-dependent splicing. RNA 7:806–818

Gudi T, Lohmann SM, Pilz RB (1997) Regulation of gene expression by cyclic GMP-dependent protein kinase requires nuclear translocation of the kinase: identification of a nuclear localization signal. Mol Cell Biol 17:5244–5254

Gui JF, Lane WS, Fu XD (1994a) A serine kinase regulates intracellular localization of splicing factors in the cell cycle. Nature 369:678–682

Gui J-F, Tronchère H, Chandler SD, Fu X-D (1994b) Purification and characterization of a kinase specific for the serine- and arginine-rich pre-mRNA splicing factors. Proc Natl Acad Sci USA 91:10824–10828

Gupta M, Fujimori A, Pommier Y (1995) Eukaryotic DNA topoisomerases I. Biochim Biophys Acta 1262:1–14

Guth S, Valcarcel J (2000) Kinetic role for mammalian SF1/BBP in spliceosome assembly and function after polypyrimidine tract recognition by U2AF. J Biol Chem 275:38059–38066

Guthrie C, Patterson B (1988) Spliceosomal snRNAs. Annu Rev Genet 22:387–419

Hanes J, Von der Kammer H, Klaudiny J, Scheit KH (1994) Characterization by cDNA cloning of two new human protein kinases: evidence by sequence comparison of a new family of mammalian protein kinases. J Mol Biol 244:665–672

Hartmann AM, Rujescu D, Giannakouros T, Nikolakaki E, Goedert M, Mandelkow EM, Gao QS, Andreadis A, Stamm S (2001) Regulation of alternative splicing of human tau exon 10 by phosphorylation of splicing factors. Mol Cell Neurosci 18:80–90

Heinrichs V, Bach M, Winkelmann G, Lührmann R (1990) U1-specific protein C needed for efficient complex formation of U1 snRNP with a 5′ splice site. Science 247:69–72

Heinrichs V, Ryner LC, Baker BS (1998) Regulation of sex-specific selection of fruitless 5′ splice sites by transformer and transformer-2. Mol Cell Biol 18:450–458

Hogges PE, Beggs JD (1994) U2 fulfills a commitment. Curr Biol 4:264–267

Howell BW, Afar DE, Lew J, Douville EM, Icely PL, Gray DA, Bell JC (1991) STY, a tyrosine-phosphorylating enzyme with sequence homology to serine/threonine kinases. Mol Cell Biol 11:568–572

Huang Y, Steitz JA (2001) Splicing factors SRp20 and 9G8 promote the nucleocytoplasmic export of mRNA. Mol Cell 7:899–905
Idriss H, Kumar A, Casas-Finet JR, Guo H, Damuni Z, Wilson SH (1994) Regulation of in vitro nucleic acid strand annealing activity of heterogeneous nuclear ribonucleoprotein protein A1 by reversible phosphorylation. Biochemistry 33:11382–11390
Jamison SF, Crow A, Garcia-Blanco MA (1992) The spliceosome assembly pathway in mammalian extracts. Mol Cell Biol 12:4279–4287
Jamison SF, Pasman Z, Wang J, Will C, Lührmann R, Manley JL, Garcia-Blanco MA (1995) U1 snRNP-ASF/SF2 interaction and 5′ splice site recognition: characterization of required elements. Nucleic Acids Res 23:3260–3267
Jensen AD, Svejstrup JQ (1996) Purification and characterization of human topoisomerase I mutants. Eur J Biochem 236:389–394
Johnson KW, Smith KA (1991) Molecular cloning of a novel human cdc2/CDC28-like protein kinase. J Biol Chem 266:3402–3407
Kanaar R, Roche SE, Beall EL, Green MR, Rio DC (1993) The conserved pre-mRNA splicing factor U2AF from *Drosophila*: requirement for viability. Science 262:569–573
Kandels-Lewis S, Séraphin B (1993) Role of U6 snRNA in 5′ splice site selection. Science 262:2035–2039
Karn J, Vidali G, Boffa LC, Allfrey VG (1977) Characterization of the non-histone nuclear proteins associated with rapidly labeled heterogeneous nuclear RNA. J Biol Chem 252:7307–7322
Kataoka N, Bachorik JL, Dreyfuss G (1999) Transportin-SR, a nuclear import receptor for SR proteins. J Cell Biol 145:1145–1152
Kohtz JD, Jamison SF, Will CL, Zuo P, Lührmann R, Garcia-Blanco MA, Manley JL (1994) Protein-protein interactions and 5′-splice-site recognition in mammalian mRNA precursors. Nature 368:119–124
Koizumi J, Okamoto Y, Onogi H, Mayeda A, Krainer AR, Hagiwara M (1999) The subcellular localization of SF2/ASF is regulated by direct interaction with SR protein kinases (SRPKs). J Biol Chem 274:11125–11131
Kojima T, Zama T, Wada K, Onogi H, Hagiwara M (2001) Cloning of human PRP4 reveals interaction with Clk1. J Biol Chem 276:32247–32256
Konarska MM, Sharp PA (1986) Electrophoretic separation of complexes involved in the splicing of precursors to mRNAs. Cell 46:845–855
Konarska MM, Sharp PA (1987) Interactions between small nuclear ribonucleoprotein particles in formation of spliceosomes. Cell 49:763–774
Konforti BB, Koziolkiewicz MJ, Konarska MM (1993) Disruption of base pairing between the 5′ splice site and the 5′ end of U1 snRNA is required for spliceosome assembly. Cell 75:863–873
Konig H, Ponta H, Herrlich P (1998) Coupling of signal transduction to alternative pre-mRNA splicing by a composite splice regulator. EMBO J 17:2904–2913
Krainer AR, Maniatis T, Ruskin B, Green MR (1984) Normal and mutant human b-globin pre-mRNAs are faithfully and efficiently spliced in vitro. Cell 36:993–1005
Krainer AR, Mayeda A, Kozak D, Binns G (1991) Functional expression of cloned human splicing factor SF2: homology to RNA-binding proteins, U1 70 K, and *Drosophila* splicing regulators. Cell 66:383–394
Krämer A (1996) The structure and function of proteins involved in mammalian pre-mRNA splicing. Annu Rev Biochem 65:367–409
Krämer A, Utans U (1991) Three protein factors (SF1, SF3 and U2AF) function in pre-splicing complex formation in addition to snRNPs. EMBO J 10:1503–1509
Krämer A, Mulhauser F, Wersig C, Groning K, Bilbe G (1995) Mammalian splicing factor SF3a120 represents a new member of the SURP family of proteins and is homologous to the essential splicing factor PRP21p of *Saccharomyces cerevisiae*. RNA 1:260–272
Krämer A, Gruter P, Groning K, Kastner B (1999) Combined biochemical and electron microscopic analyses reveal the architecture of the mammalian U2 snRNP. J Cell Biol 145:1355–1368
Kraus ME, Lis JT (1994) The concentration of B52, an essential splicing factor and regulator of splice site choice in vitro, is critical for *Drosophila* development. Mol Cell Biol 14:5360–5370

Kretzschmar M, Meisterernst M, Roeder RG (1993) Identification of human DNA topoisomerase I as a cofactor for activator-dependent transcription by RNA polymerase II. Proc Natl Acad Sci USA 90:11508–11512

Kroeger PE, Rowe TC (1989) Interaction of topoisomerase I with the transcribed region of the *Drosophila* HSP 70 heat shock gene. Nucleic Acids Res 17:8495–8509

Kuhn AN, Li Z, Brow DA (1999) Splicing factor Prp8 governs U4/U6 RNA unwinding during activation of the spliceosome. Mol Cell 3:65–75

Kumar A, Wilson SH (1990) Studies of the strand-annealing activity of mammalian hnRNP complex protein A1. Biochemistry 29:10717–10722

Kumar A, Williams KR, Szer W (1986) Purification and domain structure of core hnRNP proteins A1 and A2 and their relationship to single-stranded DNA-binding proteins. J Biol Chem 261:11266–11273

Kuroyanagi N, Onogi H, Wakabayashi T, Hagiwara M (1998) Novel SR-protein-specific kinase, SRPK2, disassembles nuclear speckles. Biochem Biophys Res Commun 242:357–364

Kuroyanagi H, Kimura T, Wada K, Hisamoto N, Matsumoto K, Hagiwara M (2000) SPK-1, a *C. elegans* SR protein kinase homologue, is essential for embryogenesis and required for germline development. Mech Dev 99:51–64

Labourier E, Rossi F, Gallouzi IE, Allemand E, Divita G, Tazi J (1998) Interaction between the N-terminal domain of human DNA topoisomerase I and the arginine-serine domain of its substrate determines phosphorylation of SF2/ASF splicing factor. Nucleic Acids Res 26:2955–2962

Labourier E, Riou JF, Prudhomme M, Carrasco C, Bailly C, Tazi J (1999a) Poisoning of topoisomerase I by an antitumor indolocarbazole drug: stabilization of topoisomerase I-DNA covalent complexes and specific inhibition of the protein kinase activity. Cancer Res 59:52–55

Labourier E, Bourbon HM, Gallouzi I, Fostier M, Allemand E, Tazi J (1999b) Antagonism between RSF1 and SR proteins for both splice site recognition in vitro and *Drosophila* development. Genes Dev 13:740–753

Labourier E, Allemand E, Brand S, Fostier M, Tazi J, Bourbon HM (1999c) Recognition of exonic splicing enhancer sequences by the *Drosophila* splicing repressor RSF1. Nucleic Acids Res 27:2377–2386

Lai MC, Lin RI, Huang SY, Tsai CW, Tarn WY (2000) A human importin-beta family protein, transportin-SR2, interacts with the phosphorylated RS domain of SR proteins. J Biol Chem 275:7950–7957

Lai MC, Lin RI, Tarn WY (2001) Transportin-SR2 mediates nuclear import of phosphorylated SR proteins. Proc Natl Acad Sci USA 98:10154–10159

Lamond AI, Konarska MM, Grabowski PJ, Sharp PA (1988) Spliceosome assembly involves the binding and release of U4 small nuclear ribonucleoprotein. Proc Natl Acad Sci USA 85:411–415

Lee K, Du C, Horn M, Rabinow L (1996) Activity and autophosphorylation of LAMMER protein kinases. J Biol Chem 271:27299–27303

Lee MP, Brawn SD, Chen A, Hsieh TS (1993) DNA topoisomerase I is essential in *Drosophila melanogaster*. Proc Natl Acad Sci USA 90:6656–6660

Legrain P, Chapon C (1993) Interaction between PRP11 and SPP91 yeast splicing factors and characterization of a PRP9-PRP11-SPP91 complex. Science 262:108–110

Legrain P, Séraphin B, Rosbash M (1988) Early commitment of yeast pre-mRNA to the spliceosome pathway. Mol Cell Biol 8:3755–3760

Legrain P, Chapon C, Galisson F (1993) Interactions between PRP9 and SPP91 splicing factors identify a protein complex required in prespliceosome assembly. Genes Dev 7:1390–1399

Lerner MR, Boyle JA, Mount SM, Wolin SL, Steitz JA (1980) Are snRNPs involved in splicing? Nature 283:220–224

Lesser CF, Guthrie C (1993) Mutations in U6 snRNA that alter splice site specificity: implications for the active site. Science 262:1982–1988

Liu LF, Miller KG (1981) Eucaryotic DNA topoisomerases: two forms of type I DNA topoisomerases from HeLa cell nuclei. Proc Natl Acad Sci USA 78:3487–3491

Liu LF, Wang JC (1987) Supercoiling of the DNA template during transcription. Proc Natl Acad Sci USA 84:7024–7027

Lührmann R (1988) snRNP proteins. In: Birnstiel ML (ed) Structure and function of major and minor small nuclear ribonucleoprotein particles. Springer, Berlin Heidelberg New York, pp 71–99

Lührmann R, Kastner B, Bach M (1990) Structure of spliceosomal snRNPs and their role in pre-mRNA splicing. Biochem Biophys Acta 1087:265–292

MacMillan AM, Query CC, Allerson CR, Chen S, Verdine GL, Sharp PA (1994) Dynamic association of proteins with the pre-mRNA branch region. Genes Dev 8:3008–3020

Madden KR, Champoux JJ (1992) Overexpression of human topoisomerase I in baby hamster kidney cells: hypersensitivity of clonal isolates to camptothecin. Cancer Res 52:525–532

Madden KR, Stewart L, Champoux JJ (1995) Preferential binding of human topoisomerase I to superhelical DNA. EMBO J 14:5399–5409

Makarova OV, Makarov EM, Luhrmann R (2001) The 65 and 110kDa SR-related proteins of the U4/U6.U5 tri-snRNP are essential for the assembly of mature spliceosomes. EMBO J 20:2553–2563

Mayeda A, Krainer AR (1992) Regulation of alternative pre-mRNA splicing by hnRNP A1 and splicing factor SF2. Cell 68:365–375

Mayeda A, Helfman DM, Krainer AR (1993) Modulation of exon skipping and inclusion by heterogeneous nuclear ribonucleoprotein A1and pre-mRNA splicing factor SF2/ASF. Mol Cell Biol 13:2993–3001

Mayeda A, Munroe SH, Càceres JF, Krainer AR (1994) Function of conserved domains of hnRNP A1 and other hnRNP A/B proteins. EMBO J 13:5483–5495

Mayrand SH, Pedersen N, Pederson T (1986) Identification of proteins that bind tightly to pre-mRNA during in vitro splicing. Proc Natl Acad Sci USA 83:3718–3722

Mayrand SH, Dwen P, Pederson T (1993) Serine/threonine phosphorylation regulates binding of C hnRNP proteins to pre-mRNA. Proc Natl Acad Sci USA 90:7764–7768

Mayrand SH, Fung PA, Pederson T (1996) A discrete 3′ region of U6 small nuclear RNA modulates the phosphorylation cycle of the heterogeneous nuclear ribonucleoprotein particle protein. Mol Cell Biol 16:1241–1246

Meier UT, Blobel G (1992) Nopp140 shuttles on tracks between nucleolus and cytoplasm. Cell 70:127–138

Menegay H, Moeslein F, Landreth G (1999) The dual specificity protein kinase CLK3 is abundantly expressed in mature mouse spermatozoa. Exp Cell Res 253:463–473

Merendino L, Guth S, Bilbao D, Martinez C, Valcarcel J (1999) Inhibition of msl-2 splicing by Sex-lethal reveals interaction between U2AF35 and the 3′ splice site AG. Nature 402:838–841

Merino A, Madden KR, Lane WS, Champoux JJ, Reinberg D (1993) DNA topoisomerase I is involved in both repression and activation of transcription. Nature 365:227–232

Mermoud JE, Cohen PTW, Lamond AI (1992) Ser/Thr-specific protein phosphatases are required for both catalytic steps of pre-mRNA splicing. Nucleic Acids Res 20:5263–5269

Mermoud JE, Calvio C, Lamond AI (1994a) Uncovering the role of Ser/Thr protein phosphorylation in nuclear pre-mRNA splicing. Adv Prot Phosphatases 8:99–118

Mermoud JE, Cohen PT, Lamond AI (1994b) Regulation of mammalian spliceosome assembly by a protein phosphorylation mechanism. EMBO J 13:5679–5688

Michaud S, Reed R (1991) An ATP-independent complex commits pre-mRNA to the mammalian spliceosome assembly pathway. Genes Dev 5:2534–2546

Michaud S, Reed R (1993) A functional association between the 5′ and 3′ splice sites is established in the earliest prespliceosome complex (E) in mammals. Genes Dev 7:1008–1020

Misteli T, Spector DL (1996) Serine/threonine phosphatase 1 modulates the subnuclear distribution of pre-mRNA splicing factors. Mol Biol Cell 7:1559–1572

Misteli T, Caceres JF, Clement JQ, Krainer AR, Wilkinson MF, Spector DL (1998) Serine phosphorylation of SR proteins is required for their recruitment to sites of transcription in vivo. J Cell Biol 143:297–307

Morham SG, Kluckman KD, Voulomanos N, Smithies O (1996) Targeted disruption of the mouse topoisomerase I gene by camptothecin selection. Mol Cell Biol 16:6804–6809

Mount SM, Salz HK (2000) Pre-messenger RNA processing factors in the *Drosophila* genome. J Cell Biol 150:37–44

Muller MT, Pfund WP, Mehta VB, Trask DK (1985) Eukaryotic type I topoisomerase is enriched in the nucleolus and catalytically active on ribosomal DNA. EMBO J 4:1237–1243

Munroe SH, Dong XF (1992) Heterogeneous nuclear ribonucleoprotein A1 catalyzes RNA:RNA annealing. Proc Natl Acad Sci USA 89:895–899

Murray HL, Jarrell KA (1999) Flipping the switch to an active spliceosome. Cell 96:599–602

Nabeshima K, Kurooka H, Takeuchi M, Kinoshita K, Nakaseko Y, Yanagida M (1995) p93dis1, which is required for sister chromatid separation, is a novel microtubule and spindle pole body-associating protein phosphorylated at the Cdc2 target sites. Genes Dev 9:1572–1585

Nagai K (1996) RNA-protein complexes. Curr Opin Struct Biol 6:53–61

Nakielny S, Dreyfuss G (1996) The hnRNP C proteins contain a nuclear retention sequence that can override nuclear export signals. J Cell Biol 134:1365–1373

Nayler O, Stamm S, Ullrich A (1997) Characterization and comparison of four serine- and arginine-rich (SR) protein kinases. Biochem J 326:693–700

Nayler O, Schnorrer F, Stamm S, Ullrich A (1998) The cellular localization of the murine serine/arginine-rich protein kinase CLK2 is regulated by serine 141 autophosphorylation. J Biol Chem 273:34341–34348

Nelson KK, Green MR (1989) Mammalian U2 snRNP has a sequence-specific RNA-binding activity. Genes Dev 3:1562–1571

Nikolakaki E, Kohen R, Hartmann AM, Stamm S, Georgatsou E, Giannakouros T (2001) Cloning and characterization of an alternatively spliced form of SR protein kinase 1 that interacts specifically with scaffold attachment factor-B. J Biol Chem 276:40175–40182

Nolen B, Yun CY, Wong CF, McCammon JA, Fu XD, Ghosh G (2001) The structure of Sky1p reveals a novel mechanism for constitutive activity. Nat Struct Biol 8:176–183

Ohkubo M, Nishimura T, Kawamoto H, Nakano M, Honma T, Yoshinari T, Arakawa H, Suda H, Morishima H, Nishimura S (2000) Synthesis and biological activities of NB-506 analogues modified at the glucose group. Bioorg Med Chem Lett 10:419–422

Ono Y, Kikkawa U, Ogita K, Fujii T, Kurokawa T, Asaoka Y, Sekiguchi K, Ase K, Igarashi K, Nishizuka Y (1987) Expression and properties of two types of protein kinase C: alternative splicing from a single gene. Science 236:1116–1120

Padgett RA, Konarska MM, Grabowski PJ, Hardy SF, Sharp PA (1984) Lariat RNAs as intermediates and products in the splicing of messenger RNA precursors. Science 225:898–903

Padmanabha R, Gehrung S, Snyder M (1991) The KNS1 gene of *Saccharomyces cerevisiae* encodes a nonessential protein kinase homologue that is distantly related to members of the CDC28/cdc2 gene family. Mol Gen Genet 229:1–9

Papoutsopoulou S, Nikolakaki E, Chalepakis G, Kruft V, Chevaillier P, Giannakouros T (1999) SR protein-specific kinase 1 is highly expressed in testis and phosphorylates protamine 1. Nucleic Acids Res 27:2972–2980

Patel NA, Chalfant CE, Watson JE, Wyatt JR, Dean NM, Eichler DC, Cooper DR (2001) Insulin regulates alternative splicing of protein kinase C beta II through a phosphatidylinositol 3-kinase-dependent pathway involving the nuclear serine/arginine-rich splicing factor, SRp40, in skeletal muscle cells. J Biol Chem 276:22648–22654

Patton JG, Mayer SA, Tempst P, Nadal-Ginard B (1991) Characterization and molecular cloning of polypyrimidine tract-binding protein: a component of a complex necessary for pre-mRNA splicing. Genes Dev 5:1237–1251

Patton JG, Porro EB, Galceran J, Tempst P, Nadal-Ginard B (1993) Cloning and characterization of PSF, a novel pre-mRNA splicing factor. Genes Dev 7:393–406

Pederson T (1974) Proteins associated with heterogeneous nuclear RNA in eukaryotic cells. J Mol Biol 83:163–183

Peled-Zehavi H, Berglund JA, Rosbash M, Frankel AD (2001) Recognition of RNA branch point sequences by the KH domain of splicing factor 1 (mammalian branch point binding protein) in a splicing factor complex. Mol Cell Biol 21:5232–5241

Petersen-Mahrt SK, Estmer C, Ohrmalm C, Matthews DA, Russell WC, Akusjarvi G (1999) The splicing factor-associated protein, p32, regulates RNA splicing by inhibiting ASF/SF2 RNA binding and phosphorylation. EMBO J 18:1014–1024
Pikielny CW, Rosbash M (1986) Specific small nuclear RNAs are associated with yeast spliceosomes. Cell 45:869–877
Pilch B, Allemand E, Facompre M, Bailly C, Riou JF, Soret J, Tazi J (2001) Specific inhibition of serine- and arginine-rich splicing factors phosphorylation, spliceosome assembly, and splicing by the antitumor drug NB-506. Cancer Res 61:6876–6884
Pinol-Roma S, Dreyfuss G (1991) Transcription-dependent and transcription-independent nuclear transport of hnRNP proteins. Science 253:312–317
Pinol-Roma S, Dreyfuss G (1992) Shuttling of pre-mRNA binding proteins between nucleus and cytoplasm. Nature 355:730–732
Pommier Y (1996) Eukaryotic DNA topoisomerase I: genome gatekeeper and its intruders, camptothecins. Semin Oncol 23:3–10
Pommier Y, Pourquier P, Fan Y, Strumberg D (1998) Mechanism of action of eukaryotic DNA topoisomerase I and drugs targeted to the enzyme. Biochim Biophys Acta 1400:83–105
Pontius BW, Berg P (1990) Renaturation of complementary DNA strands mediated by purified mammalian heterogeneous nuclear ribonucleoprotein A1 protein: implications for a mechanism for rapid molecular assembly. Proc Natl Acad Sci USA 87:8403–8407
Pontius BW, Berg P (1992) Rapid assembly and disassembly of complementary DNA strands through an equilibrium intermediate state mediated by A1 hnRNP protein. J Biol Chem 267:13815–13818
Pourquier P, Takebayashi Y, Urasaki Y, Gioffre C, Kohlhagen G, Pommier Y (2000) Induction of topoisomerase I cleavage complexes by 1-beta-D-arabinofuranosylcytosine (ara-C) in vitro and in ara-C-treated cells. Proc Natl Acad Sci USA 97:1885–1890
Prasad J, Colwill K, Pawson T, Manley JL (1999) The protein kinase Clk/Sty directly modulates SR protein activity: both hyper- and hypophosphorylation inhibit splicing. Mol Cell Biol 19:6991–7000
Query CC, Bentley RC, Keene JD (1989) A common RNA recognition motif identified within a defined U1 RNA binding domain of the 70K U1 snRNP protein. Cell 57:89–101
Query CC, Strobel SA, Sharp PA (1996) Three recognition events at the branch-site adenine. EMBO J 15:1392–1402
Rabinow L, Birchler JA (1989) A dosage-sensitive modifier of retrotransposon-induced alleles of the *Drosophila* white locus. EMBO J 8:879–889
Rabinow L, Chiang SL, Birchler JA (1993) Mutations at the Darkener of apricot locus modulate transcript levels of copia and copia-induced mutations in *Drosophila melanogaster*. Genetics 134:1175–1185
Rain JC, Tartakoff AM, Krämer A, Legrain P (1996) Essential domains of the PRP21 splicing factor are implicated in the binding to PRP9 and PRP11 proteins and are conserved through evolution. RNA 2:535–550
Rain JC, Rafi Z, Rhani Z, Legrain P, Krämer A (1998) Conservation of functional domains involved in RNA binding and protein-protein interactions in human and *Saccharomyces cerevisiae* pre-mRNA splicing factor SF1. RNA 4:551–565
Reed R (1996) Initial splice-site recognition and pairing during pre-mRNA splicing. Curr Opin Genet Dev 6:215–220
Reed R, Griffith J, Maniatis T (1988) Purification and visualization of native spliceosomes. Cell 53:949–961
Ring HZ, Lis JT (1994) The SR protein B52/SRp55 is essential for *Drosophila* development. Annu Rev Cell Biol 9:265–315
Rogers J, Wall R (1980) A mechanism for RNA splicing. Proc Natl Acad Sci USA 77:1877–1879
Roscigno RF, Garcia-Blanco MA (1995) SR proteins escort the U4/U6.U5 tri-snRNP to the spliceosome. RNA 1:692–706

Rose KM, Szopa J, Han FS, Cheng YC, Richter A, Scheer U (1988) Association of DNA topoisomerase I and RNA polymerase I: a possible role for topoisomerase I in ribosomal gene transcription. Chromosoma 96:411–416
Rossi F, Labourier E, Forné T, Divita G, Derancourt J, Riou JF, Antoine E, Cathala G, Brunel C, Tazi J (1996a) Specific phosphorylation of SR proteins by mammalian DNA topoisomerase I. Nature 381:80–82
Rossi F, Forné T, Antoine E, Tazi J, Brunel C, Cathala G (1996b) Involvement of U1 small nuclear ribonucleoproteins (snRNP) in 5′ splice site-U1 snRNP interaction. J Biol Chem 271: 23985–23991
Rossi F, Labourier E, Gallouzi IE, Derancourt J, Allemand E, Divita G, Tazi J (1998) The C-terminal domain but not the tyrosine 723 of human DNA topoisomerase I active site contributes to kinase activity. Nucleic Acids Res 26:2963–2970
Roth MB, Gall JG (1987) Monoclonal antibodies that recognize transcription unit proteins on newt lampbrush chromosomes. J Cell Biol 105:1047–1054
Roth MB, Murphy C, Gall JG (1990) A monoclonal antibody that recognizes a phosphorylated epitope stains lampbrush chromosome loops and small granules in the amphibian germinal vesicle. J Cell Biol 111:2217–2223
Roth MB, Zahler AM, Stolk JA (1991) A conserved family of nuclear phosphoproteins localized to sites of polymerase II transcription. J Cell Biol 115:587–591
Ruby SW, Abelson J (1988) An early hierarchic role of U1 small nuclear ribonucleoprotein in spliceosome assembly. Science 242:1028–1035
Rudner DZ, Kanaar R, Breger KS, Rio DC (1996) Mutations in the small subunit of the *Drosophila* U2AF splicing factor cause lethality and developmental defects. Proc Natl Acad Sci USA 93:10333–10337
Rudner DZ, Breger KS, Rio DC (1998a) Molecular genetic analysis of the heterodimeric splicing factor U2AF: the RS domain on either the large or small *Drosophila* subunit is dispensable in vivo. Genes Dev 12:1010–1021
Rudner DZ, Kanaar R, Breger KS, Rio DC (1998b) Interaction between subunits of heterodimeric splicing factor U2AF is essential in vivo. Mol Cell Biol 18:1765–1773
Ruskin B, Krainer AR, Maniatis T, Green MR (1984) Excision of an intact intron as a novel lariat structure during pre-mRNA splicing in vitro. Cell 38:317–331
Ruskin B, Zamore PD, Green MR (1988) A factor, U2AF, is required for U2 snRNP binding and splicing complex assembly. Cell 52:207–219
Ryner LC, Goodwin SF, Castrillon DH, Anand A, Villella A, Baker BS, Hall JC, Taylor BJ, Wasserman SA (1996) Control of male sexual behavior and sexual orientation in *Drosophila* by the fruitless gene. Cell 87:1079–1089
Sanford JR, Bruzik JP (1999) Developmental regulation of SR protein phosphorylation and activity. Genes Dev 13:1513–1518
Sawa H, Abelson J (1992) Evidence for a base-pairing interaction between U6 small nuclear RNA and the 5′ splice site during the splicing reaction in yeast. Proc Natl Acad Sci USA 89:11269–11273
Sawa H, Shimura Y (1992) Association of U6 snRNA with the 5′ splice site region of pre-mRNA in the spliceosome. Genes Dev 6:244–254
Schultz J, Jones T, Bork P, Sheer D, Blencke S, Steyrer S, Wellbrock U, Bevec D, Ullrich A, Wallasch C (2001) Molecular characterization of a cDNA encoding functional human CLK4 kinase and localization to chromosome 5q35 [correction of 4q35]. Genomics 71:368–370
Screaton GR, Càceres JF, Mayeda A, Bell MV, Plebanski M, Jackson DG, Bell JI, Krainer AR (1995) identification and characterization of three members of the human SR family of pre-mRNA splicing factors. EMBO J 14:4336–4349
Seghezzi W, Chua K, Shanahan F, Gozani O, Reed R, Lees E (1998) Cyclin E associates with components of the pre-mRNA splicing machinery in mammalian cells. Mol Cell Biol 18:4526–4536
Senger B, Simos G, Bischoff FR, Podtelejnikov A, Mann M, Hurt E (1998) Mtr10p functions as a nuclear import receptor for the mRNA-binding protein Npl3p. EMBO J 17:2196–2207

Séraphin B, Rosbash M (1989) Identification of functional U1 snRNA-pre-mRNA complexes committed to spliceosome assembly and splicing. Cell 59:349–358

Shannon K, Guthrie C (1991) Suppressors of a U4 snRNA mutation define a novel U6 snRNP protein with RNA-binding motifs. Genes Dev 5:773–785

Shen J, Zu K, Cass CL, Beyer AL, Hirsh J (1995) Exon skipping by overexpression of a *Drosophila* heterogeneous nuclear ribonucleoprotein in vivo. Proc Natl Acad Sci USA 92:1822–1825

Siatecka M, Reyes JL, Konarska MM (1999) Functional interactions of Prp8 with both splice sites at the spliceosomal catalytic center. Genes Dev 13:1983–1993

Siebel CW, Feng L, Guthrie C, Fu XD (1999) Conservation in budding yeast of a kinase specific for SR splicing factors. Proc Natl Acad Sci USA 96:5440–5445

Singh R, Valcarcel J, Green MR (1995) Distinct binding specificities and functions of higher eukaryotic polypyrimidine tract-binding proteins. Science 268:1173–1176

Solary E, Dubrez L, Eymin B, Bertrand R, Pommier Y (1996) Apoptosis of human leukemic cells induced by topoisomerase I and II inhibitors. Bull Cancer (Paris) 83:205–212

Sontheimer EJ, Steitz JA (1993) The U5 and U6 small nuclear RNAs as active site components of the spliceosome. Science 262:1989–1996

Staley JP, Guthrie C (1999) An RNA switch at the 5′ splice site requires ATP and the DEAD box protein Prp28p. Mol Cell 3:55–64

Steitz JA, Black DL, Gerke V, Parker KA, Krämer A, Frendewey D, Keller W (1988) Functions of the abundant U-snRNPs. In: Birnstiel M (ed) Structure and function of major and minor small nuclear ribonucleoprotein particles. Springer, Berlin Heidelberg New York, pp 115–154

Stewart AF, Herrera RE, Nordheim A (1990) Rapid induction of c-fos transcription reveals quantitative linkage of RNA polymerase II and DNA topoisomerase I enzyme activities. Cell 60:141–149

Stewart L, Ireton GC, Champoux JJ (1996) The domain organization of human topoisomerase I. J Biol Chem 271:7602–7608

Stojdl DF, Bell JC (1999) SR protein kinases: the splice of life. Biochem Cell Biol 77:293–298

Sun Q, Mayeda A, Hampson RK, Krainer AR, Rottman FM (1993) General splicing factor SF2/ASF promotes alternative splicing by binding to an exonic splicing enhancer. Genes Dev 7: 2598–2608

Swanson MS, Dreyfuss G (1988a) Classification and purification of proteins of heterogeneous nuclear ribonucleoprotein particles by RNA binding specificities. Mol Cell Biol 8:2237–2241

Swanson MS, Dreyfuss G (1988b) RNA binding specificity of hnRNP proteins: a subset bind to the 3′ end of introns. EMBO J 7:3519–3529

Tacke R, Manley JL (1999) Determinants of SR protein specificity. Curr Opin Cell Biol 11:358–362

Tacke R, Chen Y, Manley JL (1997) Sequence-specific RNA binding by an SR protein requires RS domain phosphorylation: creation of an SRp40-specific splicing enhancer. Proc Natl Acad Sci USA 94:1148–1153

Takeuchi M, Yanagida M (1993) A mitotic role for a novel fission yeast protein kinase dsk1 with cell cycle stage dependent phosphorylation and localization. Mol Biol Cell 4:247–260

Tang Z, Kuo T, Shen J, Lin RJ (2000) Biochemical and genetic conservation of fission yeast Dsk1 and human SR protein-specific kinase 1. Mol Cell Biol 20:816–824

Taylor SJ, Shalloway D (1994) An RNA-binding protein associated with Src through its SH2 and SH3 domains in mitosis. Nature 368:867–871

Tazi J, Daugeron MC, Cathala G, Brunel C, Jeanteur P (1992) Adenosine phosphorothioates (ATPaS and ATPgS) differentially affect the two steps of mammalian pre-mRNA splicing. J Biol Chem 267:4322–4326

Tazi J, Kornstädt U, Rossi F, Jeanteur P, Cathala G, Brunel C, Luhrmann R (1993) Thiophosphorylation of U1-70K protein inhibits pre-mRNA splicing. Nature 363:283–286

Tazi J, Rossi F, Labourier E, Gallouzi I, Brunel C, Antoine E (1997) DNA topoisomerase I: customs officer at the border between DNA and RNA worlds? J Mol Med 75:786–800

Teigelkamp S, Newman AJ, Beggs JD (1995a) Extensive interactions of PRP8 protein with the 5′ and 3′ splice sites during splicing suggest a role in stabilization of exon alignment by U5 snRNA. EMBO J 14:2602–2612

Teigelkamp S, Whittaker E, Beggs JD (1995b) Interaction of the yeast splicing factor PRP8 with substrate RNA during both steps of splicing. Nucleic Acids Res 23:320–326

Theissen H, Etzerodt M, Reuter R, Schneider C, Lottspeich F, Argos P, Lührmann R, Philipson L (1986) Cloning of the human cDNA for the U1 RNA-associated 70K protein. EMBO J 5:3209–3217

Tsao YP, Russo A, Nyamuswa G, Silber R, Liu LF (1993) Interaction between replication forks and topoisomerase I-DNA cleavable complexes: studies in a cell-free SV40 DNA replication system. Cancer Res 53:5908–5914

Umen JG, Guthrie C (1995a) A novel role for a U5 snRNP protein in 3′ splice site selection. Genes Dev 9:855–868

Umen JG, Guthrie C (1995b) Prp16p, Slu7p, and Prp8p interact with the 3′ splice site in two distinct stages during the second catalytic step of pre-mRNA splicing. RNA 1:584–597

Umen JG, Guthrie C (1995c) The second catalytic step of pre-mRNA splicing. RNA 1:869–885

Urasaki Y, Laco G, Takebayashi Y, Bailly C, Kohlhagen G, Pommier Y (2001) Use of camptothecin-resistant mammalian cell lines to evaluate the role of topoisomerase I in the antiproliferative activity of the indolocarbazole, NB-506, and its topoisomerase I binding site. Cancer Res 61:504–508

van der Houven van Oordt W, Diaz-Meco MT, Lozano J, Krainer AR, Moscat J, Caceres JF (2000) The MKK(3/6)-p38-signaling cascade alters the subcellular distribution of hnRNP A1 and modulates alternative splicing regulation. J Cell Biol 149:307–316

Wang C, Chua K, Seghezzi W, Lees E, Gozani O, Reed R (1998) Phosphorylation of spliceosomal protein SAP 155 coupled with splicing catalysis. Genes Dev 12:1409–1414

Wang HY, Lin W, Dyck JA, Yeakley JM, Songyang Z, Cantley LC, Fu XD (1998) SRPK2: a differentially expressed SR protein-specific kinase involved in mediating the interaction and localization of pre-mRNA splicing factors in mammalian cells. J Cell Biol 140:737–750

Wang HY, Arden KC, Bermingham JRJ, Viars CS, Lin W, Boyer AD, Fu XD (1999) Localization of serine kinases, SRPK1 (SFRSK1) and SRPK2 (SFRSK2), specific for the SR family of splicing factors in mouse and human chromosomes. Genomics 57:310–315

Wang J, Manley JL (1995) Overexpression of the SR proteins ASF/SF2 and SC35 influences alternative splicing in vivo in diverse ways. RNA 1:335–346

Wang JC, Lynch AS (1993) Transcription and DNA supercoiling. Curr Opin Genet Dev 3:764–768

Wang X, Bruderer S, Rafi Z, Xue J, Milburn PJ, Krämer A, Robinson PJ (1999) Phosphorylation of splicing factor SF1 on Ser20 by cGMP-dependent protein kinase regulates spliceosome assembly. EMBO J 18:4549–4559

Watakabe A, Tanaka K, Shimura Y (1993) The role of exon sequences in splice site selection. Genes Dev 7:407–418

Weg-Remers S, Ponta H, Herrlich P, Konig H (2001) Regulation of alternative pre-mRNA splicing by the ERK MAP-kinase pathway. EMBO J 20:4194–4203

Weighardt F, Biamonti G, Riva S (1995) Nucleo-cytoplasmic distribution of human hnRNP proteins: a search for the targeting domains in hnRNP A1. J Cell Sci 108:545–555

Wilusz J, Shenk T (1990) A uridylate tract mediates efficient heterogeneous nuclear ribonucleoprotein C protein-RNA cross-linking and functionally substitutes for the downstream element of the polyadenylation signal. Mol Cell Biol 10:6397–6407

Wilusz J, Feig DI, Shenk T (1988) The C proteins of heterogeneous nuclear ribonucleoprotein complexes interact with RNA sequences downstream of polyadenylation cleavage sites. Mol Cell Biol 8:4477–4483

Woppmann A, Will CL, Konstädt U, Zuo P, Manley JM, Lührmann R (1993) Identification of an snRNP-associated kinase activity that phosphorylates arginine/serine-rich domains typical of splicing factors. Nucleic Acids Res 21:2815–2822

Wu JY, Maniatis T (1993) Specific interactions between proteins implicated in splice site selection and regulated alternative splicing. Cell 75:1061–1070

Wu S, Romfo CM, Nilsen TW, Green MR (1999) Functional recognition of the 3′ splice site AG by the splicing factor U2AF35. Nature 402:832–835

Wyatt J, Sontheimer EJ, Steitz JA (1992) Site-specific cross-linking of mammalian U5 snRNP to the 5′ splice site before the first step of pre-mRNA splicing. Genes Dev 6:2542–2553

Xiao SH, Manley JL (1997) Phosphorylation of the ASF/SF2 RS domain affects both protein-protein and protein-RNA interactions and is necessary for splicing. Genes Dev 11:334–344

Xiao SH, Manley JL (1998) Phosphorylation-dephosphorylation differentially affects activities of splicing factor ASF/SF2. EMBO J 17:6359–6367

Yang X, Bani MR, Lu SJ, Rowan S, Ben-David Y, Chabot B (1994) The A1 and A1b proteins of heterogeneous nuclear ribonucleoparticles modulate 5′ splice site selection in vivo. Proc Natl Acad Sci USA 91:6924–6928

Yeakley JM, Tronchere H, Olesen J, Dyck JA, Wang HY, Fu XD (1999) Phosphorylation regulates in vivo interaction and molecular targeting of serine/arginine-rich pre-mRNA splicing factors. J Cell Biol 145:447–455

Yoshinari T, Matsumoto M, Arakawa H, Okada H, Noguchi K, Suda H, Okura A, Nishimura S (1995) Novel antitumor indolocarbazole compound 6-N-formylamino-12,13-dihydro-1,11-dihydroxy-13-(beta-D-glucopyranosyl)-5H-indolo[2,3-a]pyrrolo[3,4-c]carbazole-5,7(6H)-dione (NB-506): induction of topoisomerase I-mediated DNA cleavage and mechanisms of cell line-selective cytotoxicity. Cancer Res 55:1310–1315

Yun B, Farkas R, Lee K, Rabinow L (1994) The Doa locus encodes a member of a new protein kinase family and is essential for eye and embryonic development in *Drosophila melanogaster*. Genes Dev 8:1160–1173

Yun B, Lee K, Farkas R, Hitte C, Rabinow L (2000) The LAMMER protein kinase encoded by the Doa locus of *Drosophila* is required in both somatic and germline cells and is expressed as both nuclear and cytoplasmic isoforms throughout development. Genetics 156:749–761

Zahler AM, Lane WS, Stolk JA, Roth MB (1992) SR proteins: a conserved family of pre-mRNA splicing factors. Genes Dev 6:837–847

Zahler AM, Neugebauer KM, Lane WS, Roth MB (1993) Distinct functions of SR proteins in alternative pre-mRNA splicing. Science 260:219–222

Zamore PD, Green MR (1989) Identification, purification, and biochemical characterization of U2 small nuclear ribonucleoprotein auxiliary factor. Proc Natl Acad Sci USA 86:9243–9247

Zamore PD, Green MR (1991) Biochemical characterization of U2 snRNP auxiliary factor: an essential pre-mRNA splicing factor with a novel intranuclear distribution. EMBO J 10:207–214

Zamore PD, Patton JG, Green MR (1992) Cloning and domain structure of the mammalian splicing factor U2AF. Nature 335:609–614

Zhang H, Wang JC, Liu LF (1988) Involvement of DNA topoisomerase I in the transcription of human ribosomal RNA genes. Cell 85:1060–1064

Zhang M, Zamore PD, Carmo-Fonseca M, Lamond AI, Green MR (1992) Cloning and intracellular localization of the U2 small nuclear ribonucleoprotein auxiliary factor small subunit. Proc Natl Acad Sci USA 89:8769–8773

Zhu J, Krainer AR (2000) Pre-mRNA splicing in the absence of an SR protein RS domain. Genes Dev 14:3166–3178

Zhu J, Mayeda A, Krainer AR (2001) Exon identity established through differential antagonism between exonic splicing silencer-bound hnRNP A1 and enhancer-bound SR proteins. Mol Cell 8:1351–1361

Zhuang Y, Weiner AM (1986) A compensatory base change in U1 snRNA suppresses a 5′ splice site mutation. Cell 46:827–835

Zorio DA, Blumenthal T (1999) Both subunits of U2AF recognize the 3′ splice site in *Caenorhabditis elegans*. Nature 402:835–838

Zuo P, Maniatis T (1996) The splicing factor U2AF35 mediates critical protein-protein interactions in constitutive and enhancer-dependent splicing. Genes Dev 10:1356–1368

Zuo P, Manley JL (1993) Functional domains of the human splicing factor ASF/SF2. EMBO J 12:4727–4737

Splicing Regulation in *Drosophila* Sex Determination

P. Förch and J. Valcárcel[1]

Posttranscriptional regulation is of fundamental importance for establishing the gene expression programs that determine sexual identity in the fruitfly *Drosophila melanogaster*. The protein Sex-lethal acts as a master regulatory switch, being expressed exclusively in female flies and inducing female-specific patterns of alternative splicing on target genes. As a consequence, other regulatory factors are expressed in a sex-specific manner, and these factors control somatic and germline sexual differentiation, sexual behavior and X chromosome dosage compensation. Here, we review the molecular mechanisms responsible for splicing regulation in *Drosophila* sexual determination.

1 Introduction

Soon after the discovery of introns in eukaryotic genes (Berget et al. 1977; Chow and Broker 1978), examples of the selective use of splice sites in different tissues were reported (Breathnach et al. 1980; DeNoto et al. 1981; Mariman et al. 1983; King and Piatigorsky 1983). The extent to which alternative splicing contributes to the regulation of gene expression, however, has only recently been fully appreciated (Lander et al. 2001; Venter et al. 2001). Current estimates indicate that 45–60% of human genes are alternatively spliced (Modrek and Lee 2002), providing a means to generate literally thousands of mRNA isoforms from a single primary transcript (Schmucker et al. 2000). Given the combinatorial potential and the prevalence of alternative splicing, the study of the molecular mechanisms underlying splice site selection has become the focus of much interest (Hastings and Krainer 2001). Progress has been particularly appreciable in systems for which genetic work laid the foundations for identifying regulators of specific patterns of RNA processing. These advances also helped to establish several early paradigms in the field. One paradigm was the existence of tissue- or stage-specific factors that can determine patterns of alternative splicing on several target genes. Another was that the same regulator could influence a diverse array of patterns of splicing in different genes

[1] Gene Expression Programme, European Molecular Biology Laboratory, Meyerhofstrasse 1, 69117 Heidelberg, Germany

Progress in Molecular and Subcellular Biology, Vol. 31
Philippe Jeanteur (Ed.)

(e.g., 5′ and 3′ splice site selection, exon skipping, intron retention). A third paradigm was that regulators often influence the pattern – and thereby the activity – of RNA processing factors, which leads to the establishment of regulatory cascades that amplify, diversify or autoregulate the initial signal. Perhaps the most relevant realization coming from these studies was that regulation of alternative RNA processing is of fundamental importance for complete programs of cell differentiation and development.

One of the best-studied examples of such a program occurs during sexual determination of the fruit fly *Drosophila melanogaster*, and this will be the topic of this chapter. As a number of excellent and comprehensive reviews have covered this issue in recent years (MacDougall et al. 1995; Cline and Meyer 1996; López 1998; Schütt and Nöthiger 2000), we will focus here on the molecular mechanisms of splicing regulation.

2
Establishment and Maintenance of Sexual Identity

The key genetic difference between sexes in many species, including *Drosophila*, is the presence of a Y chromosome in males and two X chromosomes in females (reviewed by Cline and Meyer 1996). Although the Y chromosome does harbor genes involved in male-specific functions, the primary signal that triggers different sex determination programs in *Drosophila* does not depend on the presence or absence of this chromosome, but rather on the ratio between the number of X chromosomes and autosomes (X:A = 1 for females, X:A = 0.5 for males). This signal is interpreted through a complex interplay of transcriptional activators encoded by genes located in the X chromosome (the *numerator* genes *sisterless* A B, C and *runt*) and at least one transcriptional repressor located in an autosome (the *denominator* gene *deadpan*; reviewed by Cline 1993; Schütt and Nöthiger 2000), acting through multiple redundant elements in the promoter (Estes et al. 1995).

In adult flies the output of gene expression from X chromosomes is equalized in both sexes by the process known as dosage compensation (see below). This process, however, is not operative prior to the blastoderm stage of embryogenesis, the time at which sex identity is defined in every somatic cell of the organism. As a consequence, the amounts of *numerator* transcription factors in females are double those in males, and this results in activation of an early or "establishment" promoter of the gene *Sex-lethal* (*Sxl*) in females, but not in males (Keyes et al. 1992).

Sxl plays a key role in orchestrating the changes in gene expression responsible for all aspects of sexual determination, including somatic and germ line differentiation, dosage compensation, sexual behavior and the maintenance of sexual identity throughout the life of the animal (Fig. 1). *Sxl* encodes a protein with RNA binding domains that modulates alternative splicing and translation

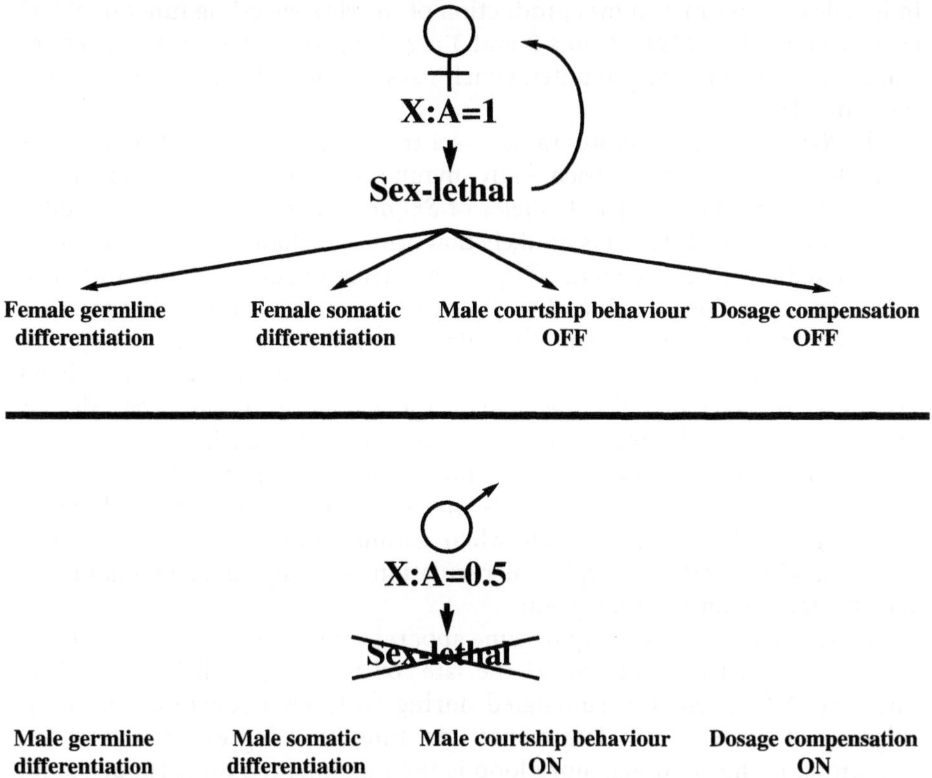

Fig. 1. The protein Sex-lethal plays a central role in *Drosophila* sex determination. Expression of the protein Sex-lethal (SXL) depends on the ratio of X chromosomes to autosomes (X:A) In female flies (X:A = 1) SXL is expressed, and its presence induces female germline and somatic differentiation, while it represses male courtship behavior and inhibits dosage compensation. SXL is also required for maintenance of the female differentiation state through an autoregulatory loop. In male flies (X:A = 0.5) SXL is not expressed, and this results in male germline and somatic differentiation, while it allows the activation of male courtship behavior and dosage compensation

of target pre-mRNAs and mRNAs (Bell et al. 1988; reviewed by Schütt and Nöthiger 2000).

One of the most striking aspects of *Sxl* regulation is that after the critical period when the X:A signal is read out and the establishment promoter is activated in females, a constitutive promoter is switched on in both males and females. Despite the presence of *Sxl* transcripts in both sexes, however, Sex-lethal (SXL) protein is only produced in female flies. This is due to a critical difference between the transcripts in the two sexes: exon 3 is included in male flies and skipped in females (Salz et al. 1989; Bell et al. 1991). This exon contains stop codons in frame with the initiator ATG located in exon 2, thereby generating mRNAs unable to code for full-length SXL in males. Exon skipping in females, in contrast, allows production of mRNAs encoding functional SXL (Fig. 3; Bopp et al. 1991). Exons 2 and 3 are skipped in transcripts generated from the establishment promoter, which uses a different exon 1 with its own initiator ATG.

The *Sxl* switch is therefore framed as a transcriptional regulation problem in early embryos and subsequently maintained by a posttranscriptional regulation mechanism. As null alleles of *Sxl* do not compromise the viability of male flies (Cline 1993), it is unlikely that exon 3-including transcripts, or the truncated SXL proteins generated by their translation, have any function. This raises the intriguing question of why posttranscriptional control has arisen at all when transcriptional regulation already establishes proper sex-specific expression early in development. The answer to that question most likely resides in the fact that the mechanism for activation of the establishment promoter relies on the absence of dosage compensation during early embryogenesis, a process that needs to be implemented to ensure viability at later stages of development. In other words, the establishment promoter is designed to read the X:A ratio at the time when chromosome dosage effects can be detected, while posttranscriptional regulation is designed to maintain sex-specific expression of SXL protein.

The rationale for this set up became apparent when SXL protein was shown to control alternative splicing of the late *Sxl* transcripts (Bell et al. 1991). The early SXL protein, accumulated during early embryogenesis, promotes skipping of exon 3 from late transcripts, thus inducing expression of late SXL protein. This autoregulatory loop is then maintained throughout the life of female flies. In summary, while a transcriptional regulatory network first activates SXL expression, maintenance relies on the posttranscriptional activities of the protein acting on its own pre-mRNA.

Another curious aspect of the *Sxl* switch is that although transcripts that include exon 3 contain stop codons more than 70 nucleotides upstream from the last exon-exon junction, these transcripts accumulate and are apparently not subject to nonsense-mediated decay (NMD; reviewed by Maquat and Carmichael 2001).

3 Sex-Lethal Proteins

The domain structure of late SXL protein is shown in Fig. 2. It is composed of an amino-terminal region of 95 amino acids rich in glycine and asparagine (GN domain), two RNA recognition motifs (RRMs; Lee et al. 1994; reviewed by Varani and Nagai 1998) and an 80 amino acid carboxy-terminal region without apparent sequence similarities to known protein domains. Detailed analysis of late transcripts revealed the existence of several mRNA isoforms expressed from mid-embryogenesis through adulthood (Salz et al. 1989; Samuels et al. 1991). They differ by the use of four alternative polyadenylation sites and by

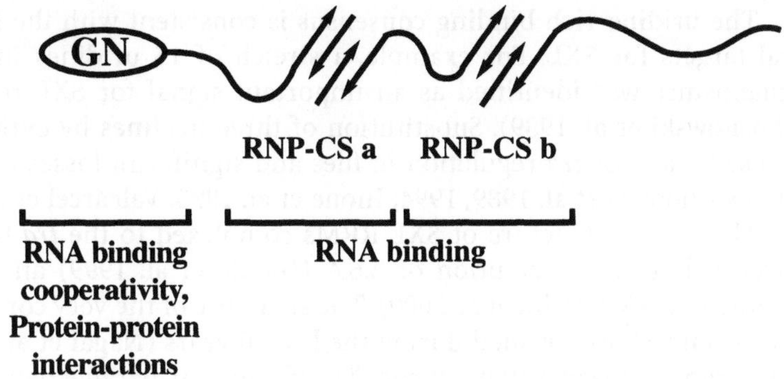

Fig. 2. Domain structure of the SXL protein. RRM type of RNA binding domains are indicated by *antiparallel arrows*, which represent the characteristic antiparallel β-sheets that provide a binding platform for RNA. The amino-terminal domain is rich in glycine and asparagine residues, and is important for cooperative RNA binding and protein-protein interactions

two alternative splicing events (in addition to the sex-specific inclusion/skipping of exon 3). These mRNA variants can encode up to six protein isoforms that differ in amino acid sequences outside of the RRM motifs. Although males do not express the major SXL forms, smaller versions of the protein seem to be present in the head and thorax of males (Bopp et al. 1991). At present, it is unclear whether the different mRNA and protein isoforms of SXL have biologically distinct functions.

The RNA binding properties of SXL have been extensively studied. Identification of *cis*-acting elements important for SXL regulation present in target genes like *transformer*, *Sex-lethal* and *male-specific lethal-2* strongly suggested that SXL binds uridine-rich sequences (Sosnowski et al. 1989; Bell et al. 1991; Kelley et al. 1995; Zhou et al. 1995; Bashaw and Baker 1997). Biochemical experiments further confirmed this expectation (Inuoe et al. 1990; Valcárcel et al. 1993; Samuels et al. 1994; Wang and Bell 1994; Singh et al. 1995), and established that both RRMs are required for specific recognition of SXL targets (Kanaar et al. 1995; Samuels et al. 1998). Iterative selection of randomized RNAs (SELEX) showed that high affinity binding sites for SXL contain stretches of 13 or more uridines (Singh et al. 1995). These stretches can be interrupted by guanidine residues without significant loss of affinity, while adenosine or cytidine residues are more disruptive. Independent SELEX experiments derived a similar consensus with an additional AG dinucleotide 3′ to the uridine-rich stretch, thus resembling the structure of 3′ splice site regions (Sakashita and Sakamoto 1994). The ability of Sex-lethal to interact with sequences resembling 3′ splice site AG was also confirmed by structural NMR studies of the second RRM (Chi et al. 1999), but was not observed in other studies (Samuels et al. 1994). Still a third set of SELEX experiments suggested that in addition to uridine stretches, RNAs without uridine runs but particular secondary structures can also be bound by SXL (Wang et al. 1997).

The uridine-rich binding consensus is consistent with the known biological targets for SXL. For example, a stretch of 15 uridines interrupted by 2 guanosines was identified as an important signal for SXL regulation in *tra* (Sosnowski et al. 1989). Substitution of three uridines by cytidines, however, caused a loss of *tra* regulation in flies and significant losses in binding affinity (Sosnowski et al. 1989, 1994; Inoue et al. 1990; Valcárcel et al. 1993).

The crystal structure of SXL RRMs complexed to the *tra* binding site has been solved at a resolution of 2.6 Å (Handa et al. 1999) and has also been studied by NMR (Kim et al. 2000). The structure of the very common RRM fold has been well established during the last 10 years (Nagai et al. 1990; reviewed by Varani and Nagai 1998). It consists of four anti-parallel beta sheets forming an extended surface that interacts with the RNA, where two particularly well conserved motifs of six and eight amino acids, essential for RNA binding, are located adjacent to each other in the two central beta sheets. The two RRMs of SXL are required for specific binding (Kanaar et al. 1995), form a V-shaped cleft when bound to RNA (Handa et al. 1999), and each of the motifs appear to establish nonanalogous points of contact (Lee et al. 1997). A striking conclusion of these structural analyses was that the protein establishes numerous contacts with many residues and chemical groups. This was also confirmed by detailed RNA interference and chemical protection studies (Singh et al. 2000). Consistent with this notion, many mutations in the SXL binding site in *tra* alter regulation by SXL in transgenic flies (Sosnowski et al. 1994).

A possible rationale for the establishment of extensive interactions with the target RNA is that uridine-rich sequences are common at the 3′ end of introns where they play an important role in the splicing process (see below). Therefore, a high level of discrimination is required to avoid general repression of 3′ splice sites (Singh et al. 1995).

Regions outside the RRMs influence the RNA binding properties of the protein (Wang and Bell 1994; Wang et al. 1997). For example, the amino-terminal GN domain has been shown to facilitate cooperative interactions between Sex-lethal proteins bound to adjacent uridine-rich stretches (Wang and Bell 1994), through direct protein-protein interactions mediated by this domain and possibly the RRM motifs as well (Wang et al. 1997; Samuels et al. 1998). Consistent with this notion, the presence of proteins containing glycine-rich domains alters the RNA binding properties of SXL in in vitro assays (Wang et al. 1997), an observation that may have physiological relevance.

A fascinating issue regarding SXL proteins is the evolution of their function. The protein is very conserved, particularly in the RRM domains, between Diptera that have diverged for hundreds of millions of years. Remarkably, however, the function that the SXL homologues carry out in these species is not conserved. Indeed, while SXL-mediated sex-specific splicing is conserved among different species of the *Drosophila* genus (Bopp et al. 1996), SXL protein is not expressed in a sex-specific fashion in other diptera, and *Sxl* alternative splicing is related to expression in different tissues (Meise et al. 1998; Saccone et al. 1998; Sievert et al. 2000). This raises the question of what is the ancestral function of SXL-like proteins. One idea that has been put forward is that an ancestral function of SXL, possibly on germline determination and/or neural development, was adapted in *Drosophila melanogaster* to achieve sex-specific splicing (for more extensive discussion, see Schütt and Nöthiger 2000).

4 Somatic Sexual Differentiation

4.1 Regulation of *transformer*

As *Sxl*, the gene *transformer* (*tra*) is transcribed in both male and female flies. Two alternative 3′ splice sites are present in intron 1 (Boggs et al. 1987; Nagoshi et al. 1988). Males use exclusively the upstream 3′ splice sites, while both the upstream and downstream 3′ splice sites are used in females (Fig. 3). For this reason, the upstream site is known as "non-sex-specific", and the downstream one as "female-specific". Also similar to *Sxl*, transcripts present in male flies have limited coding capacity because they contain early stop codons which also seem to escape nonsense-mediated decay. Use of the female-specific 3′ splice site allows skipping of the premature stop codons and synthesis of full length TRA protein, which is functional to trigger female differentiation (McKeown et al. 1987).

Genetic data and experiments carried out in transgenic flies, cells in culture and in vitro assays have established that the switch in 3′ splice site utilization is controlled by SXL (Nagoshi et al. 1988; Sosnowski et al. 1989; Inoue et al. 1990; Valcárcel et al. 1993). SXL binds to the pyrimidine-rich (Py)tract preceding the non-sex-specific site, and as described in detail in the previous section, mutations that reduce the uridine content of this tract can block both SXL binding and function (Sosnowski et al. 1989, 1994; Inoue et al. 1990; Valcárcel et al. 1993).

Py-rich tracts are a common feature of the 3′ end of introns in higher eukaryotes and they are important for early events in the assembly of spliceosomal complexes (reviewed by Reed 2000). These sequences serve as binding sites for the splicing factor U2AF^{65} (Zamore et al. 1992), which is important for the recruitment of U2 snRNP (Ruskin et al. 1988; Zamore et al. 1992). One simple model to explain the regulation of alternative splicing of *tra* is that binding of SXL prevents association of U2AF^{65} to the Py-tract of the non-sex-specific site, thus inhibiting its use and allowing the use of the female-specific site. Indeed, evidence for repression of the non-sex-specific site by SXL was obtained in transgenic fly experiments (Sosnowski et al. 1989), and results from in vitro splicing assays using human nuclear extracts indicated that SXL can block access of U2AF^{65} to the non-sex-specific site (Valcárcel et al. 1993).

Two factors may contribute to the exclusive use of the non-sex-specific site in the absence of SXL. First, U2AF^{65} has 100-fold higher affinity for the uridine-

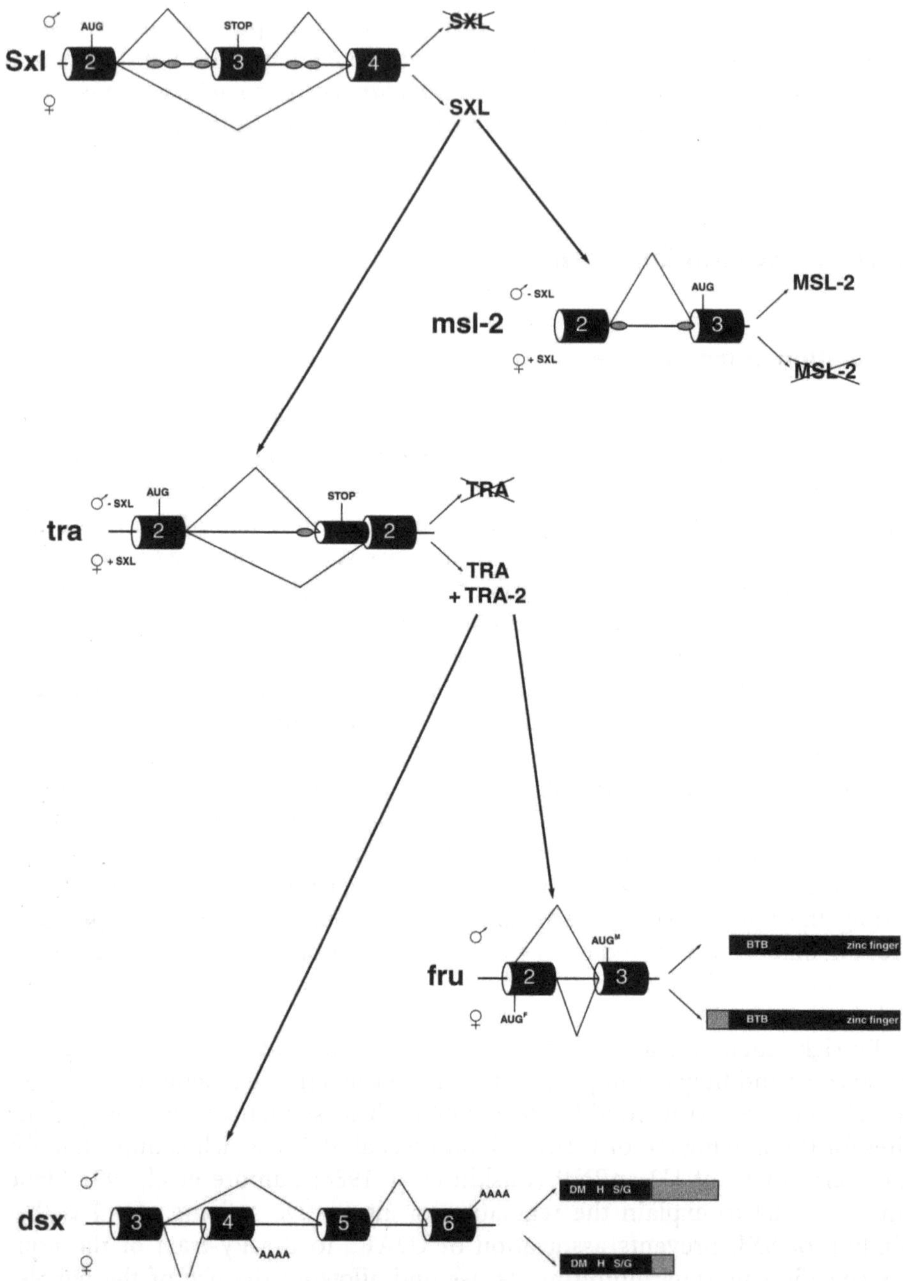
Sxl
AUG
STOP
2
3
4
SXL
SXL
msl-2
♂ - SXL
♀ + SXL
AUG
2
3
MSL-2
MSL-2
tra
♂ - SXL
♀ + SXL
AUG
STOP
2
2
TRA
TRA
+ TRA-2
fru
AUG[M]
AUG[F]
2
3
BTB
zinc finger
BTB
zinc finger
dsx
3
4
5
6
AAAA
AAAA
DM H S/G
DM H S/G

rich Py-tract of the non-sex-specific 3′ splice site than for the less uridine-rich Py-tract associated with the female-specific site (Valcárcel et al. 1993). Second, the closer proximity of the non-sex-specific site to the 5′ splice site also promotes its use (Valcárcel et al. 1997). Thus, accessibility of $U2AF^{65}$ to the competing 3′ splice sites can play a role both in the selection of the non-sex-specific site in males and in the selection of the female-specific site in the presence of SXL in females.

The domain structure of $U2AF^{65}$ consists of an amino-terminal arginine and serine-rich (RS) motif and three RRMs. While RRMs are required for recognition of Py-tracts, the RS domain and part of the first RRM are important to facilitate the recruitment of U2 snRNP to the pre-mRNA (Zamore et al. 1992; Fleckner et al. 1997). Interestingly, when the "effector" domain of $U2AF^{65}$ was fused to SXL RNA binding domains, the chimeric protein activated, rather than repressed the use of the non-sex-specific site (Valcárcel et al. 1993). This result indicates that providing $U2AF^{65}$ activity to SXL converts the protein from a splicing repressor to an activator. As this suffices to disrupt its regulatory function on *tra* splicing both in vitro and in transgenic flies (Granadino et al. 1997), these data argue that $U2AF^{65}$ is the main target of SXL in the regulation of *tra* splicing.

A different mechanism needs to be invoked, however, to explain results obtained in transgenic flies expressing a fusion protein between the amino-terminal 99 amino acids of SXL and beta-galactosidase (Deshpande et al. 1999). Expression of this protein caused activation of the female-specific site in male flies and was even able to cause feminization, thus proving the biological activity of the protein in somatic differentiation. Also consistent with a role for the amino-terminal domain in *tra* regulation, deletion of the amino-terminal 40 residues of SXL compromised the ability of the protein to regulate *tra* and caused sex transformation when this deletion mutant was expressed in male flies (Yanowitz et al. 1999). A similar conclusion could be derived from the

Fig. 3. Cascade of regulated splicing in *Drosophila* sex determination. SXL protein promotes skipping of exon 3 from its own pre-mRNA, thus removing premature stop codons present within the exon which are in frame with the initiator ATG present in exon 2 and allowing the synthesis of more SXL. In addition SXL regulates alternative splicing of *male-specific lethal-2* (*msl-2*) and *transformer* (*tra*) SXL-induced retention of an intron at the 5′ UTR of *msl-2* transcripts results in translational inhibition of the mRNA in females. SXL represses the use of a 3′ splice site in *tra*, which allows skipping of stop codons in frame with the initiator ATG and production of TRA protein in female flies. TRA itself is a regulator of alternative splicing of the genes *doublesex* (*dsx*) and *fruitless* (*fru*). TRA acts in combination with TRA-2, which is expressed in both sexes, to induce the use of a female-specific 3′ splice site and polyadenylation signal in *dsx*, and the use of a downstream 5′ splice site in exon 2 of *fru*. Sex-specific DSX and FRU isoforms that differ in their carboxy or amino-terminal domains are generated as a result. Male and female patterns of splicing pattern are indicated by *male and female symbols*. Binding sites for SXL are represented by *striped ovals*. Exons are represented by *cylinders* with indication of their number in the pre-mRNA, introns by *thin lines*. AAA represents polyadenylation sites. AUG and STOP represent initiator and stop codons

observation that expression in transgenic *Drosophila* flies of the SXL proteins from *Musca domestica* or *Ceratitis capitata* – which are very similar in the RNA binding domains to *Drosophila* SXL – does not cause feminization of male flies (Meise et al. 1998; Saccone et al. 1998). In addition, *Drosophila virilis* SXL is expressed both in male and female flies, where both proteins bind to similar loci in polytene chromosomes, but only the female protein elicits female-specific splicing (Bopp et al. 1996). The two proteins only differ in their 25 amino-terminal residues, suggesting again a critical role of the amino-terminal domain in somatic differentiation.

A mutant harboring a deletion of the first 94 amino acids of SXL, however, was able to induce the use of the female-specific site in similar experiments (Granadino et al. 1997), and the amino-terminal domain of SXL was also dispensable to regulate *tra* splicing in vitro (Valcárcel et al. 1997). Although the discrepancies between the results obtained with different amino-terminal deletion mutants remain to be explained, the activity of the amino-terminal domain fused to beta-galactosidase does argue that recognition of the Py-tract by SXL is not required for regulation of *tra* splicing. This is in contrast to the effect of mutations in the Py-tract of the non-sex-specific site, which disrupt both SXL binding and splice site regulation, in vitro and in vivo (Sosnowski et al. 1989; Inoue at al 1992; Valcárcel et al. 1993). In addition, it is difficult to envision how the regulatory function of the amino-terminal domain can be directed to specific transcripts and splice sites in the absence of targeting RNA binding motifs (i.e., why it does not affect splicing in general). Further work will be required to understand in molecular terms the effect of the beta-galactosidase-SXL fusion on *tra* splicing.

4.2 Regulation of *doublesex*

The gene *doublesex* (*dsx*) is alternatively spliced in a sex-specific fashion to generate transcripts encoding transcription factors that differ in their carboxy-terminal domains (reviewed by López 1995). Although these different amino acid sequences do not affect the DNA binding properties of the factors, sex-specific activation of transcription of distinct sets of genes is achieved through differential expression of the DSX isoforms (Burtis and Baker 1989; Waterbury et al. 2000). Sex-specific expression of DSX protein homologues appears to be conserved in a variety of insects (reviewed in Ohbayashi et al. 2001). Genetic and molecular data indicated that *dsx* splicing is under the control of the products of the genes *tra* and *tra-2* (Boggs et al. 1987; Nagoshi et al. 1988; Hedley and Maniatis 1991; Hoshijima et al. 1991; Ryner and Baker 1991; Tian and Maniatis 1992). TRA-2 protein is expressed in both sexes, while TRA is expressed exclusively in females, as described in the previous section. Therefore, regulation of *tra* splicing by SXL results in accumulation of TRA protein and this, in turn, triggers female-specific splicing of *dsx* transcripts.

Both TRA and TRA-2 contain protein domains characteristic of splicing factors, in particular RS-rich domains. These are signature features of a large family of proteins (SR proteins) involved in RNA processing (reviewed by Fu 1995; Manley and Tacke 1996; Hastings and Krainer 2001). SR proteins play multiple functions in splicing (reviewed by Valcárcel and Green 1996), and they are thought to stimulate various steps in the assembly of splicing complexes through the establishment of RNA-protein and protein-protein interactions, the latter mediated by the RS domains (Wu and Maniatis 1993). While one of the two RS domains present in TRA-2 is dispensable, the other is required for both protein-protein and RNA-protein interactions, and the RRM domain is required for all the functions of TRA-2 in either splicing activation or repression (see below; Amrein et al. 1994). The essential RS domain of TRA-2 can be replaced by RS domains from other splicing factors with different functional efficiency, suggesting specific interactions mediated by each RS domain in vivo (Dauwalder and Mattox 1998).

One well-established function of SR proteins is to associate to purine-rich exonic enhancers and to promote the use of weak 3′ splice sites (Lavigueur et al. 1993; Sun et al. 1993; Watakabe et al. 1993; Xu et al. 1993). Different enhancers bind a distinct set of SR proteins (Ramchatesingh et al. 1995).

Dsx splicing regulation also involves enhancement of a weak, female-specific 3′ splice site (Hedley and Maniatis 1991; Hoshijima et al. 1991; Ryner and Baker 1991). TRA and TRA-2 also associate with exonic sequences, which are composed of six 13-nucleotide repeats and a purine-rich element (Hedley and Maniatis 1991; Tian and Maniatis 1992). TRA-2 binds to the 3′ end of the repeats and, through cooperative interactions with TRA, stabilizes the interaction of specific SR proteins with the repeats 5′ end. A similar effect has the association of TRA-2/TRA with the purine-rich element, although a different SR protein is recruited (Tian and Maniatis 1993; Lynch and Maniatis 1995, 1996). The main difference between the *dsx* enhancer and purine-rich enhancers frequently present in exons downstream from constitutive and regulated 3′ splice sites (Schaal and Maniatis 1999a,b) is that the dsx enhancer is at a significant distance from the female-specific site. When the purine-rich element embedded within the repeats is located close to the 3′ splice site, it can then exert enhancer functions in the absence of TRA and TRA-2 (Tian and Maniatis 1994). Regulation is therefore achieved by making a 3′ splice site dependent on an enhancer sequence that requires TRA and TRA-2 to recruit SR proteins and act at a distance to stimulate splicing.

Several mechanisms have been proposed for how exonic enhancers can work. Stabilization of $U2AF^{65}$ binding to the weak Py-tract through bridging interactions mediated by the RS domain of $U2AF^{35}$ and that of SR proteins bound to the enhancer has been documented in a reconstituted system using purified components (Zuo and Maniatis 1996). Analysis in complete extracts, however, has suggested alternative mechanisms (Kan and Green 1999; Blencowe 2000). In addition, deletion of the RS domain of $U2AF^{35}$ in transgenic flies did not have any effect on sex determination, even in a sensitized genetic background and despite the fact that the mutation did cause other developmental defects (Rudner et al. 1998). The fact that $U2AF^{35}$ can both facilitate $U2AF^{65}$ binding and other events in spliceosome assembly (Guth et al. 2001) suggests that various mechanisms can act to stimulate 3′ splice site utilization depending on the specific sequence context and the concentration of different factors involved.

One factor particularly important for activation of the *dsx* female-specific site in the SR protein is RBP1. RBP1 associates to the TRA/TRA-2 repeats and, in addition, also binds to the weak, relatively purine-rich Py-tract (Heinrichs and Baker 1995). Overexpression of RBP1 (but not other SR proteins) can activate female-specific splicing of *dsx* in transient transfection assays (Heinrichs and Baker 1995), suggesting that RBP1 can play a critical role in recruiting (or replacing) U2AF.

Human homologues of TRA-2 have been identified (Dauwalder et al. 1996; Tacke et al. 1998). They not only show striking sequence similarity, but also one of the human TRA-2 isoforms is able to replace the biological function of the *Drosophila* protein (Dauwalder et al. 1996). Human TRA-2 proteins also stimulate 3′ splice site utilization acting through exonic enhancers in a sequence-specific fashion (Tacke et al. 1998).

The state of phosphorylation of SR proteins regulates their functions in vitro (reviewed by Stojdl and Bell 1999). In vivo evidence for their involvement in splicing regulation was obtained when mutants in the *Drosophila* LAMMER kinase (*Doa*) gene were found to alter sexual differentiation and *dsx* alternative splicing acting synergistically with *tra* and *tra-2* mutations (Du et al. 1998). An important implication of these results is that the effects of an SR protein kinase can be substrate-dependent and affect very specific programs of gene regulation.

5 Dosage Compensation

A number of mechanisms have evolved to balance the expression of X-linked genes between males and females (reviewed by Cline and Meyer 1996; Marin et al. 2000). In *Drosophila* the problem has been solved by doubling the transcription rate in males of genes associated with the X chromosome. This effect is mediated by a chromatin remodeling complex, the male-specific lethal (MSL) complex, that binds to multiple sites along the X chromosome and enhances transcription twofold (reviewed by Amrein 2000; Pannuti and Lucchesi 2000). The MSL complex is composed of several proteins that associate with two non-coding RNAs, roX1 and roX2, that associate with specific nucleation sites on the X chromosome and then spread from these sites to remodel the whole chromosome (reviewed by Kelley and Kuroda 2000). One of the components of the complex, the protein MSL-2, is expressed only in male flies. Absence of MSL-2 in females prevents MSL complex assembly and therefore hypertranscription of the two X chromosomes, which would be lethal.

Molecular cloning and RNA expression analysis indicated that an intron present at the 5′ untranslated region (UTR) of the *msl-2* gene was spliced in males, but retained in females (Kelley et al. 1995; Zhou et al. 1995; Bashaw and Baker 1996). Interestingly, uridine-rich sequences were present within the intron and also at the 3′ UTR of *msl-2* transcripts, suggesting that SXL could bind to these sequences to inhibit MSL-2 protein expression in females. Results of experiments in transgenic flies and in cells in culture indicated that SXL carries out two different functions in the regulation of *msl-2* expression (Kelley et al. 1997; Bashaw and Baker 1997; Gebauer et al. 1998). Binding of SXL to uridine-rich sequences within the 5′ UTR inhibits splicing and this allows the unspliced RNA to be exported to the cytoplasm, where SXL inhibits translation (reviewed by Gebauer et al. 1997). Translational inhibition requires SXL binding to uridine-rich sequences at both the 5′ and 3′ UTRs (Bashaw and Baker 1997; Kelley et al. 1997; Gebauer et al. 1999).

In vitro experiments using HeLa nuclear extracts indicated that efficient inhibition of splicing of the 5′ UTR intron requires SXL binding to uridine-rich sequences present immediately downstream from the 5′ splice site and at the Py-tract region (Förch et al. 2001). Binding to the site close to the 5′ splice site interferes with the assembly of U1 snRNP through an indirect mechanism: SXL antagonizes the binding of a protein (TIA-1 in human extracts) required for U1 snRNP binding to the weak 5′ splice site present in this intron (Förch et al. 2000, 2001). A homologue of TIA-1 exists in *Drosophila* (Brand and Bourbon 1993), although its function has not yet been characterized.

Binding of SXL to the Py-tract blocks the binding of $U2AF^{65}$, and consequently of U2 snRNP to the 3′ splice site region (Merendino et al. 1999), as was the case, at least in vitro, for the regulation of the non-sex-specific site of *tra* (see Sect. 5.1). Interestingly, inhibition of *msl-2* 3′ splice site depended upon an uncommon feature of the pre-mRNA: an unusually long distance of 13 nucleotides between the Py-tract and the 3′ splice site AG. A long Py-tract/AG distance prevents efficient interaction between the 3′ splice site AG and $U2AF^{35}$, which cannot contribute to stabilize $U2AF^{65}$ binding, thus allowing SXL to displace the U2AF heterodimer (Merendino et al. 1999).

In summary, inhibition of *msl-2* splicing requires blockage of both the 5′ and 3′ splice sites, and these splice sites harbor sequence features that make them susceptible to regulation by SXL. Regulation of 3′ splice site choice in *tra* did not require a weakened $U2AF^{35}$-3′ splice site interaction, suggesting that regulation of intron retention in *msl-2* is exerted through a tighter level of control of early events in spliceosome assembly.

6 Sexual Behavior

6.1 Fruitless

An additional target of TRA/TRA-2 is the gene *fruitless* (*fru*), which regulates sexual behavior in *Drosophila* (Ito et al. 1996; Ryner et al. 1997). *Fruitless* loss of function mutants show male-to-male courtship behavior and malformation of the male-specific muscle of Lawrence (MOL), which is important for copulation. The gene is expressed in a small number (about 0.5%) of neurons of the central nervous system (CNS). The gene contains four promoters and transcripts derived from promoter P1 are alternatively spliced in a sex-specific manner (Fig. 3; Anand et al. 2001)). Two 5′ splice sites 1.5 kb apart compete for splicing to a common 3′ splice site 70 kb downstream (Ryner et al. 1997). The upstream 5′ splice site is used in males, while the downstream site is used in females. As a consequence of this alternative splicing event, different start codons are used, and male transcripts encode a protein with 101 additional amino acids at the amino terminus.

FRU proteins are transcription factors containing BR-C, ttk and bab (BTB) and zinc-finger domains, and it was assumed that the different amino termini contributed different biological functions. Recent work, however, has shown that the female protein is not expressed in the CNS, and that expression of either the male or the female proteins induce formation of Muscle of Lawrence (MOL; Usui-Aoki et al. 2000). Therefore, it is not completely clear at present how sex-specific splicing of *fru* contributes to regulation of the function of the gene.

Genetic data indicated that *fru* is under the control of *tra* and *tra-2*. Sequence analyses revealed the presence of three repeats characteristic of cis-acting TRA/TRA-2 responsive elements located 38 nucleotides upstream of the female-specific 5′ splice site (Ryner et al. 1997). These sequences are bound by TRA/TRA-2 in vitro, and expression of TRA in Schneider cells induced activation of the female 5′ splice site even in the absence of a male site, while it did not repress the use of the male site in the absence of the female site (Heinrichs et al. 1998). These data indicate that TRA/TRA-2 (and most likely other SR proteins associated to the complex) act, as in *doublesex*, to activate female-specific splicing. The interesting mechanistic difference, however, is that a 5′ splice site is activated in *fru*, while a 3′ splice site is activated in *dsx*. Given the long distance between the 5′ and 3′ splice sites in *fru* (more than 70 kb), it seems unlikely that the TRA/TRA-2 enhancer carries out its function indirectly, acting on the 3′ splice site. It seems more likely that the enhancer targets early events in 5′ splice site recognition. Thus, an enhancer can act on different molecular machines depending on its location on the pre-mRNA. Another striking example of this phenomenon was found in an adenovirus pre-mRNA containing an enhancer bound by SR proteins upstream from the branchpoint region. This enhancer, which has a stimulatory effect on weak 3′ splice sites when positioned downstream from the regulated site, is inhibitory for 3′ splice sites when located upstream from the branchpoint (Kanopka et al. 1998).

7 Transformer-2 Autoregulation

Apart from the control of *dsx* and *fru*, TRA-2 proteins control sex-specific splicing and polyadenylation of the gene *exuperantia* (Hazelrig et al. 1990; Hazelrig and Tu 1994) and its own splicing in the male germ line. Four different alternatively spliced transcripts are generated from the *tra-2* gene to give rise to three protein isoforms (Amrein et al. 1988; Goralski et al. 1989; Mattox et al. 1990). While the different isoforms have redundant functions in female-specific splicing of *dsx*, the TRA-2^{226} isoform is essential for spermatogenesis. Excessive levels of the protein, however, result in abnormal spermatids and male sterility (McGuffin et al. 1998). Appropriate levels of TRA-2^{226} are achieved through splicing repression of *tra-2* pre-mRNA intron M1 by the TRA-2^{226} protein (Mattox et al. 1990; Mattox and Baker 1991). As the TRA-2^{226} initiator codon is contained in part at the 3′ end of the M1 intron, splicing of the intron prevents TRA-2^{226} expression (Fig. 4; Mattox et al. 1996).

Mutation of conserved features in the M1 intron indicate that regulation depends on the presence of a suboptimal, nonconsensus 3′ splice site. Substitution by other weak splice sites maintains regulation, but replacement by a strong one prevents inhibition by TRA-2^{226}. An additional intronic enhancer is necessary to promote splicing of the M1 intron in the soma and is important to overcome the weakness of the 3′ splice site. Therefore, there is a balance between the strength of the splice site and additional enhancer sequences to allow regulation in different tissues.

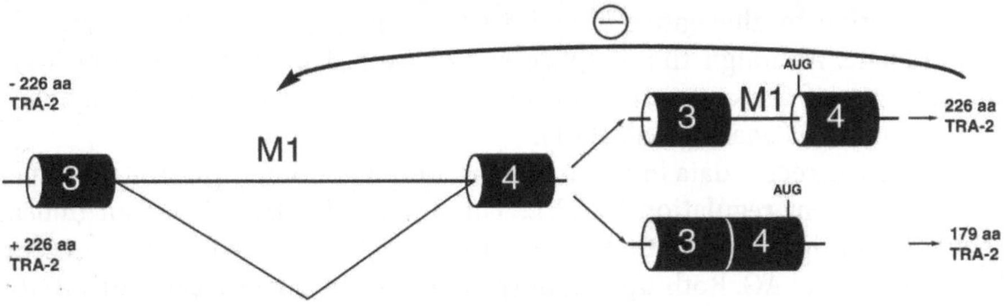

Fig. 4. Negative feedback regulation of tra-2 pre-mRNA splicing by the 226 isoform of the TRA-2 protein. Tra-2^{226} represses splicing of the intron between exons 3 and 4 (intron M1) Splicing of intron M1 eliminates the initiator codon for the 226 amino acids isoform. Use of a downstream initiator generates the 179 amino acids isoform of TRA-2. Exons and introns are represented as in Fig. 3

Contrary to factors at higher levels in the regulatory cascade, the expression strategy and the activity of TRA-2 proteins is conserved among different *Drosophila* species (Chandler et al. 1997).

8
Sex-lethal Autoregulation

As mentioned in Section 3, SXL autoregulates its own expression by promoting skipping of exon 3 from its pre-mRNA (Fig. 3; Bell et al. 1991). Multiple uridine-rich sequences located in the introns upstream and downstream from exon 3 bind SXL and contribute to regulation (Sakamoto et al. 1992; Horabin and Schedl 1993b; Wang and Bell 1994; Penalva et al. 2001). The regulatory effects of the different uridine-rich sequences seem to be additive, and removal or mutation of any single individual site still allows SXL function (Sakamoto et al. 1992). Cooperative RNA binding between adjacent – and perhaps also between more distant – uridine-rich stretches is important for regulation, because deletion of the amino-terminal glycine-rich domain responsible for these interactions compromises *Sxl* autoregulation (Wang and Bell 1994). Binding sites located in the downstream intron seem to make a stronger contribution to regulation (Horabin and Schedl 1993b). As most of the regulatory signals are remote from the splice sites, steric block of the access of splicing factors to the splice sites – as suggested for the regulation of *tra* and *msl-2* – seems an unlikely mechanism for *Sxl* autoregulation.

Several pieces of data suggested that the 5′ splice site of exon 3 is a target for SXL regulation. A minigene harboring *Sxl* genomic sequences from exon 2 to 4, in which the 3′ splice site of exon 3 was deleted, was still subject to regulation in transgenic female flies (Horabin and Schedl 1993b). Similarly, transient transfection of a minigene in which the length of intron 2 was reduced to a size that precludes splicing between exons 2 and 3 still responded to SXL (Penalva et al. 2001). The absence of a functional 3′ splice site, or the impossibility of intron 2 removal, switched the regulatory decision from exon skipping/retention to alternative 5′ splice site competition, which SXL was able to modulate. Although the 5′ splice site of exon 3 is predicted to be weak, the strength of the competing sites does not seem to play a critical role in autoregulation (Penalva et al. 2001).

In contrast, recent data indicate that the identity of the 3′ splice site is important for efficient regulation by SXL. The 3′ end of intron 2 has an unusual sequence configuration in that the Py-tract and AG dinucleotide are preceded by an additional AG. Both upstream (proximal) and downstream (distal) AGs can be used for catalysis (Bell et al. 1991), and quantitative analysis showed that the proximal site is preferentially used (Penalva et al. 2001). The integrity of the Py-tract and distal AG are, however, important for exon 3 definition, because their mutation results in constitutive exon 3 skipping. Mutation of the proximal site results in catalysis at the distal site and, intriguingly, in loss of

SXL regulation. These observations suggest that a switch in 3′ splice site AG recognition between early events of exon definition and later events leading to catalysis is important for efficient SXL function.

A model has been proposed for how SXL can promote exon 3 skipping based on genetic and physical interactions between SXL and a protein component of both U1 and U2 snRNPs. This protein, the product of the gene *sans-fille* (*snf*), is the functional homologue in *Drosophila* of the vertebrate U1A and U2B" polypeptides (Flickinger and Salz 1994; Polycarpou-Schwartz et al. 1996). SXL has been shown to associate with SNF in RNase-sensitive complexes (Deshpande et al. 1996), an interaction that involves the RRM1 of SXL (Samuels et al. 1998), but that appears to be disrupted by an excess of the amino terminal domain (Deshpande et al. 1999). Based on these interactions and on the synergistic effects that *snf* and *Sxl* mutations have on *Sxl* autoregulation (Salz 1992; Albrecht and Salz 1993; Salz and Flickinger 1996), it has been proposed that SXL binds to U1 and/or U2 snRNP bound to the 5′ and 3′ splice sites of exon 3 and somehow inactivates them to proceed further in the splicing pathway (Deshpande et al. 1996). One curious aspect of the observed genetic interactions is that both viable loss-of-function and gain-of-function mutations have the same effect on *Sxl* autoregulation. In addition, the alleles of *snf* are predicted to fail to bind RNA, and also show higher levels of the protein in the cytoplasm (Stitzinger et al. 1999a). For these reasons, the molecular basis for the genetic interactions observed between *snf* and *Sxl*, and the significance of the RNA-mediated complexes that their protein products form, remain to be clarified.

Recent results suggest that the function of a splicing factor involved in the second catalytic step of the reaction, SPF45 (Neubauer et al. 1998), is important for SXL autoregulation. RNA interference of SPF45 in cultures of *Drosophila* cells reduces the activity of SXL, concomitant with a switch in the use of the alternative 3′ splice site AGs from the proximal to the distal site. SPF45 binds directly to the proximal AG and promotes its utilization. SXL interacts with SPF45 through the amino-terminal glycine-rich domain and inhibits splicing at the second step of the reaction (Lallena et al. 2002).

Further work will be required to assess the mechanism(s) that operate in *Sxl* autoregulation in vivo.

Another aspect that remains obscure is the function of the products of other genes that show genetic interactions with Sxl. The genes *virilizer* (*vir*) and *female-lethal (2)d* (*fl(2)d*) were identified as *Sxl* modifiers important for proper Sxl autoregulation (Granadino et al. 1990; Hilfiker et al. 1995). Both genes are also important for regulation of *tra* (Hilfiker et al. 1995; Granadino et al. 1997), and have additional non-sex-specific functions at least in part related to alternative splicing regulation of the homeotic gene *Ultrabithorax* (Burnette et al. 1999). Unfortunately, cloning of *vir* and *fl(2)d* has not provided new insights into how their products may operate to modulate SXL activity (Penalva et al. 2000; Niessen et al. 2001).

It is important to point out, however, that a variety of indirect effects can compromise SXL activity, as illustrated by the observation that the levels of tRNA synthetases can act as dosage-sensitive modifiers of *Sxl* (Stitzinger et al. 1999b). It is conceivable that the levels and activity of SXL are particularly sensitive to changes in the overall efficiency of translation. However, this is in contrast to the existence of a feedback regulatory loop by which binding of SXL to uridine-rich sequences at the 3′ UTR of its own mRNA controls its own translation (Yanowitz et al. 1999).

Finally, it is remarkable that the complexities associated with the regulation of exon 3 skipping, both in terms of the multiple *cis*-acting sequences and *trans*-acting factors involved, do not seem to affect proper splicing of transcripts derived from the early promoter. These transcripts skip exons 2 and 3, thus joining the early promoter-specific exon 1 directly to exon 4. This pattern of splicing is associated with the sequences around the 5′ splice site of exon 1 (Horabin and Schedl 1993a; Zhu et al. 1997).

9 Perspectives

The examples shown in this review illustrate two important principles of gene regulation. First, they show how regulators are multifunctional, being able to affect various patterns of alternative processing in different genes. This implies the possibility of amplification, autoregulation and diversification of signals, and supports the notion that complete cellular programs of gene expression can be modulated through combinatorial control of the expression (or modification) of a limited set of splicing factors and regulators (for review see Smith and Valcárcel 2000).

A second principle illustrated by these examples is that the same regulator can act through diverse mechanisms, depending on the location of its binding sites, context of regulatory signals, strength of the splice sites, etc. Regulation can also occur at different steps of the splicing pathway.

Taken together with the prevalence of alternative splicing in higher eukaryotes, and its capacity to significantly amplify the information content of genomes, it seems clear that splicing regulation is a major and versatile contributor to the transcriptome and proteome complexity that shapes programs of cell differentiation and development.

References

Albrecht EB, Salz HK (1993) The *Drosophila* sex determination gene snf is utilized for the establishment of the female-specific splicing pattern of Sex-lethal. Genetics 134:801–807

Amrein H (2000) Multiple RNA-protein interactions in *Drosophila* dosage compensation. Genome Biol 1:1030.1–1030.5

Amrein H, Gorman M, Nothiger R (1988) The sex-determining gene tra-2 of *Drosophila* encodes a putative RNA binding protein. Cell 55:1025–1035

Amrein H, Hedley M, Maniatis T (1994) The role of specific protein-RNA and protein-protein interactions in positive and negative control of pre-mRNA splicing by Transformer-2. Cell 76:735–746

Anand A, Villella A, Ryner LC, Carlo T, Goodwin SF, Song HJ, Gailey DA, Morales A, Hall JC, Baker BS, Taylor BJ (2001) Molecular genetic dissection of the sex-specific and vital functions of the *Drosophila melanogaster* sex determination gene fruitless. Genetics 158:1569–1595

Bashaw GJ, Baker BS (1996) Dosage compensation and chromatin structure in *Drosophila*. Curr Opin Genet Dev 6:496–501

Bashaw G, Baker B (1997) The regulation of the *Drosophila* msl-2 gene reveals a function for Sex-lethal in translational control. Cell 89:789–798

Bell LR, Maine EM, Schedl P, Cline TW (1988) Sex-lethal, a *Drosophila* sex determination switch gene, exhibits sex-specific RNA splicing and sequence similarity to RNA binding proteins. Cell 55:1037–1046

Bell LR, Horabin JI, Schedl P, Cline TW (1991) Positive autoregulation of sex-lethal by alternative splicing maintains the female determined state in *Drosophila*. Cell 65:229–39

Berget SM, Moore C, Sharp PA (1977) Spliced segments at the 5′ terminus of adenovirus 2 late mRNA. Proc Natl Acad Sci USA 74:3171–3175

Blencowe BJ (2000) Exonic splicing enhancers: mechanism of action, diversity and role in human genetic diseases. Trends Biochem Sci 25:106–110

Boggs R, Gregor P, Idriss, S, Belote, J, McKeown M (1987) Regulation of sexual differentiation in *D. melanogaster* via alternative splicing of RNA from the transformer gene. Cell 50:739–747

Bopp D, Bell LR, Cline TW, Schedl P (1991) Developmental distribution of female-specific Sex-lethal proteins in *Drosophila melanogaster*. Genes Dev 5:403–415

Bopp D, Calhoun G, Horabin JI, Samuels M, Schedl P (1996) Sex-specific control of Sex-lethal is a conserved mechanism for sex determination in the genus *Drosophila*. Development 122:971–982

Brand S, Bourbon H (1993) The developmentally regulated *Drosophila* gene rox8 encodes an RRM-type RNA binding protein structurally related to human TIA-1-type nucleolysins. Nucleic Acids Res 21:3699–3704

Breathnach R, Mantei N, Chambon P (1980) Corrected splicing of a chicken ovalbumin gene transcript in mouse L cells. Proc Natl Acad Sci USA 77:740–744

Burnette JM, Hatton AR, Lopez AJ (1999) Trans-acting factors required for inclusion of regulated exons in the Ultrabithorax mRNAs of *Drosophila melanogaster*. Genetics 151:1517–1529

Burtis K, Baker B (1989) *Drosophila* double-sex gene controls somatic sexual differentiation by producing alternatively spliced mRNAs encoding related sex-specific polypeptides. Cell 56:997–1010

Chandler D, McGuffin ME, Piskur J, Yao J, Baker BS, Mattox W (1997) Evolutionary conservation of regulatory strategies for the sex determination factor transformer-2. Mol Cell Biol 17:2908–2919

Chi SW, Muto Y, Inoue M, Kim I, Sakamoto H, Shimura Y, Yokoyama S, Choi BS, Kim H (1999) Chemical shift perturbation studies of the interactions of the second RNA-binding domain of the *Drosophila* sex-lethal protein with the transformer pre-mRNA polyuridine tract and 3′ splice-site sequences. Eur J Biochem 260:649–660

Chow LT, Broker TR (1978) The spliced structures of adenovirus 2 fiber message and the other late mRNAs. Cell 15:497–510

Cline TW (1993) The *Drosophila* sex determination signal: how do flies count to two? Trends Genet 9:385–390

Cline TW, Meyer BJ (1996) Vive la difference: males vs. females in flies vs. worms. Annu Rev Genet 30:637–702

Dauwalder B, Mattox W (1998) Analysis of the functional specificity of RS domains in vivo. EMBO J 17:6049–6060

Dauwalder B, Amaya-Manzanares F, Mattox W (1996) A human homologue of the *Drosophila* sex determination factor transformer-2 has conserved splicing regulatory functions. Proc Natl Acad Sci USA 93:9004–9009

DeNoto FM, Moore DD, Goodman HM (1981) Human growth hormone DNA sequence and mRNA structure: possible alternative splicing. Nucleic Acids Res 9:3719–3730

Deshpande G, Samuels M, Schedl P (1996) Sex-lethal interacts with splicing factors in vitro and in vivo. Mol Cell Biol 16:5036–5047

Deshpande G, Calhoun G, Schedl PD (1999) The N-terminal domain of Sxl protein disrupts Sxl autoregulation in females and promotes female-specific splicing of tra in males. Development 126:2841–2853

Du C, McGuffin ME, Dauwalder B, Rabinow L, Mattox W (1998) Protein phosphorylation plays an essential role in the regulation of alternative splicing and sex determination in *Drosophila*. Mol Cell 2:741–750

Estes PA, Keyes LN, Schedl P (1995) Multiple response elements in the Sex-lethal early promoter ensure its female-specific expression pattern. Mol Cell Biol 15:904–917

Fleckner J, Zhang M, Valcarcel J, Green MR (1997) U2AF^{65} recruits a novel human DEAD box protein required for the U2 snRNP-branchpoint interaction. Genes Dev 11:1864–1872

Flickinger TW, Salz HK (1994) The *Drosophila* sex determination gene snf encodes a nuclear protein with sequence and functional similarity to the mammalian U1A snRNP protein. Genes Dev 8:914–925

Förch P, Puig O, Kedersha N, Martinez C, Granneman S, Seraphin B, Anderson P, Valcarcel J (2000) The apoptosis-promoting factor TIA-1 is a regulator of alternative pre-mRNA splicing. Mol Cell 6:1089–1098

Förch P, Merendino L, Martinez C, Valcarcel J (2001) Modulation of msl-2 5′ splice site recognition by Sex-lethal. RNA 7:1185–1191

Fu X-D (1995) The superfamily of arginine/serine-rich splicing factors. RNA 1:663–680

Gebauer F, Merendino L, Hentze MW, Valcarcel J (1997) Novel functions for 'nuclear factors' in the cytoplasm: the Sex-lethal paradigm. Semin Cell Dev Biol 8:561–566

Gebauer F, Merendino L, Hentze M, Valcárcel J (1998) The *Drosophila* splicing regulator Sex-lethal directly inhibits translation of male-specific-lethal 2 mRNA. RNA 4:142–150

Gebauer F, Corona DF, Preiss T, Becker PB, Hentze MW (1999) Translational control of dosage compensation in *Drosophila* by Sex-lethal: cooperative silencing via the 5′ and 3′ UTRs of msl-2 mRNA is independent of the poly(A) tail. EMBO J 18:6146–6154

Goralski TJ, Edstrom JE, Baker BS (1989) The sex determination locus transformer-2 of *Drosophila* encodes a polypeptide with similarity to RNA binding proteins. Cell 56:1011–1018

Granadino B, Campuzano S, Sánchez L (1990) The *Drosophila melanogaster* fl(2)d gene is needed for the female-specific splicing of Sex-lethal. EMBO J 9:2597–2602

Granadino B, Penalva LO, Green MR, Valcarcel J, Sanchez L (1997) Distinct mechanisms of splicing regulation in vivo by the *Drosophila* protein Sex-lethal. Proc Natl Acad Sci USA 94:7343–7348

Guth S, Tange TO, Kellenberger E, Valcarcel J (2001) Dual function for U2AF(35) in AG-dependent pre-mRNA splicing. Mol Cell Biol 21:7673–7681

Handa N, Nureki O, Kurimoto K, Kim I, Sakamoto H, Shimura Y, Muto Y, Yokoyama S(1999) Structural basis for recognition of the tra mRNA precursor by the Sex-lethal protein. Nature 398:579–585

Hastings ML, Krainer AR (2001) Pre-mRNA splicing in the new millennium. Curr Opin Cell Biol 13:302–309

Hazelrigg T, Tu C (1994) Sex-specific processing of the *Drosophila exuperantia* transcript is regulated in male germ cells by the tra-2 gene. Proc Natl Acad Sci USA 91:10752–10756

Hazelrigg T, Watkins WS, Marcey D, Tu C, Karow M, Lin XR (1990) The exuperantia gene is required for *Drosophila* spermatogenesis as well as anteroposterior polarity of the developing oocyte, encodes overlapping sex-specific transcripts. Genetics 126:607–617

Hedley M, Maniatis T (1991) Sex-specific splicing and polyadenylation of dsx pre-mRNA requires a sequence that binds specifically to tra-2 protein in vitro. Cell 65:579–586

Heinrichs V, Baker BS (1995) The *Drosophila* SR protein RBP1 contributes to the regulation of doublesex alternative splicing by recognizing RBP1 RNA target sequences. EMBO J 14:3987–4000

Heinrichs V, Ryner LC, Baker BS (1998) Regulation of sex-specific selection of fruitless 5′ splice sites by transformer and transformer-2. Mol Cell Biol 18:450–458

Hilfiker A, Amrein H, Dubendorfer A, Schneiter R, Nothiger R (1995) The gene virilizer is required for female-specific splicing controlled by Sxl, the master gene for sexual development in *Drosophila*. Development 121:4017–4026

Horabin J, Schedl P (1993a) Sex-lethal autoregulation requires multiple cis-acting elements upstream and downstream of the male exon and appears to depend largely on controlling the use of the male exon 5′ splice site. Mol Cell Biol 13:7734–7746

Horabin JI, Schedl P (1993b) Regulated splicing of the *Drosophila* sex-lethal male exon involves a blockage mechanism. Mol Cell Biol 13:1408–1414

Hoshijima K, Inoue K, Higuchi I, Sakamoto H, Shimura Y (1991) Control of doublesex alternative splicing by transformer and transformer-2 in *Drosophila*. Science 252:833–836

Inoue K, Hoshijima H, Sakamoto H, Shimura Y (1990) Binding of the *Drosophila* Sex-lethal gene product to the alternative splice site of transformer primary transcript. Nature 344:461–463

Inoue K, Hoshijima K, Higuchi I, Sakamoto H, Shimura Y (1992) Binding of the *Drosophila* transformer and transformer-2 proteins to the regulatory elements of doublesex primary transcript for sex-specific RNA processing. Proc Natl Acad Sci USA 89:8092–8096

Ito H, Fujitani K, Usui K, Shimizu-Nishikawa K, Tanaka S, Yamamoto D (1996) Sexual orientation in *Drosophila* is altered by the satori mutation in the sex-determination gene fruitless that encodes a zinc finger protein with a BTB domain. Proc Natl Acad Sci USA 93:9687–9692

Kan J, Green M (1999) Pre-mRNA splicing of IgM exons M1 and M2 is directed by a juxtaposed splicing enhancer and inhibitor. Genes Dev. 13:462–471

Kanaar R, Lee A, Rudner D, Wemmer D, Rio D (1995) Interaction of the Sex-lethal RNA binding domains with RNA. EMBO J 14:4530–4539

Kanopka A, Muhlemann O, Petersen-Mahrt S, Estmer C, Ohrmalm C, Akusjarvi G (1998) Regulation of adenovirus alternative RNA splicing by dephosphorylation of SR proteins. Nature 393:185–187

Kelley R, Solovyeva I, Lyman L, Richman R, Solovyev V, Kuroda M (1995) Expression of Msl-2 causes assembly of dosage compensation regulators on the X chromosomes and female lethality in *Drosophila*. Cell 81:867–877

Kelley R, Wang J, Bell L, Kuroda M (1997) Sex-lethal controls dosage compensation in *Drosophila* by a non-splicing mechanism. Nature 387:195–199

Kelley RL, Kuroda MI (2000) The role of chromosomal RNAs in marking the X for dosage compensation. Curr Opin Genet Dev 10:555–561

Keyes LN, Cline TW, Schedl P (1992) The primary sex determination signal of *Drosophila* acts at the level of transcription. Cell 68:933–943

Kim I, Muto Y, Watanabe S, Kitamura A, Futamura Y, Yokoyama S, Hosono K, Kawai G, Takaku H, Dohmae N, Takio K, Saskamoto H, Shimura Y (2000) Interactions of a didomain fragment of the *Drosophila* sex-lethal protein with single-stranded uridine-rich oligoribonucleotides derived from the transformer and Sex-lethal messenger RNA precursors: NMR with residue-selective [5-2H]uridine substitutions. J Biomol NMR 17:153–165

King CR, Piatigorsky J (1983) Alternative RNA splicing of the murine alpha A-crystallin gene: protein-coding information within an intron. Cell 32:707–712

Lallena MJ, Chalmers KS, Leamazares S, Lamand AI, Valcárcel J (2002) Splicing regulation at the second catalytic step by sex-lethal involves 3′ splice site recognition by SPF45. Cell 109:285–296

Lander ES, Linton LM, Birren B, Nusbaum C, Zody MC, Baldwin J, Devon K, Dewar K, Doyle M, FitzHugh W, Funke R, Gage D, Harris K, Heaford A, Howland J, Kann L, Lehoczky J, LeVine R, McEwan P, McKernan K, Meldrim J, Mesirov JP, Miranda C, Morris W, Naylor J, Raymond C, Rosetti M, Santos R, Sheridan A, Sougnez C, Stange-Thomann N, Stojanovic N, Subramanian

A, Wyman D, Rogers J, Sulston J, Ainscough R, Beck S, Bentley D, Burton J, Clee C, Carter N, Coulson A, Deadman R, Deloukas P, Dunham A, Dunham I, Durbin R, French L, Grafham D, Gregory S, Hubbard T, Humphray S, Hunt A, Jones M, Lloyd C, McMurray A, Matthews L, Mercer S, Milne S, Mullikin JC, Mungall A, Plumb R, Ross M, Shownkeen R, Sims S, Waterston RH, Wilson RK, Hillier LW, McPherson JD, Marra MA, Mardis ER, Fulton LA, Chinwalla AT, Pepin KH, Gish WR, Chissoe SL, Wendl MC, Delehaunty KD, Miner TL, Delehaunty A, Kramer JB, Cook LL, Fulton RS, Johnson DL, Minx PJ, Clifton SW, Hawkins T, Branscomb E, Predki P, Richardson P, Wenning S, Slezak T, Doggett N, Cheng JF, Olsen A, Lucas S, Elkin C, Uberbacher E, Frazier M et al. (2001) Initial sequencing and analysis of the human genome. Nature 409:860–921

Lavigueur A, LaBranche H, Kornblihtt A, Chabot B (1993) A splicing enhancer in the human fibronectin alternate ED1 exon interacts with SR proteins and stimulates U2 snRNP binding. Genes Dev 7:2405–2417

Lee AL, Kanaar R, Rio DC, Wemmer DE (1994) Resonance assignments and solution structure of the second RNA-binding domain of sex-lethal determined by multidimensional heteronuclear magnetic resonance. Biochemistry 33:13775–13786

Lee AL, Volkman BF, Robertson SA, Rudner DZ, Barbash DA, Cline TW, Kanaar R, Rio DC, Wemmer DE. (1997) Chemical shift mapping of the RNA-binding interface of the multiple-RBD protein sex-lethal. Biochemistry 36:14306–14317

Lopez AJ (1995) Developmental role of transcription factor isoforms generated by alternative splicing. Dev Biol 172:396–411

Lopez AJ (1998) Alternative splicing of pre-mRNA: developmental consequences and mechanisms of regulation. Annu Rev Genet 32:279–305

Lynch KW, Maniatis T (1995) Synergistic interactions between two distinct elements of a regulated splicing enhancer. Genes Dev 9:284–293

Lynch KW, Maniatis T (1996) Assembly of specific SR protein complexes on distinct regulatory elements of the *Drosophila* doublesex splicing enhancer. Genes Dev 10:2089–20101

MacDougall C, Harbison D, Bownes M (1995) The developmental consequences of alternate splicing in sex determination and differentiation in *Drosophila*. Dev Biol 172:353–376

Manley J, Tacke R (1996) SR proteins and splicing control. Genes Dev 10:1569–1579

Maquat LE, Carmichael GG (2001) Quality control of mRNA function. Cell 104:173–176

Mariman EC, van Beek-Reinders RJ, van Venrooij WJ (1983) Alternative splicing pathways exist in the formation of adenoviral late messenger RNAs. J Mol Biol 163:239–256

Marin I, Siegal ML, Baker BS (2000) The evolution of dosage-compensation mechanisms. Bioessays 22:1106–1114

Mattox W, Baker BS (1991) Autoregulation of the splicing of transcripts from the transformer-2 gene of *Drosophila*. Genes Dev 5:786–796

Mattox W, Palmer MJ, Baker BS (1990) Alternative splicing of the sex determination gene transformer-2 is sex-specific in the germ line but not in the soma. Genes Dev 4:789–805

Mattox W, McGuffin ME, Baker BS (1996) A negative feedback mechanism revealed by functional analysis of the alternative isoforms of the *Drosophila* splicing regulator transformer-2. Genetics 143:303–314

McGuffin ME, Chandler D, Somaiya D, Dauwalder B, Mattox W (1998) Autoregulation of transformer-2 alternative splicing is necessary for normal male fertility in *Drosophila*. Genetics 149:1477–1486

McKeown M, Belote JM, Baker BS (1987) A molecular analysis of transformer, a gene in *Drosophila melanogaster* that controls female sexual differentiation. Cell 48:489–499

Meise M, Hilfiker-Kleiner D, Dubendorfer A, Brunner C, Nothiger R, Bopp D (1998) Sex-lethal, the master sex-determining gene in *Drosophila*, is not sex-specifically regulated in *Musca domestica*. Development 125:1487–1494

Merendino L, Guth S, Bilbao D, Martinez C, Valcarcel J (1999) Inhibition of msl-2 splicing by Sex-lethal reveals interaction between $U2AF^{35}$ and the 3′ splice site AG. Nature 402:838–841

Modrek B, Lee C (2002) A genomic view of alternative splicing. Nat Genet 30:13–19

Nagai K, Oubridge C, Jessen T, Li J, Evans P (1990) Crystal structure of the RNA-binding domain of the U1 small nuclear ribonucleoprotein A. Nature 348:515–520
Nagoshi R, McKeown M, Burtis K, Belote J, Baker B (1988) The control of alternative splicing at genes regulating sexual differentiation in *D. melanogaster*. Cell 53:229–236
Neubauer G, King A, Rappsilber J, Calvio C, Watson M, Ajuh P, Sleeman J, Lamond A, Mann M (1998) Mass spectrometry and EST-database searching allows characterization of the multi-protein spliceosome complex. Nat Genet 20:46–50
Niessen M, Schneiter R, Nothiger R (2001) Molecular identification of virilizer, a gene required for the expression of the sex-determining gene Sex-lethal in *Drosophila melanogaster*. Genetics 157:679–688
Ohbayashi F, Suzuki MG, Mita K, Okano K, Shimada T (2001) A homologue of the *Drosophila* doublesex gene is transcribed into sex-specific mRNA isoforms in the silkworm, *Bombyx mori*. Comp Biochem Physiol B Biochem Mol Biol 128:145–158
Pannuti A, Lucchesi JC (2000) Recycling to remodel: evolution of dosage-compensation complexes. Curr Opin Genet Dev 10:644–650
Penalva LO, Ruiz MF, Ortega A, Granadino B, Vicente L, Segarra C, Valcarcel J, Sanchez L (2000) The *Drosophila* fl(2)d gene, required for female-specific splicing of Sxl and tra pre-mRNAs, encodes a novel nuclear protein with a HQ-rich domain. Genetics 155:129–139
Penalva LO, Lallena MJ, Valcarcel J (2001) Switch in 3′ splice site recognition between exon definition and splicing catalysis is important for sex-lethal autoregulation. Mol Cell Biol 21:1986–1996
Polycarpou-Schwarz M, Gunderson SI, Kandels-Lewis S, Seraphin B, Mattaj IW (1996) *Drosophila* SNF/D25 combines the functions of the two snRNP proteins U1A and U2B′ that are encoded separately in human, potato, yeast. RNA 2:11–23
Ramchatesingh J, Zahler A, Neugebauer K, Roth M, Cooper T (1995) A subset of SR proteins activates splicing of the cardiac troponin T alternative exon by direct interactions with an exonic enhancer. Mol Cell Biol. 15:4898–4907
Reed R (2000) Mechanisms of fidelity in pre-mRNA splicing. Curr Opin Cell Biol 12:340–345
Rudner D, Kanaar R, Breger K, Rio D (1998) Interaction between subunits of heterodimeric splicing factor U2AF is essential in vivo. Mol Cell Biol 18:1765–1773
Ruskin B, Zamore P, Green M (1988) A factor, U2AF, is required for U2 snRNP binding and splicing complex assembly. Cell 52:207–219
Ryner L, Baker B (1991) Regulation of doublesex pre-mRNA processing occurs by 3′ splice site activation. Genes Dev 5:2071–2085
Ryner L, Goodwin S, Castrillon D, Anand A, Villella A, Baker B, Hall J, Taylor B, Wasserman S (1997) Control of male sexual behavior and sexual orientation in *Drosophila* by the fruitless gene. Cell 87:1079–1089
Saccone G, Peluso I, Artiaco D, Giordano E, Bopp D, Polito LC (1998) The *Ceratitis capitata* homologue of the *Drosophila* sex-determining gene sex-lethal is structurally conserved, but not sex-specifically regulated. Development 125:1495–1500
Sakamoto H, Inoue K, Higuchi I, Ono Y, Shimura Y (1992) Control of *Drosophila* Sex-lethal pre-mRNA splicing by its own female-specific product. Nucleic Acids Res 20:5533–5540
Sakashita E, Sakamoto H (1994) Characterization of RNA binding specificity of the *Drosophila* sex-lethal protein by in vitro ligand selection. Nucleic Acids Res 22:4082–4086
Salz HK (1992) The genetic analysis of snf: a *Drosophila* sex determination gene required for activation of Sex-lethal in both the germline and the soma. Genetics 130:547–554
Salz HK, Flickinger TW (1996) Both loss-of-function and gain-of-function mutations in snf define a role for snRNP proteins in regulating Sex-lethal pre-mRNA splicing in *Drosophila* development. Genetics 144:95–108
Salz HK, Maine EM, Keyes LN, Samuels ME, Cline TW, Schedl P (1989) The *Drosophila* female-specific sex-determination gene, Sex-lethal, has stage-, tissue-, sex-specific RNAs suggesting multiple modes of regulation. Genes Dev 3:708–719

Samuels ME, Schedl P, Cline TW (1991) The complex set of late transcripts from the *Drosophila* sex determination gene sex-lethal encodes multiple related polypeptides. Mol Cell Biol 11:3584–3602

Samuels M, Bopp D, Colvin R, Roscigno R, Gracia-Blanco M, Schedl P (1994) RNA binding by Sxl proteins in vitro and in vivo. Mol Cell Biol 14:4975–4990

Samuels M, Deshpande G, Schedl P (1998) Activities of the Sex-lethal protein in RNA binding and protein:protein interactions. Nucleic Acids Res 26:2625–2637

Schaal T, Maniatis T (1999a) Selection and characterization of pre-mRNA splicing enhancers: identification of novel sr protein-specific enhancer sequences. Mol Cell Biol 19:1705–1719

Schaal TD, Maniatis T (1999b) Multiple distinct splicing enhancers in the protein-coding sequences of a constitutively spliced pre-mRNA. Mol Cell Biol 19:261–273

Schmucker D, Clemens JC, Shu H, Worby CA, Xiao J, Muda M, Dixon JE, Zipursky SL (2000) *Drosophila* Dscam is an axon guidance receptor exhibiting extraordinary molecular diversity. Cell 101:671–684

Schütt C, Nöthiger R (2000) Structure, function and evolution of sex-determining systems in Dipteran insects. Development 127:667–677

Sievert V, Kuhn S, Paululat A, Traut W (2000) Sequence conservation and expression of the sex-lethal homologue in the fly *Megaselia scalaris*. Genome 43:382–390

Singh R, Valcárcel J, Green M (1995) Distinct binding specificities and functions of higher eukaryotic polypyrimidine tract-binding proteins. Science 268:1173–1176

Singh R, Banerjee H, Green MR (2000) Differential recognition of the polypyrimidine-tract by the general splicing factor $U2AF^{65}$ and the splicing repressor sex-lethal. RNA 6:901–911

Smith CW, Valcarcel J (2000) Alternative pre-mRNA splicing: the logic of combinatorial control. Trends Biochem Sci 25:381–388

Sosnowski BA, Belote JM, McKeown M (1989) Sex-specific alternative splicing of RNA from the transformer gene results from sequence-dependent splice site blockage. Cell 58:449–459

Sosnowski B, Davis D, Boggs R, Madigan S, McKeown M (1994) Multiple portions of a small region of the *Drosophila* transformer gene are required for efficient in vivo sex-specific regulated RNA splicing and in vitro Sex-lethal binding. Dev Biol 161:302–312

Stitzinger SM, Pellicena-Palle A, Albrecht EB, Gajewski KM, Beckingham KM, Salz HK (1999a) Mutations in the predicted aspartyl tRNA synthetase of *Drosophila* are lethal and function as dosage-sensitive maternal modifiers of the sex determination gene Sex-lethal. Mol Gen Genet 261:142–151

Stitzinger SM, Conrad TR, Zachlin AM, Salz HK (1999b) Functional analysis of SNF, the *Drosophila* U1A/U2B" homolog: identification of dispensable and indispensable motifs for both snRNP assembly and function in vivo. RNA 5:1440–1450

Stojdl DF, Bell JC (1999) SR protein kinases: the splice of life. Biochem Cell Biol 77:293–298

Sun Q, Mayeda A, Hampson R, Krainer A, Rottman F(1993) General splicing factor SF2/ASF promotes alternative splicing by binding to an exonic splicing enhancer. Genes Dev 7:2598–2608

Tacke R, Tohyama M, Ogawa S, Manley JL (1998) Human Tra2 proteins are sequence-specific activators of pre-mRNA splicing. Cell 93:139–148

Tian M, Maniatis T (1992) Positive control of pre-mRNA splicing in vitro. Science 256:237–240

Tian M, Maniatis T (1993) A splicing enhancer complex controls alternative splicing of doublesex pre-mRNA. Cell 74:105–114

Tian M, Maniatis T (1994) A splicing enhancer exhibits both constitutive and regulated activities. Genes Dev 8:1703–1712

Usui-Aoki K, Ito H, Ui-Tei K, Takahashi K, Lukacsovich T, Awano W, Nakata H, Piao ZF, Nilsson EE, Tomida J, Yamamoto D. (2000) Formation of the male-specific muscle in female *Drosophila* by ectopic fruitless expression. Nat Cell Biol 8:500–506

Valcárcel J, Green M (1996) The SR protein family: pleiotropic effects in pre-mRNA splicing. Trends Biochem. Sci. 21:296–301

Valcárcel J, Singh R, Zamore P, Green M (1993) The protein Sex-lethal antagonizes the splicing factor U2AF to regulate alternative splicing of transformer pre-mRNA. Nature 362:171–175

Valcárcel J, Martínez C, Green M (1997) Functional analysis of splicing factors and regulators. In: Richter JD (ed) mRNA formation and function. Academic Press, New York, pp 31–53

Varani G, Nagai K (1998) RNA recognition by RNP proteins during RNA processing. Annu Rev Biophys Biomol Struct 27:407–445

Venter JC, Adams MD, Myers EW, Li PW, Mural RJ, Sutton GG, Smith HO, Yandell M, Evans CA, Holt RA, Gocayne JD, Amanatides P, Ballew RM, Huson DH, Wortman JR, Zhang Q, Kodira CD, Zheng XH, Chen L, Skupski M, Subramanian G, Thomas PD, Zhang J, Gabor Miklos GL, Nelson C, Broder S, Clark AG, Nadeau J, McKusick VA, Zinder N, Levine AJ, Roberts RJ, Simon M, Slayman C, Hunkapiller M, Bolanos R, Delcher A, Dew I, Fasulo D, Flanigan M, Florea L, Halpern A, Hannenhalli S, Kravitz S, Levy S, Mobarry C, Reinert K, Remington K, Abu-Threideh J, Beasley E, Biddick K, Bonazzi V, Brandon R, Cargill M, Chandramouliswaran I, Charlab R, Chaturvedi K, Deng Z, Di Francesco V, Dunn P, Eilbeck K, Evangelista C, Gabrielian AE, Gan W, Ge W, Gong F, Gu Z, Guan P, Heiman TJ, Higgins ME, Ji RR, Ke Z, Ketchum KA, Lai Z, Lei Y, Li Z, Li J, Liang Y, Lin X, Lu F, Merkulov GV, Milshina N, Moore HM, Naik AK, Narayan VA, Neelam B, Nusskern D, Rusch DB, Salzberg S, Shao W, Shue B, Sun J, Wang Z, Wang A, Wang X, Wang J, Wei M, Wides R, Xiao C, Yan C et al. (2001) The sequence of the human genome. Science 291:1304–1351

Wang J, Bell L (1994) The Sex-lethal amino terminus mediates cooperative interactions in RNA binding and is essential for splicing regulation. Genes Dev 8:2072–2085

Wang J, Dong Z, Bell LR (1997) Sex-lethal interactions with protein and RNA. Roles of glycine-rich and RNA binding domains. J Biol Chem 272:22227–22235

Watakabe A, Tanaka K, Shimura Y (1993) The role of exon sequences in splice site selection. Genes Dev 7:407–418

Waterbury JA, Horabin JI, Bopp D, Schedl P (2000) Sex determination in the *Drosophila* germline is dictated by the sexual identity of the surrounding soma. Genetics 155:1741–1756

Wu J, Maniatis T (1993) Specific interactions between proteins implicated in splice site selection and regulated alternative splicing. Cell 75:1061–1070

Xu R, Teng J, Cooper T (1993) The cardiac troponin T alternative exon contains a novel purine-rich positive splicing element. Mol Cell Biol 13:3660–3674

Yanowitz JL, Deshpande G, Calhoun G, Schedl PD (1999) An N-terminal truncation uncouples the sex-transforming and dosage compensation functions of sex-lethal. Mol Cell Biol 19:3018–3028

Zamore P, Patton J, Green M (1992) Cloning and domain structure of the mammalian splicing factor U2AF. Nature 355:609–614

Zhou S, Yang Y, Scott M, Pannuti A, Fehr K, Eisen A, Koonin E, Fouts D, Wrightsman R, Manning J, Lucchesi J (1995) Male-specific lethal 2: a dosage compensation gene of *Drosophila*, undergoes sex-specific regulation and encodes a protein with RING finger and a metallothionein-like cystein cluster. EMBO J 14:2884–2895

Zhu C, Urano J, Bell L (1997) The Sex-lethal early splicing pattern uses a default mechanism dependent on the alternative 5′ splice sites. Mol Cell Biol 17:1674–1681

Zuo P, Maniatis T (1996) The splicing factor $U2AF^{35}$ mediates critical protein-protein interactions in constitutive and enhancer-dependent splicing. Genes Dev 10:1356–1368

Alternative Pre-mRNA Splicing and Regulation of Programmed Cell Death

J.Y. Wu, H. Tang, and N. Havlioglu[1]

Apoptosis or programmed cell death (PCD) is critical for development and homeostasis of multi-cellular organisms. It also plays an important role in pathogenesis of a large number of human diseases. PCD is under complex regulation at multiple levels. An important level of this regulation is at the level of pre-mRNA splicing because a number of PCD regulatory genes are expressed as functionally distinct or even antagonistic isoforms as a result of alternative splicing. In this chapter, we review recent molecular and biochemical studies on alternative splicing of these PCD genes, ranging from extracellular signals, death receptors to intracellular components of the death machinery. As a major mechanism to generate transcript variability, alternative splicing affects not only intracellular distribution, but also functional activities of these PCD genes, modulating a presumably tightly controlled process of cell death. In some cases, alternative splicing is involved in the life-or-death decision. Studies with molecular and biochemical approaches have begun to reveal mechanisms underlying alternative splicing regulation of PCD gene expression and function. Many important questions remain to be addressed by future studies.

1 Introduction

Since the original description of apoptosis (Kerr et al. 1972), also known as programmed cell death (PCD), significant progress has been made in understanding this form of cell death (for recent reviews see Horvitz 1999; Korsmeyer et al. 1999; Rathmell and Thompson 1999). At the cellular level, prominent morphological features of PCD include cell shrinkage and nuclear condensation, usually without dramatic inflammatory responses (Kerr et al. 1972). It is now widely accepted that PCD is critical for development and homeostasis of multicellular organisms (Jacobson et al. 1997; Horvitz 1999; Korsmeyer et al. 1999; Rathmell and Thompson 1999). For example, PCD is essential for deleting autoreactive cells during lymphoid development and

[1]Department of Pediatrics and Department of Molecular Biology and Pharmacology, MPRB Rm3107, Washington University School of Medicine, St. Louis, Missouri 63110, USA; e-mail: jwu@molecool.wustl.edu

Progress in Molecular and Subcellular Biology, Vol. 31
Philippe Jeanteur (Ed.)

eliminating excessive neurons during neural development. Insufficient or excessive cell death also plays an important role in the pathogenesis of a variety of human diseases, including cancer, autoimmune diseases, viral infections and neurodegenerative disorders (Thompson 1995; Nijhawan et al. 2000).

With the rapid progress in developmental biology, genetics and molecular cellular biology in the last two decades, a large number of genes involved in controlling PCD have been identified and characterized. We have now begun to understand molecular mechanisms underlying PCD regulation. Genetic studies on developmental cell death in *C. elegans* have delineated critical pathways controlling PCD and revealed basic players in PCD, *ced* genes (reviewed in Horvitz et al. 1994). *ced-3* and *ced-4* genes act to turn on the cell death program, whereas *ced-9* suppresses PCD. The genetic program controlling PCD is highly conserved through evolution. Vertebrate homologues of *ced-9* are anti-apoptotic members of the bcl-2 superfamily. Apoptosis protease activating factor 1 (apaf-1; Zou et al. 1997) is a vertebrate homologue of *ced-4*. *ced-3* gene product is a cysteine protease (Yuan et al. 1993) related to members of vertebrate ICE (interleukin-1β-converting enzyme) family now named caspases (cysteine aspases; Alnemri et al. 1996). Both genetic and biochemical studies significantly advanced our understanding of signal transduction pathways and functional mechanisms of different cell death genes (see references in Horvitz 1999; Korsmeyer et al. 1999; Talapatra and Thompson 2001). A number of different signals including growth factor deprivation, activation of certain cell surface receptors (death receptors) and metabolic or cell cycle perturbation may activate cell death machinery via divergent signal transduction pathways. These divergent death signals appear to converge into a common final pathway. This involves a cascade of activation of caspases that in turn results in the terminal events leading to the ultimate demise of the cell. For example, activation of receptors of the tumor necrosis factor receptor (TNFR) family, including CD95 (Fas/APO-1) and TNFR-1, leads to the recruitment of death domain (DD)-containing proteins (such as the Fas-associated DD), which interact with and activate caspase-8. This in turn initiates a cascade of caspase activation. Different stimuli that cause perturbations in mitochondrial homeostasis and cytochrome C release from the inner mitochondrial space and into the cytosol can also commit cells to cytochrome c-APAF1-caspase 9-mediated death pathway. Members of the Bcl-2 superfamily are known to regulate cytochrome C release and mitochondrial function (Korsmeyer et al. 1999; Talapatra and Thompson 2001). It is the cascade of caspase activation that initiates the final degradation events of cell death.

With sequencing of the entire human genome nearly complete, it is now known that the vast majority of human genes contain introns and that more than 50% of human genes undergo alternative splicing to produce distinct products from the same genetic locus. The most recent development in genome sequencing projects has also provided a more comprehensive cataloguing of genes that may play a role in PCD (Aravind et al. 2001). It is clear that mammalian animals develop much more sophisticated and complex regulatory mechanisms at multiple levels to regulate this crucial process. At each level of regulation, there appears to be gene amplification to generate a number of gene family members with structural similarities (and perhaps distinct physiological activities). It is particularly interesting to note that critical regulatory PCD gene families such as "death domain"(DD)-containing, "death effector domain" (DED)-containing protein families and bcl-2 family have all evolved into multi-member superfamilies (Aravind et al. 1999, 2001). There is only one gene of bcl-2 family identified in *C. elegans*, whereas the mammalian bcl-2 superfamily contains multiple anti-apoptotic and multiple pro-apoptotic members (Korsmeyer et al. 1999; Aravind et al. 2001). It is clear that expression and function of the large number of cell death genes are regulated both transcriptionally and post-transcriptionally. An important form of posttranscriptional regulation occurs at the step of pre-mRNA splicing. Alternative pre-mRNA splicing is a major mechanism to generate transcript variability from individual genetic loci. Perhaps this form of posttranscriptional regulation via alternative splicing provides a single most powerful mechanism to contribute to genetic and functional diversity. It also has an additional advantage of being efficient and versatile. This type of alternative splicing regulation of PCD gene expression and function is the focus of this chapter. We will review recent advances in functional and mechanistic studies of alternative splicing of PCD genes. Because a large number of PCD genes undergo alternative splicing and generate an enormous number of different gene products, it is difficult to cover all aspects and all of the PCD genes published within this short chapter. Instead, we are only using a few examples from each gene family to demonstrate the great potential and extreme complexity of alternative splicing in regulation of PCD gene expression and function. We will focus on cell death machinery components and the genes containing domains directly implicated in apoptosis regulation. Although a number of genes encoding transcription factors, translational regulators, or cell cycle proteins are also implicated in cell death processes, those genes that are primarily involved in cell cycle control, transcriptional regulation as well as other processes not directly linked to apoptosis regulation are not included in this review because of limited space.

2
Alternative Splicing of PCD Genes

Analysis of the genome sequencing data has revealed more than 200 genes that are implicated in cell death regulation (Aravind et al. 2001). Some of these genes have been identified because of their functional activities in cell death, and others are categorized based on their predicted peptide sequences with similarity to known PCD proteins. It is clear that the expression and function of genes acting at every step of cell death regulatory pathways are influenced by alternative splicing. These genes range from extracellular signals, membrane-associated death signal receptors, intracellular signal transducing genes to death executor genes. Some examples of alternatively spliced programmed cell death genes are listed in Table 1. Alternative splicing has a dramatic impact on the expression and function of a large number of PCD genes.

2.1 Death Signals and Death Receptors: Modulation of Both Ligands and Receptors by Alternative Splicing

The initial signal or stimulus that triggers cellular responses which ultimately lead to cell death often involves deprivation of growth or trophic factors. One of the best-characterized signals for neuronal cell death during development is, for example, deprivation of trophic factors such as nerve growth factor (NGF). Alternative splicing events that cause a loss of functional activity of growth or trophic factors may also, in theory, result in apoptosis. Persephin is a trophic factor that has sequence similarity with the neurturin (NTN) and glial cell line-derived neurotrophic factor (GDNF), of the TGF-beta superfamily. Persephin promotes the survival of ventral midbrain dopaminergic neurons in cell culture and prevents their degeneration after 6-hydroxydopamine treatment in vivo. Persephin also supports the survival of motor neurons (Milbrandt et al. 1998). Posttranscriptional processing of persephin genes in both humans and mice involves the removal of an 88-nucleotide intron from the region encoding the prodomain of the protein. This splicing event has been shown to be rather inefficient, leaving a significant fraction, if not majority, of RNA exported into cytoplasm retaining this intron (Milbrandt et al. 1998). It is interesting to note that the prodomains of both the NTN and GDNF genes are also interrupted by an intron (Milbrandt et al. 1998). It remains unclear how this intron removal is regulated during development. However, it is likely that this splicing event is under regulation to generate a functionally active form under certain physiological conditions. It is conceivable that activation of this splicing event in response to certain stimuli could provide an efficient post-transcriptional mechanism controlling the production of functional mRNAs for such trophic factors.

Many cytokine genes undergo alternative splicing to produce functionally different proteins (see references in Atamas 1997; Youn et al. 1998, Robinson and Stringer 2001). In some cases, it is not clear why cells express different cytokine isoforms with apparently similar activities (for example, human beta-chemokines CKbeta8 and CKbeta8-1; see Youn et al. 1998). In other cases, tissue-specific alternative splicing produces different isoforms in different tissues (for example, stromal cell derived factor-1, SDF-1; see Gleichmann et al. 2000). The expression of different alternative splicing isoforms of SDF-1 appears to be differentially regulated in the nervous system (Gleichmann et al. 2000). Another recent example is found in vascular endothelial growth factor

Table 1. Multiple genes acting at different steps of cell death pathway undergo alternative splicing with significant functional impact

Family	Gene	Domain motif	Alternative splicing consequences
Extracellular signals and receptors			
Growth factors, ligands			
TGFbeta	Persephin, GDNF, NTN		Intron retention in predomain, inactive
Cytokine	SDF-1		Tissue-specific alternative splicing
	VEGF		Different activities
Death receptors			
TNFR	Fas (CD95), LARD	DD	Soluble versus membrane-associated receptor
Intracellular signal transduction			
Mitochondrial membrane proteins of Bcl2 superfamily			
Bcl-2	Bcl2	BH	Membrane-bound versus soluble
	Bcl-x	BH	Active or inactive in preventing PCD
Adaptor proteins			
	MADD	DD	Susceptible or resistant to PCD
TIR	MyD88	DD, TIR	Distinct activities predicted
Pyrin	Marenostrin	Pyrin	Differential subcellular localization
BIR activation	TRAF2A	BIR; RING	Stimulating versus inhibiting NF-kappaB activation
	Livin	BIR; RING	Distinct activities
	Survivin	BIR; coiled-coil domain	Distinct activities predicted
Other metastasis suppressor	CC3	?	Induce or prevent apoptosis
Effector or regulatory enzymes and their regulatory proteins			
Ap-ATPase	APAF1	ATPase, CARD	Unclear in mammals, antagonistic activities in worm
Kinase	IRAK1	Serine-threonine kinase	Active and inactive kinase
Caspases	casp-2	Caspase	Functionally antagonistic isoforms
	casp-8	Caspase, DED	Active or inactive enzyme
	casp-9	Caspase	Functionally antagonistic isoforms
Caspases regulator	cFLIP	Caspase, DED	Functionally antagonistic isoforms
Nuclear effectors			
ICAD	DFF45	ICAD regulatory domain	Differential subcellular localization

TNFR, tumor necrosis factor receptor; TIR, Toll-interleukin receptor; DD, death domain; DED, death effector domain; CARD, caspase recruitment domain; TRAF, tumor necrosis factor receptor (TNFR)-associated factor; IAP, inhibitor of apoptosis protein; BIR, baculovirus inhibitor of apoptosis repeats; ICAD, inhibitor of caspase-activated DNase; MADD, MAPK-activating death domain-containing protein

(VEGF) gene. Several different isoforms with different expression patterns and biological properties are produced via alternative splicing (Robinson and Stringer 2001). Understanding this regulation event will have an impact on human diseases involving angiogenesis and vasculogenesis, especially the processes associated with neovascularization during tumorigenesis and associated with resistance of tumor cell death in response to chemotherapeutic drugs.

The expression of many receptors for trophic or growth factors is also under regulation by alternative splicing to generate functionally distinct isoforms. For example, a TrkA alternative splicing isoform of nerve growth factor receptor shows enhanced responses to neurotrophin 3 (Clary and Reichhardt 1994). A number of genes encoding cytokine receptors use alternative splicing to generate functionally diverse receptor isoforms. For example, alternative splicing generates truncated interleukin 7 (IL-7) receptor isoforms that have been implicated in the pathogenesis of leukemia (Korte et al. 2000).

A large number of membrane receptors have been identified that transduce extracellular death signals to intracellular components of the cell death machinery. Tumor necrosis factor receptor 1 (TNF-R1) and Fas are both cysteine-rich cell surface receptors related to the low-affinity nerve growth factor family. Binding of their corresponding ligands, FasL and TNF, to Fas and TNF-R1 induces apoptosis. The extracellular region of these TNF receptor family members contains two to six repeats of a cysteine-rich motif. These receptors also share homology in their cytoplasmic domain. This domain, which is sufficient to trigger apoptosis when overexpressed alone, has thus been named the death domain (DD). Several other members of this family have been identified including lymphocyte-associated receptor of death (LARD, also named as DR3, wsl, Apo-3 or TRAMP), death receptor-4 (DR4), and TRAIL receptor inducer of cell killing-2 (Trick-2; Pan et al. 1997; Screaton et al. 1997a,b). Many members of the death receptor family have multiple isoforms produced as a result of alternative splicing. Human-activated peripheral mononuclear cells, EBV-infected B cells, hematopoietic and nonhematopoietic tumor cell lines express several Fas mRNA variants in addition to the predominant Fas mRNA species which generates a membrane-bound form of the functional receptor (Ruberti et al. 1996). With the exception of FasEx06Del, which is characterized by an in frame deletion of exon 6 encoded region, in all of these variants alternative splicing leads to translation in a different reading frame with premature termination codons, resulting in the formation of soluble receptor proteins lacking the transmembrane domain. Interestingly, in FasEX08Del, which was predominantly found in tumor-resistant clones, alternative splicing leads to the skipping of exon 8, and therefore, a change in the downstream reading frame with a premature termination codon and elimination of the cytoplasmic death domain. Two TRICK2 mRNA isoforms, TRICK2α and TRICK2β, are generated by alternative splicing (Cascino et al. 1995, 1996). It is possible these soluble receptors or the receptor lacking the death domain may function as dominant inhibitory forms of the full-length receptor, contributing to insufficient cell death or tumor resistance. Alternative splicing of LARD pre-mRNA by skipping one or more exons generates at least 11 distinct isoforms (Screaton et al. 1997a). The full-length LARD-1, isoforms LARD-1b and LARD-8 encode proteins containing both the transmembrane domain and death domain, whereas the other isoforms generate secreted molecules without the transmembrane domain. TRICK2 can transduce the cytotoxic signal from TRAIL/APO-2L alternative splicing and differ by 29 amino acid residues extending into the extracellular domain (Screaton et al. 1997b).

Toll-like receptors (TLRs) are important for host defense. Mice deficient in TLR4 are hyporesponsive to lipopolysaccharide (LPS). The inclusion of an additional exon between the second and third exon of mouse TLR4 converts the membrane-bound receptor into a soluble form because of an in-frame stop codon. This soluble TLR4 inhibits LPS-mediated TNF-alpha production and NF-kappaB activation (Iwami et al. 2000).

It is obvious from the few examples described above that alternative splicing can generate different isoforms of these death receptors and, in some cases, convert a membrane-associated receptor into soluble proteins that inhibit the receptor signal transduction function. It is completely unknown how such alternative splicing events are regulated.

2.2 Cell Death Signal Transducing Adaptor Genes: Multiple Isoforms with Distinct Activities

Several families of cell death adaptor proteins are responsible for transducing extracellular death signals to intracellular cell death machinery. Many of these adaptor proteins contain distinct modular domains (Aravind et al. 1999, 2001; Fesik 2000). Bcl-2 family was the first cell death protein family identified, and this family has now expanded into a superfamily containing many other proteins with Bcl-2 homology (BH). Several other types of protein modules or motifs have been identified in PCD proteins, including DD (death domain), DED (death effector domain), CARD (caspase recruitment domain), and amino-terminal regulatory domain of ICAD (inhibitor of caspase-activated DNase, also named DNA fragmentation factor 45 or DFF45) as well as pyrin, TIR (Toll-interleukin receptor), MATH (meprin and the TRAF-homology) domains. In mammals, especially human, each family of these death adaptor genes has significantly expanded in numbers. In addition to gene family expansion, alternative splicing further increases the functional diversity of these cell death adaptor genes. Although in most cases, we do not fully understand the functional differences of different splicing isoforms yet, it is likely that such extraordinary transcript complexity generated by alternative splicing allows cells to respond to different death signals under different conditions.

APAF-1, a mammalian homologue of *C. elegans* Ced-4 protein, contains both a CARD domain and an AP-ATPase domain. APAF1/Ced4 proteins serve as key regulators of cell death by interacting with anti-apoptotic Bcl-2 family members and controlling cytochrome C-dependent caspase activation (Chinnaiyan et al. 1997; Spector et al. 1997; Wu Det al. 1997; W. Wu et al 1997; Zou et al. 1997; also see references in Nijhawan et al. 2000). In the worm, two forms of Ced-4, Ced-4L and Ced-4S, are generated by alternative selection of the two 3′ splice sites (Shaham and Horvitz 1996). Ced-4S is expressed as a predominant form. Ced-4L transcript contains an in-frame insertion of 72 nucleotides and encodes a protein with a 24-amino acid insertion relative to Ced-4S. Ced-4S induces cell death, whereas Ced-4L can protect cells from developmental cell death. In animals carrying the ced-4 n2273 mutant allele which has a mutation at the Ced-4S 3′ splice site (AG to AA), the Ced-4L level is increased and the anterior pharynx of the mutant animals has extra cells that fail to undergo PCD (Shaham and Horvitz 1996). A human protein Apaf-1 (apoptotic protease activating factor-1) has been identified as a vertebrate homologue of ced-4 (Zou et al. 1997). The *Drosophila* apaf1 gene is alternatively spliced to produce at least two isoforms with different activity in cell death. The human apaf-1 gene also undergoes alternative splicing to generate multiple isoforms that may have different function activities (Benedict et al. 2000; Walke and Morgan 2000; H. Tang and J.Y. Wu, unpubl. results). Further investigation is necessary to define a regulatory role of alternative splicing in APAF-1 function.

Downstream of tumor necrosis factor receptors are TRAF (TNFR-associated factor) family members. Several members of this family exhibit alternative splicing (Brink and Lodish 1998; Gamper et al. 2001). TRAF2, for example, has an alternative splicing variant, TRAF2A, that contains a 7-amino acid insert within its RING finger domain. TRAF2A mRNA is expressed at a high level in the spleen and at the lowest level in the brain. Unlike TRAF2, TRAF2A is incapable of stimulating NF-kappaB activity when overexpressed in transfected cells and acts as a dominant inhibitor of TNFR2-dependent NF-kappaB activation. (Brink and Lodish 1998). Another interesting adapter gene is MADD (MAPK-activating death domain-containing protein). Two alternative splicing isoforms of this gene have been reported, DENN-SV (differentially expressed in normal and neoplastic cells) and IG20. These two isoforms in cell culture show distinct activities, rendering cells either more resistant or susceptible to tumor necrosis factor alpha (TNF-alpha)-induced apoptosis (Al-Zoubi et al. 2001).

A tumor metastasis suppressor gene for small cell lung carcinoma, CC3, has been implicated in cell death. CC3 shows some sequence similarity to bacterial short-chain dehydrogenases/reductases. Alternative splicing of the CC3 gene results in two products, CC3 and TC3. These two isoforms show antagonistic activity in cell death. Overexpression of CC3 causes massive death of rodent fibroblasts, whereas TC3 protects against death induced by overexpression of BAX or CC3, or by TNF treatment (Whitman et al. 2000). MyD88 was previously characterized as a myeloid differentiation response gene in mice. MyD88 proteins appear to be cytoplasmic proteins containing an N-terminal death domain and C-terminal TIR domain (Hardiman et al. 1996; Du et al. 2000). Their function in cell death is not clear yet. Both human and mouse MyD88 have multiple splicing variants that are predicted to have different functional properties (Hardiman et al. 1996).

Bcl-2 gene family proteins are well-studied adaptor proteins that mediate a complex network of interactions among different anti-/pro-apoptotic proteins as well as downstream molecules (Korsmeyer et al. 1999). Bcl-2 proteins play important roles in controlling mitochondrial permeability, cytochrome C release and subsequent caspase activation. In the human, this family has multiple members containing both anti- and pro-apoptotic genes. Prototypical anti-apoptotic members usually contain four BH (BH1–4) domains, and they include Bcl-2, Bcl-w, Bfl-1, Brag-1, Mcl-1, Boo/DIVA and A1. Pro-apoptotic members, on the other hand, contain either three BH domains (BH1-3) Bax, Bak, Mtd/Bok or only the BH3 domain (Bad, Bid, Bik, Bim, Blk, Hrk, Noxa and Rad9). In addition, several genes of this superfamily including Bcl-x and Bak encode for both anti-apoptotic (Bcl-xL, N-Bak) and pro-apoptotic (Bcl-xS, Bak) as a result of alternative splicing (Boise et al. 1993; Sun et al. 2001).

Many members of Bcl-2 superfamily contain a carboxyl-terminal transmembrane domain that anchors the protein to the mitochondrion and other intracellular membrane structures. Several proteins of this family, including Bcl-2, Bcl-x and Bax, have both membrane-bound and soluble isoforms generated by alternative splicing. Protein-protein interaction among different Bcl-2 family members as well as their interactions with other PCD regulatory proteins is an important mechanism for their function. It has been proposed that the relative ratio of death-preventing and death-promoting proteins and selective interactions among different Bcl-2 family members determine whether the cell is to survive or to die. From the few examples described below, it is clear that alternative splicing plays a significant role in generating functional diversity and complexity for these Bcl-2 superfamily members, providing an additional level of fine-tuning regulation.

The Bcl-2 gene produces at least two different transcripts. Bcl-2α is produced by splicing within the first exon and joining to the second exon, whereas Bcl-2β mRNA is transcribed from the first exon (Tsujimoto and Croce 1986). Bcl-2α is the predominant isoform containing a carboxyl-terminal transmembrane domain and has potent death-inhibitory activity. Bcl-2β lacks the transmembrane domain and has reduced (Hockenbery et al. 1993) or no (Tanaka et al. 1993) anti-apoptotic function.

The Bcl-x gene is critical for development of both hematopoietic tissues and the central nervous system. Bcl-x-deficient mice exhibit massive cell death of immature hematopoietic cells and neurons (Motoyama et al. 1995). Bcl-x has several transcripts with different biological activities. The mouse Bcl-x gene exhibits a three-exon structure with the first exon containing a 5′ untranslated region (Grillot et al. 1997). The coding region of Bcl-xL is generated by a splicing reaction that ligates exons 2 and 3, whereas Bcl-xS is produced by the use of an alternative 5′ splice site located within exon 2 with the BH1/BH2 domain deleted. Bcl-xL is similar to Bcl-2 in inhibiting cell death, whereas Bcl-xS antagonizes the anti-apoptotic action of bcl-2 or bcl-xL (Boise et al. 1993). Other murine Bcl-x isoforms include Bcl-xΔTM (Fang et al. 1994), Bcl-xβ (Gonzalez-Garcia et al. 1994; Shiraiwa et al. 1996) and Bcl-xγ (Yang et al. 1997). Bcl-xΔTM deletes the carboxyl-terminal transmembrane domain of Bcl-x via alternative splicing. The involved 5′ splice donor site is in the exon 3 and located 12 nucleotides downstream of the 3′ splice acceptor site for Bcl-xs. Bcl-xΔTM is possibly a product of an intraexonic splicing using the noncanonical splice donor and acceptor sites, because both the 5′ and 3′ splice sites reside in the exon 3 of the Bcl-x gene. Bcl-xΔTM can inhibit apoptosis in B-cells. The Bcl-xβ mRNA is generated by ligation of exon 1 and exon 2 with the intron 2 retained. Bcl-xβ is expressed in embryonic and postnatal tissues and can inhibit apoptosis in neurons. Bcl-xγ carries a unique carboxyl-terminus and is expressed primarily in thymocytes and mature T-cells (Yang et al. 1997). Expression of Bcl-xγ in mature T-cells is associated with ligation of the T-cell receptor (TCR). Unlike other Bcl-x isoforms, Bcl-xγ is specifically associated with TCR ligation and is important for resistance to TCR-dependent apoptosis. Activated T-cells which express Bcl-xγ are resistant to PCD after CD3 ligation of TCR. Generation of Bcl-xγ involves alternative splicing which ligates exon 1 and exon 2 with exon 2 spliced to an exon different from the exon 3. The exon specific to Bcl-xγ has not been well characterized. It is not known how cells regulate the production of these Bcl-x isoforms.

Similar to Bcl-2 and Bcl-xL, the Bcl-w protein promotes cell survival (Gibson et al. 1996). The Bcl-w gene contains at least four closely spaced exons with the coding region split between exon 3 and exon 4. Bcl-w-rox corresponds to transcripts spliced from exon 3 of bcl-w gene to an exon of the adjacent rox gene. The biological significance of such "gene fusion" as a result of alternative splicing remains to be investigated.

Bax shows extensive amino acid homology with Bcl-2 and forms homodimers and heterodimers with Bcl-2 in vivo. When Bax predominates, the death repressor activity of Bcl-2 is inhibited, and cell death is accelerated (Oltvai et al. 1993). The Bax gene consists of six exons and produces several different transcripts. Baxα encodes a protein containing a transmembrane domain. The retention of intron 5 results in the Baxβ mRNA that generates a product lacking the predicted transmembrane domain. The predicted membrane form α and cytosolic β forms of Bax are similar to the α integral membrane form of Bcl-2 and predicted β cytosolic form. The γ forms of Bax mRNA species lack the small exon 2, shifting the reading frame in exon 3. As a result, Baxγ utilizes a stop codon in exon 3 and is a much smaller peptide compared to Baxα and Baxβ. It is possible that these different Bax isoforms have distinct activities in inhibiting Bcl-2 function.

Bim (Bcl-2 interacting mediator of cell death) is a new member of Bcl-2 family which can promote apoptosis (O'Connor et al. 1998). It has a BH3 region that is the only domain with sequence homology to Bcl-2 family members. This

BH3 region is required for its Bcl-2 binding and for most of its cytotoxicity. At least seven Bim alternative splicing isoforms have been identified (O'Connor et al. 1998; Miyashita et al. 2001). Five of these Bim isoforms, BimEL, BimL, BimS, Bim-alpha1 and Bim-alpha2, have a BH3 region and can induce apoptosis. Bim-alpha1, alpha2, and beta1-beta4 isoforms do not contain a C-terminal hydrophobic region. The expression profiles of Bim isoforms are different among normal human tissues, suggesting a tissue-specific regulation of bim gene function (Miyashita et al. 2001).

Another BH3 domain-containing protein was recently reported, PUMA (P53 upregulated modulator of apoptosis; Nakano and Vousden 2001). The PUMA gene undergoes alternative splicing to generate isoforms containing or lacking BH3 domain. PUMA-alpha and PUMA-beta proteins contain BH3 domain, interact with Bcl-2 and induce cytochrome C release with subsequent cell death. The function of other PUMA isoforms that do not contain BH3 domain is not clear.

A recent study has identified a neuron-specific splicing isoform of Bak, N-Bak. Inclusion of a small 20-base pair exon leads to the formation of an anti-apoptotic BH3-only protein in neurons, whereas skipping of this exon results in pro-apoptotic Bak in non-neuronal cells (Sun et al. 2001).

Pyrin (also named marenostrin) is a protein encoded by familial Mediterranean fever (MEFV) gene. The pyrin domain consists of approximately 100 amino acid residues in the highly conserved N-terminal region of pyrin genes of different species. This pyrin domain has now been found, not only in proteins implicated in inflammation, but also in proteins involved in apoptosis (Bertin and DiStefano 2000). The human MEFV gene has at least two different products, full-length (MEFV-fl) and MEFV-d2, generated by in-frame alternative splicing of exon 2. MEFV-fl protein appears to be a cytoplasmic protein, whereas MEFV-d2 is concentrated in the nucleus (Papin et al. 2000). Thus, these two alternative splicing isoforms may have distinct functional activities, although the physiological function of these proteins remains unknown.

From these examples it is clear that, in addition to the mechanism of gene amplification, alternative splicing of different genes in these cell death adaptor gene families provides tightly controlled and fine-tuned regulation at multiple levels, perhaps thereby ensuring proper responses of cells to different death or survival signals.

2.3 Caspase Family and Caspase Regulatory Proteins: Fine-Tuned Dosage Control

Activation of caspases appears to be a central final common pathway for different death signals. Some caspases, such as caspase-8, contain a DED domain in addition to caspase enzyme domain. These caspases may have additional regulatory roles in cell death. Activated caspases act as the executive killer of apoptosis. To date, at least 14 homologues of human origin have been published, and they have now been named caspases 1–14 (Alnemri et al. 1996; Alnemri 1997; Nicholson and Thornberry 1997; Nunez et al. 1998). Adding further to the complexity, many members of this family have different isoforms generated as a result of alternative splicing.

Several alternatively spliced isoforms of ICE (caspase-1) have been described (Alnemri et al. 1995). All the alternative splicing events are within the coding sequence of ICE. Four ICE isoforms maintain open reading frames and are designated as ICEβ, γ, δ, and ε. ICEα and ICEβ have partial intramolecular auto-processing activity. However, most of the propeptide is deleted in ICEγ, and the region containing the cleavage sites between the p20 and p10 subunits of ICE is deleted in ICEδ. ICEε is homologous to the P10 subunit of active ICE. It can form a heterodimer with P20 and may regulate its activity in vivo by forming an inactive complex. Only the ICEα, β and γ isoforms can induce apoptosis.

Alternative splicing of caspase-2 gives rise to at least three transcripts, caps-2L, casp-2S and casp-2β. A previous study (Wang et al. 1994) suggested that production of casp-2L and casp-2S resulted from the alternative use of two competing 5′ splice sites. However, our analyses of the genomic DNA and cDNA sequences indicate that production of casp-2L and casp-2S transcripts results from alternative exclusion or inclusion of a 61-base-pair (bp) exon, respectively (Jiang et al. 1998). Inclusion of the 61-bp exon results in translation termination using a more upstream stop codon in casp-2S. Therefore, a longer casp-2S transcript encodes a shorter protein product. The sequence of the 61-bp exon which is present in caps-2S, but absent from casp-2L is identical between human and mouse casp-2 sequence (Wang et al. 1994). Casp-2L contains sequence homologous to both P20 and P10 subunits of ICE, whereas casp-2S is a truncated version of casp-2L, containing only the P20 region. Casp-2L and casp-2S have antagonistic activities in cell death. A protein specifically interacting the casp-2S isoform has been identified that may be a mediator of anti-apoptotic function of casp-2S (Ito et al. 2000). Another casp-2 isoform, casp-2β, contains a deletion downstream from the first potential aspatic proteolytic cleavage site between the large and small subunits. The deletion alters the coding frame and causes a termination 10-amino acid residues downstream of the aspatic acid residue, essentially generating a functional large subunit. Unlike casp-2S, casp-2β may act as a positive regulator of casp-2 activity (Alnemri 1997).

Similar to casp-2, casp-3 gene also uses alternative exon inclusion/skipping to generate active versus inactive isoforms (either containing or lacking conserved QACRG sequence at the catalytic site; Huang et al. 2001). In addition, the two casp-3 isoforms respond to proteasome inhibition to a different degree (Huang et al. 2001).

Caspase-6 (Mch-2) gene has two transcripts, Mch-2α and Mch-2β. Mch-2β has one-half of its P20 subunit coding region deleted and therefore encodes an inactive enzyme (Fernandes-Alnemri et al. 1995b). Alternative splicing of caspase-7 (Mch-3) pre-mRNA generates two isoforms, Mch3α and Mch-3β. Mch3α encodes a peptide which can be processed into an active caspase, whereas Mch-3β is produced by alternative splicing which leads to the deletion of the region encoding the caspase active site with QACRG pentapeptide sequence (Fernandes-Alnemri et al. 1995b).

Human caspase-8 has multiple isoforms as a result of alternative splicing. Retention of intron 8 leads to a 136-bp insertion and frame shift of the transcript. This casp-8L encodes the N-terminal two repeats of death effector domain (DED) of caspase-8, but lacks the carboxyl-terminal half of the proteolytically active domain (Horiuchi et al. 2000). The ratio between the enzymatically active casp-8 and inactive caspase-8L product appears to be important for lymphocyte function, as suggested by the observation of imbalanced expression of caspase-8L transcript in patients with systemic lupus erythematosus (Horiuchi et al. 2000). Functionally antagonistic alternative splicing isoforms of caspase-9 have also been described (Srinivasula et al. 1999).

More recent analysis of different members of the human caspase family indicates that the majority of these caspase genes undergo alternative splicing at or near the regions encoding the enzyme active site (Hicks, Tang and Wu, unpubl. observ.). The delicate balance of different caspase isoforms, especially that between the enzymatically active and enzymatically inactive products, is likely to be an important factor in cell death regulation or execution.

Some caspase genes have alternative splicing outside the mature protease region or in the untranslated regions. Caspase-3 (CPP32, also named Yama or apopain), caspase-4 (Mih-1, also known as TX, ICErel-II or Ich-2) and caspase-10 (Mch-4) have isoforms which contain deletions in their propeptide domain or 5′ nontranslated region (Alnemri 1997, Fernandes-Alnemri et al. 1995a). Such alternative splicing may influence translation or processing of their pro-caspases.

A number of inhibitory proteins antagonize caspase activities, including proteins containing baculovirus inhibitor of apoptosis protein (IAP) repeats (BIR). These proteins also have alternative splicing isoforms with distinct properties. A death regulator Usurpin, also named CASH, Casper, CLARP, FLAME-1, FLIP, I-FLICE or MRIT, has been identified as a protein with sequence features similar to pro-caspase-8 and pro-caspase-10. This protein is ubiquitously expressed and has at least three splicing variants (Rasper et al. 1998). The functional differences among these isoforms remain to be defined. Cellular FADD-like interleukin-1-converting enzyme (FLICE) inhibitory protein (cFLIP) has two isoforms cFLIP-L and cFLIP-S that show opposite activities in mediating Fas ligand- and TRAIL-induced upregulation of c-Fos in Jurkat cells, with cFLIP-L inhibiting and cFLIP-S stimulating such death receptor-induced and caspase-8-dependent responses (Siegmund et al. 2001). Survin is a gene involved in both oncogenesis and apoptosis. Both human and mouse survin genes produce at least three alternative splicing isoforms that are differentially expressed among different tissues (Conway et al. 2000). Full-length survin contains a single IAP repeat and a carboxyl-terminal coiled-coil domain, whereas survin-121 isoform lacks the coiled-coil domain as a result of the retention of intron 3. The third isoform, survin-40 lacks both domains and is generated because of the skipping of its exon 2 (Conway et al. 2000). It is reasonable to predict that these different survin isoforms have different functional activities. Human Livin gene, also of the inhibitor of apoptosis (IAP) family, has two splicing variants, Livin alpha and beta, and both contain a single copy of baculovirus IAP repeat and RING domain. Livin alpha and beta have a distinct tissue distribution and anti-apoptotic properties (Ashhab et al. 2001).

The currently identified alternative splicing events in different genes encoding caspases and caspase inhibitory proteins suggest that alternative splicing may be used by cells as a general mechanism to modulate caspase activity. It is obvious that the dosage of active caspases has to be tightly controlled in response to different death signals. By generating enzymatically active versus inactive isoforms and different caspase regulatory gene products, alternative splicing is an excellent mechanism of fine-tuning the level of active caspases at the post-transcriptional level. It remains to be investigated how such alternative splicing events are regulated and whether they use similar or distinct regulatory mechanisms.

2.4 Caspase-Activated DNase and Inhibitor of Caspase-Activated DNase

Downstream of caspases, DNA fragmentation and degradation of chromosomal DNA are mediated by a caspase-activated DNase (CAD), DFF40 (Liu et al. 1998). The activity of DFF40 is regulated by its interaction with inhibitor of caspase-activated DNase (ICAD). DFF40 gene contains at least seven exons, although its alternative splicing has not been reported. ICAD has two isoforms, ICAD-L/DFF45 and ICAD-S/DFF35, produced by alternative splicing. In ICAD-S, alternative splicing removes a nuclear localization signal present at the carboxyl terminus of ICAD-L. Therefore, ICAD-L is localized to the nucleus, whereas ICAD-S lacks the nuclear localization signal and is distributed throughout the cell. It is possible that ICAD-L and ICAD-S have different properties in modulating DFF40 nuclease activity because of their different intracellular distribution (Samejima and Earnshaw 2000; Masuoka et al. 2001).

3 The Role of Alternative Splicing in the Regulation of Functional Activities of PCD Genes

Accumulating evidence suggests that alternative pre-mRNA splicing may play an important role in the regulation of programmed cell death. First, as described above, a large number of genes involved in PCD undergo alternative splicing in a tissue- or development stage-specific manner. For example, naive B- and T-cells express very little LARD-1, but express combinations of the other isoforms. Upon T-cell activation, a programmed change in alternative splicing occurs so that the full-length, membrane-band LARD-1 predominates. Adult neural tissue predominantly expresses the bcl-xL mRNA, whereas immature thymocytes that are in the process of undergoing selection in the thymus express a relatively high level of bcl-xS transcript (Boise et al. 1993). Expression of casp-2S is highest in the embryonic brain, and only casp-2L is detected in the adult thymus and spleen (Wang et al. 1994; Jiang et al. 1998). Second, a number of PCD genes utilize alternative splicing to generate functionally distinct gene products from the same genes. Adding further to the complexity, several critical PCD regulators have been found to use alternative splicing to generate gene products that have antagonistic activities in cell death (Fig. 1). It is clear that alternative splicing plays an important role in regulating expression and function of these PCD genes. The mechanism by which alternative splicing may regulate function of these PCD genes can be summarized in at least three aspects: (1) regulating subcellular localization of PCD gene products; (2) modulating functional activity of PCD gene products; and (3) influencing stability of mRNA and/or translational control.

3.1
Regulation of Subcellular Localization

Several members of the death receptor family and of Bcl-2 superfamily have both membrane-associated as well as soluble forms (see review by Jiang and Wu 1999). Cellular and subcellular localization of these death regulatory molecules are determined by including or excluding transmembrane domain as a result of alternative splicing. These different isoforms may very likely have distinct functions in PCD. Expression of two isoforms of TRICK2 that have different sequences between the transmembrane domain and the TRAIL-binding domain may be involved in the regulation of responses to ligand binding. Several human Fas splicing variant mRNAs encode soluble proteins. In vitro apoptosis inhibition assays have shown that the soluble Fas isoforms block Fas-mediated apoptosis induced by agonistic antibody (Cheng et al. 1994; Cascino et al. 1995) and by its natural ligand (Papoff et al. 1996). These soluble Fas isoforms retain the ability to interact with Fas, but they are not able to form active trimers because they lack other domains including the death domain. As a consequence, the signal provided by FasL or by Fas antibody is blocked, resulting in inhibition of apoptosis. Another Fas isoform, FasEX08Del, is an alternative splicing product with skipping of exon 8, and therefore, contains a premature termination codon with elimination of the entire intracytoplasmic death domain, although it retains the transmembrane domain. FasEX08Del is only expressed in tumor-resistant clones (Cascino et al. 1995). It is possible that

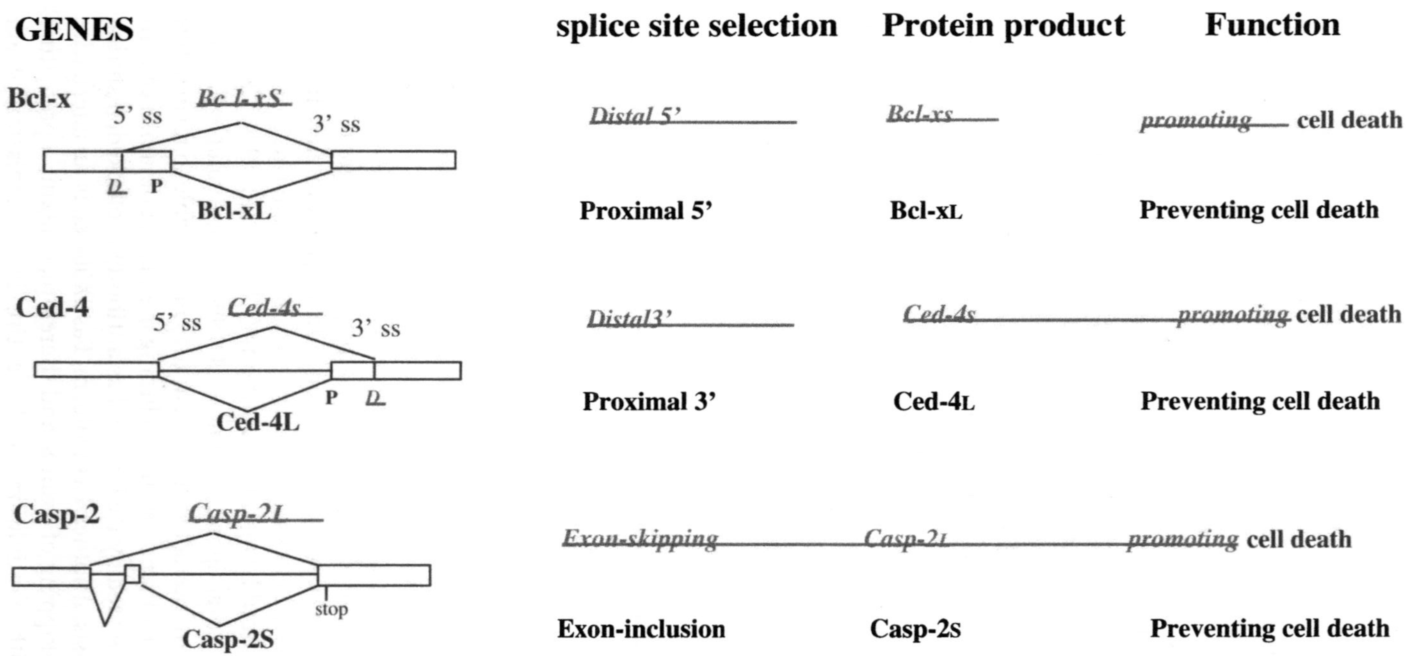

Fig. 1. Examples of alternative splicing events that generate functionally antagonistic products in cell death

heterotrimers containing FasEX08Del isoform may render the receptor incapable of coupling with relevant death domain-binding effectors, or unable to displace death domain-binding effectors. The inability of the receptor to couple with relevant intracellular signal transducers may silence a specific pathway. These studies suggest that different alternative splicing products of the death receptor may have distinct roles in the propagation of the cell death or survival signals.

A number of genes of the Bcl-2 family undergo alternative splicing to generate different isoforms with distinct intracellular localization. For example, both membrane-bound and soluble isoforms of Bcl-2 gene family members are produced as a result of alternative splicing. The cytosolic Bcl-2 family members may provide an additional level of regulation by homodimerizing or heterodimering with the integral membrane forms.

Even the terminal events such as chromosome fragmentation and DNA degradation may involve regulation by alternative splicing. The removal of the nuclear localization signal at the carboxyl terminus of ICAD-L as a result of alternative splicing leads to the cytoplasmic distribution of ICAD-S. Such an alternative splicing event is likely to influence the ability of ICAD to interact with CAD and therefore, its ability to regulate the DNase activity.

3.2
Modulation of Functional Activities

The complex patterns of alternative splicing of different members of Bcl-2 superfamily members provide another excellent example how alternative splicing contributes to the functional diversity of these genes. In addition to membrane-bound versus soluble isoforms generated by alternative splicing, many members of Bcl-2 superfamily have other types of alternatively spliced transcripts encoding isoforms with different peptide sequences outside the transmembrane domain. Because protein-protein interactions among different domains of different Bcl-2 family members and with downstream signal transducing molecules are important for function of Bcl-2 family members, it is likely that different isoforms containing distinct domains have different properties in PCD regulation. For example, a simple switch in the alternative use of two competing 5′ splice sites (Fig. 1) results in the formation of Bcl-xL and Bcl-xS, which have different protein sequences. Bcl-xL contains a BH1/BH2 domain that is deleted in Bcl-xS. Association of pro-survival with pro-apoptotic proteins, such as Bcl-2/Bcl-xL with Bax/Bad, requires the BH1 and BH2 domains of the former and the BH3 domain of the later protein-protein interaction partner. Bcl-xS contains only a BH3 domain and lacks the conserved BH1/BH2 domain. Bcl-xS prevents the stably expressed Bcl-2 or Bcl-xL from inhibiting apoptosis upon growth factor deprivation. Bcl-xS may act as a dominant inhibitor of Bcl-2/Bcl-xL by binding though the BH3 domain. Therefore, the two alternative splicing products, Bcl-xL and Bcl-xS, have antagonistic activities in cell death. The delicate balance of different pro-apoptotic and anti-apoptotic products of the members of Bcl-2 superfamily is likely to play a critical role in determining the cell's susceptibility to death signals.

Alternative splicing events which generate truncated peptides may provide an additional mechanism to quantitatively regulate gene expression. For example, production of Baxγ via alternative splicing may provide a post-transcriptional regulatory mechanism to avoid making excessive the death promoting full-length Bax product.

Alternative splicing events in Bcl-2 and APAF1/Ced-4 gene families may also regulate interactions between members in these two families. Some Bcl-2 family members prevent apoptosis induced by a variety of stimuli, whereas others promote apoptosis. Competitive dimerization between Bcl-2 and APAF1/Ced-4 family members is thought to regulate their function. There is a complex network of protein-protein interactions among different Bcl-2 superfamily members and other cell death signal transducing proteins. APAF1/Ced-4 proteins act as adapter molecules between Bcl2/Ced-9 and caspases/Ced-3. The two splicing variants of Ced-4, Ced-4L and Ced-4S, have opposite functions in PCD, suggesting the important role of alternative splicing regulating PCD. Consistent with this, in the worm carrying a mutation at the Ced-4S 3′ splice site, Ced-4L expression was increased and extra cells were found due to decreased cell death (Shaham and Horvitz 1996).

Alternative splicing is an important mechanism of regulating the enzymatic activity of caspases. Active caspases consist of a p10 plus p20 heterodimer produced by cleavage of the precursor protein at aspartate residues. Alternative splicing isoforms that produce peptides truncated upstream or at the caspase enzyme-active domain, such as Mch2β, Mch3β, casp-2S and casp-8L all encode inactive enzymes. Active caspases can induce apoptosis, whereas inactive isoforms may be capable of forming a heterodimer with active enzymes and regulate their activity in vivo. More recent analyses indicate that the majority of the known caspase family members have alternative splicing at or near the regions encoding peptide sequences contributing to their active sites for enzyme activity (Hicks, H. Tang and J.Y. Wu, unpubl. results).

From the few examples described above, it is clear that alternative splicing can regulate each step of cell death induction by generating distinct gene products with different or even antagonistic activities in cell death (Table 1, Fig. 1). Such examples can be found in almost every family of genes important for PCD regulation.

3.3
Influencing mRNA Stability or Translational Regulation

It is known that alternative splicing can generate mRNAs with differential turnover rates or with differential properties in translational control (reviewed in Smith et al. 1989; Curtis et al. 1995). For example, genes encoding growth factors such as colony-stimulating factor-1 have alternatively spliced transcripts which have different stability (Shaw and Kamen, 1986; Ladner et al. 1987). It has been observed that many alternative splicing events occur at either 5′ or 3′ untranslated regions of cell death genes. For example, CPP32β contains a deletion of most of its 5′ untranslated region. CPP32α and CPP32β transcripts may have a difference in the stability and/or ability in regulating translation. A conserved AU-rich element has been identified in the 3′ untranslated region of bcl-2 mRNA, and this element interacts with AU-rich element binding proteins and is associated with bcl-2 downregulation during apoptosis (Schiavone et al. 2000; Donnini et al. 2001). IL-1RI-associated kinase-1 (IRAK1) is a membrane proximal serine-threonine kinase important for IL-1 signaling. IRAK1 gene produces at least two alternative splicing isoforms. Compared to IRAK1, the IRAK1b isoform is kinase inactive and more stable (Jensen and Whitehead 2001). It is not proven yet whether this enhanced stability is a result of an increase in mRNA stability or occurs at the protein level.

Although it is not clear yet how the stability of the alternatively spliced transcripts of the above-mentioned PCD genes are regulated, it is obvious that mRNA stability of different PCD gene isoforms may have a significant impact on their functional activity in cell death.

4 Molecular Mechanism Underlying the Regulation of Alternative Splicing of PCD Genes

4.1 Splicing Machinery and Alternative Splicing Regulators

Since the discovery of pre-mRNA splicing more than 20 years ago, tremendous effort has been put into identifying molecular components of mammalian splicing machinery and understanding regulatory mechanisms controlling alternative splicing (Moore et al. 1993; Black 1995; Krämer 1996; Reed 1996; Manley and Tacke 1996; Wang and Manley 1997; Konarska 1998; Smith and Valcarcel 2000; Goldstrohm et al. 2001; Hastings and Krainer 2001). Similar to yeast splicing machinery, the mammalian spliceosome is assembled in an orderly pathway containing, in addition to pre-mRNA, several snRNPs and a number of accessory protein factors . The assembly and activation of the multi-component spliceosome involves a complex array of RNA, protein-RNA and protein-protein interactions in a highly dynamic fashion (Staley and Guthrie 1998).

A critical issue in pre-mRNA splicing and alternative splicing regulation is specific recognition of splice sites and proper association between appropriate 5′ and 3′ splice sites. Biochemical studies using vertebrate splicing substrates as well as genetic studies using yeast systems provide compelling evidence that snRNPs including U1 (or U11), U2 (or U12) and U5 play essential roles in splice site recognition and association. Studies using mammalian and viral splicing model substrates indicate that several families of proteins play essential roles in spliceosome assembly and in regulating alternative splicing. These include the heteronuclear ribonucleoproteins family (hnRNP proteins), proteins containing serine-arginine-rich sequences (SR proteins) and specific RNA-binding proteins (such as proteins binding to polypyrimidine tract and branch site; Krämer 1996). Splicing factors of the SR and hnRNP families are emerging as important alternative splicing regulators (Berget 1995; Black 1995; Fu 1995; Krämer 1996; Wang and Manley 1997; Boucher et al. 2001; see also Sanford, Longman and Cáceres, this Vol.). Several studies have suggested that SR proteins and an hnRNP protein, hnRNPA1, have antagonistic effects on splice site selection. In several model substrate systems, SR proteins have been found to act as factors interacting with splicing enhancers, whereas polypyrimidine-tract-binding protein (PTB), a protein of the hnRNP family, acts as a splicing repressor in the splicing of several genes (reviewed in Grabowski and Black 2001). Several other proteins including KH-domain-containing and CUGBP proteins have also been shown to play regulatory roles in alternative splicing (reviewed in Grabowski and Black 2001). Disruption of or defects in alternative splicing regulation has been implicated in the pathogenesis of a number of human diseases, including those associated with either excessive or insufficient cell death (Cooper and Mattox 1997; Blencowe 2000; Philips and Cooper 2000; Dredge et al. 2001; Hastings and Krainer 2001).

4.2
Splicing Signals and *cis*-Regulatory Elements for Splicing

The basic splicing signals include the 5′ splice site, branch site and polypyrimidine track-AG at the 3′ splice site. These signals are initially recognized by the U1 snRNP, U2 snRNP and U2 snRNP auxiliary factor (U2AF), respectively. In mammals, these basic splicing signals tend to be degenerate and are not sufficient by themselves to confer the specificity required to achieve accurate splice site selection. A number of other factors also contribute to splice site recognition and influence splicing efficiency. These include the distance between two splice sites, the size of an exon, as well as local secondary structures in the pre-mRNAs. In addition, various types of exonic and intronic elements have been identified that modulate the use of nearby splice sites. For example, exonic splicing enhancers (ESEs) in different genes have been described, in particular, of purine-rich sequences (for example, Blencowe 2000). These splicing enhancers can promote the use of either an upstream 3′ splice site (Lavigueur et al. 1993; Watakabe et al. 1993; Xu et al. 1993) or downstream 5′ splice site (Humphrey et al. 1995; Elrick et al. 1998; Heinrichs et al. 1998; Bourgeois et al. 1999; Cote et al. 1999). Most of these prototypical ESEs bind different subsets of SR proteins, a family of highly conserved nuclear phosphoproteins that are essential for splicing and crucial for regulating alternative splicing of many pre-mRNAs (reviewed in Fu 1995; Chabot 1996; Valcarcel and Green 1996; Tacke and Manley 1999; Hastings and Krainer 2001).

Enhancer elements residing in the intron regions, usually downstream of alternative exons, are generally more diverse in sequence (Balvay et al. 1992; Huh and Hynes 1994; Zhao et al. 1994; Black 1995; Del Gatto and Breathnach 1995; Lou et al. 1995; Sirand-Pugnet et al. 1995; Carlo et al. 1996; Ryan and Cooper 1996; Zhang et al. 1996; McCullough and Berget 1997; Wei et al. 1997; Haut and Pintel 1999; Pret et al. 1999). Although the factors interacting with these elements are beginning to be identified (Min et al. 1995, 1997; Gallego et al. 1997; Chou et al. 1999), mechanisms involved in the splicing stimulation are still largely unknown.

Splicing repressors or silencers have also been found in exons (Chew et al. 2000 and references within) or introns (Gontarek et al. 1993; Chan and Black 1995; Kanopka et al. 1996; Zhang et al. 1996; Carstens et al. 1998; Blanchette and Chabot 1999; Jin et al. 1999; Cote et al. 2001a,b). However, in most cases, the molecular mechanism underlying the splicing inhibition is not clear.

Although the involvement of alternative splicing in modulating functional activities of different PCD genes has been studied for several years, the underlying molecular mechanisms still remain to be elucidated. To our knowledge, the casp-2 system is the only mini-gene model system that has been established to characterize alternative splicing of PCD genes and to dissect underlying molecular mechanisms using biochemical assays (Jiang et al. 1998; Cote et al. 2001a,b). We have begun to examine the cis-elements and trans-acting factors involved in modulating the formation of anti-apoptotic and pro-apoptotic casp-2 isoforms by alternative splicing. Using the casp-2 model system, we have demonstrated that several splicing factors can regulate casp-2 alternative splicing of both the cotransfected minigene and endogenous gene. SR proteins including SC35 and ASF/SF2 promote exon skipping to produce casp-2L, whereas, hnRNPA1 facilitates exon inclusion to produce casp-2S (Jiang et al. 1998). These effects of SR proteins and hnRNP A1 on caspase-2 splicing are opposite to what were predicted from the effects of corresponding splicing regulators on other model splicing substrates used in previous studies. In addition to relative concentrations of SR proteins and their antagonistic hnRNP proteins, other mechanisms could regulate functional activities through modification of these proteins such as reversible protein phosphorylation. Several kinases capable of phosphorylating SR proteins have been identified and phosphatase activities associated with the spliceosome have also been reported (reviewed in Fu 1995). Another potential link between SR proteins and apoptosis regulation is association of phosphorylated SR proteins including SC35 with U1-snRNP complex in cells undergoing apoptotic cell death (Utz et al. 1998).

Biochemical studies and experiments using cell cultures have revealed an evolutionarily conserved 100-base-pair intronic element, In100, in casp-2 gene that specifically inhibits its exon 9 splicing. As described previously, alternative splicing of this 61-bp exon 9 leads to the formation of two functionally antagonistic products (casp-2L and casp-2S; Fig. 1). The In100 element contains a decoy 3′ splice site juxtaposed to a PTB-binding domain, both of which contribute to the full activity of In100 in inhibiting exon 9 inclusion (Cote et al. 2001a,b). Based on a number of observations, we proposed a model for In100 function in regulating caspase-2 alternative splicing (Fig. 2). The upstream portion of In100 contains a sequence with features of an authentic 3′ splice site

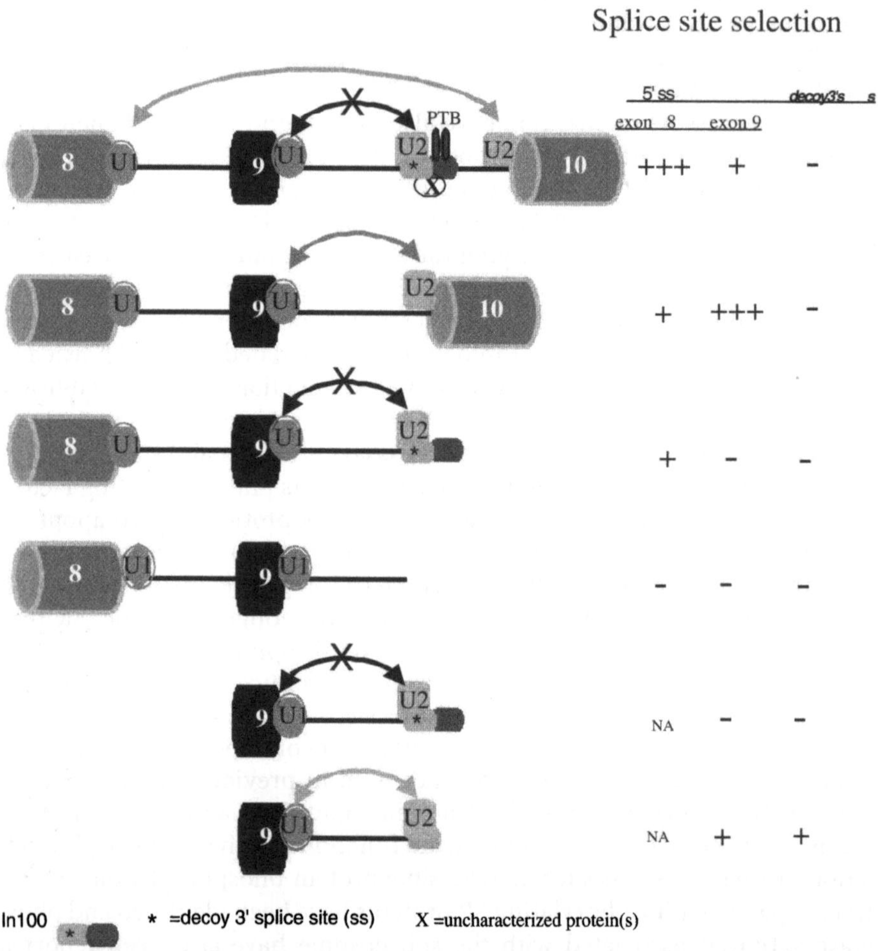

Fig. 2. A model for the function of an intronic repressor element that regulates caspase-2 exon 9 alternative splicing. The splicing of exon 9 (*black*) in caspase-2 gene is repressed by an intronic element, In100, that contains a decoy 3′ splice site (*ss*) domain and a PTB-binding domain. The interactions of some yet-to-be characterized factors (*X*) and of PTB with In100 contribute to the nonproductive interaction with the 5′ ss of exon 9 blocking the selection of this 5′ss even in the presence of a high level of U1 snRNP binding. The decoy 3′ss inside In100 is only utilized when it is isolated and with the downstream PTB-binding domain removed. Molecular mechanisms underlying such differential recognition of decoy versus authentic splice sites remain to be elucidated

(including a branch site, a polypyrimidine tract and AG dinucleotide). However, this site is not used under normal conditions, unless it is isolated and the downstream PTB-binding domain of In100 is deleted. Thus, we call this site a decoy 3′ splice site. Our biochemical and cell culture experiments show that this decoy 3′ acceptor site interacts nonproductively with the 5′ splice site of the alternative exon 9, thus repressing the efficient use of the 5′ splice site of exon 9 despite a high level of U1 snRNP binding to this 5′ splice site. Downstream of the decoy 3′ splice site resides a second functionally distinct domain that interacts with PTB. The binding of PTB to CU-rich motifs within this downstream domain juxtaposed to the decoy 3′ acceptor site correlates well with the repressor activity of this domain. Interestingly, this negative effect of the PTB binding site-containing downstream domain can also be observed in the context of competing 5′ splice sites. PTB can modulate recognition of the adjacent decoy 3′ acceptor site. In addition to factors interacting with an authentic 3′ splice site (such as U2AF or U2 snRNP), PTB as well as at least two novel proteins (yet to be further characterized; J. Cote, H. Tang and J.Y. Wu, unpubl. results) appear to interact with In100 and contribute to the recognition of the In100 decoy 3′ splice site by spliceosome as an intronic repressor element, rather than an authentic 3′ splice site (illustrated in Fig. 2). PTB has been implicated in the alternative splicing of several other genes (reviewed in Grabowski and Black 2001). However, the regulatory role of PTB in casp-2 alternative splicing and its mechanism of action appear to be distinct from previously described systems (see discussion in Cote et al. 2001b). Furthermore, our recent survey of known human genes involved in cell death regulation suggests that In100-like intronic elements (i.e., 3′ splice site juxtaposed to PTB-binding domains) may represent a general intronic splicing repressor motif and that such elements may play a role in the regulation of alternative splicing of other cell death genes (Hicks, Tang and Wu, unpubl. observ.). It remains to be investigated how this decoy 3′ splice site inside In100 is differentially recognized by splicing machinery as a regulatory element rather than being used as an authentic 3′ cleavage site.

Molecular dissection of cis-acting elements and *trans*-acting splicing regulators involved in caspase-2 alternative splicing has provided a good entry site to understanding mechanisms underlying the complex regulation of alternative splicing of important PCD genes. The involvement of such a pseudo- or decoy splice site in alternative splicing regulation may provide an explanation for the phylogenic conservation of certain pseudo-splicing signals in mammalian introns. Further investigation is necessary to prove the generality of such splicing regulatory motifs.

4.3
Changes in Expression of Splicing Regulators During Cell Death

Efforts have been made to examine the changes in gene expression or protein localization during induction of programmed cell death. A recent study has investigated gene expression changes during Fas-induced apoptosis of Jurkat T-cells. In this study, subtractive 2-D gel patterns and MALDI-MS (matrix-assisted laser desorption/ionization mass spectrometry) were used to examine proteins that were modified or changed during apoptosis. Some of the identified molecules are involved in alternative splicing or other aspects of RNA processing. These include hnRNP A, K and R; the splicing factors p54, SRp30c, ASF2 and KH-type splicing regulatory proteins, alpha NAC, NS-1 associated protein-1 and poly(A) binding protein 4 (Thiede et al. 2001). Splicing factor PSF has been reported to undergo significant changes during apoptosis, including protein phosphorylation and protein-protein interaction profiles (Shav-Tal et al. 2001).

Another family of proteins have also been identified that are involved in both cell death and alternative splicing, TIA-1 and the related protein TIAR. TIA-1 was first identified as a protein localized to the granules of cytolytic lymphocytes that could induce DNA fragmentation (Tian et al. 1991). Both TIA-1 and TIAR appear to promote cell death in thymocytes, and they contain RNA recognition motifs (RRMs). TIA-1 has been shown to play a role in the regulation of splicing (Förch et al. 2000). It has been implicated in both alternative splicing and translational regulation (for references see the review by Förch and Valcarcel 2001). Among the potential target substrates of alternative splicing by TIA-1 are the fibroblast growth factor receptor 2 and the Fas receptor (reviewed in Förch and Valcarcel 2001).

4.4 Tissue- or Cell-Type Specificity of Alternative Splicing of Programmed Cell Death Genes

A number of PCD genes undergo alternative splicing in a tissue- or cell type-specific manner. It is not clear how such tissue specificity is achieved. So far, no splicing factors involved in PCD gene splicing regulation have been shown to be expressed with strict tissue specificity. Our studies suggest that multiple factors contribute to the complex and dynamic alternative splicing pattern of these PCD genes (Tang, Cote and Wu, unpubl. results).

A few splicing regulators that are enriched or specifically expressed in the nervous system have been reported, including nPTB, NOVA and elav (reviewed in Dredge et al. 2001; Grabowski and Black 2001; J.D. Keene 2001). It is not known whether these RNA binding proteins or splicing regulators are involved in alternative splicing of PCD genes.

Our study on molecular mechanisms regulating alternative splicing of PCD genes is only beginning. A number of important questions remain to be addressed. We do not understand why so many different isoforms of PCD genes are generated and how they function in vivo to control cell death. Our knowledge is extremely limited in terms of mechanisms controlling alternative splicing of critical PCD genes. In fact, the casp-2 system is the only model system established, to our knowledge, for biochemical characterization of alternative splicing regulation of cell death genes. It is likely that different PCD gene alternative splicing events are regulated by distinct molecular mechanisms. It is also conceivable that alternative splicing events of different PCD genes are coordinated to a certain extent during cell death. It is not known how splicing machinery respond to death signals and what signal transduction pathways are utilized to transduce death signals into nuclear spliceosomal components. Almost nothing is known about potential splicing regulators involved in controlling cell death under physiological or pathological conditions. More systematic studies using combined biochemical, molecular, cell biological and bioinformatic approaches on alternative splicing of critical PCD genes are necessary to obtain a more comprehensive picture of alternative splicing regulation of PCD gene function and to elucidate mechanisms underlying alternative splicing regulation in cell death.

5 Conclusion

A large number of genes involved in PCD undergo alternative splicing to produce multiple isoforms with distinct functional activities in cell death. Alternative splicing not only generates products with different subcellular localization (membrane-associated versus soluble proteins; nuclear versus cytoplasmic), but also produces proteins with different functional activities. Molecular analyses of expression and regulation of these genes have suggested a very important role of alternative splicing in regulating PCD. So far, very little is known about molecular mechanisms controlling alternative splicing of these PCD genes. Our work using caspase-2 as a model system has initiated molecular dissection of the link between alternative splicing and PCD regulation. Further study is necessary to elucidate the role of alternative splicing in modulation of PCD and mechanisms underlying this important alternative splicing regulation. Studying molecular mechanisms underlying functionally important alternative splicing events not only helps us understand pathogenetic mechanisms of human diseases caused by splicing defects, but may also provide insights into designing new therapies for such diseases (for example, Puttaraju et al. 2001; Sazani and Kole, this Vol.).

Acknowledgement. The authors' work was supported by grants from the NIH, a scholarship from the Leukemia Society of America and funding from the Society for Progress Supranuclear Palsy and Muscular Dystrophy Association. We would like to thank members of the Wu lab for critical reading of the manuscript. We thank Angie Hantak for excellent assistance. We apologize to colleagues for using reviews in many places rather than citing original research articles due to space limitations.

References

Alnemri ES (1997) Mammalian cell death proteases: a family of highly conserved aspartate specific cysteine proteases. J Cell Biochem 64:33–42

Alnemri ES, Fernandes-Alnemri T, Litwack G (1995) Expression of four novel isoforms of human interleukin-1converting enzyme with different apoptotic activities. J Biol Chem 270:4312–4317

Alnemri ES, Livingston DJ, Nicholson DW, Salvesen G, Thornberry NA, Wong WW, Yuan J (1996) Human ICE/CED-3 protease nomenclature. Cell 87:171

Al-Zoubi AM, Efimova EV, Kaithamana S, Martinez O, El-Idrissi Mel-A, Dogan RE, Prabhakar BS (2001) Contrasting effects of IG20 and its splice isoforms, MADD and DENN-SV, on tumor necrosis factor alpha-induced apoptosis and activation of caspase-8 and -3. J Biol Chem 276:47202–47211

Aravind L, Dixit VM, Koonin EV (1999) The domains of death: evolution of the apoptosis machinery. Trends Biochem Sci 24(2):47–53

Aravind L, Dixit VM, Koonin EV (2001) Apoptotic molecular machinery: vastly increased complexity in vertebrates revealed by genome comparisons. Science 291(5507):1279–1284

Ashhab Y, Alian A, Polliack A, Panet A, Yehuda DB (2001) Two splicing variants of a new inhibitor of apoptosis gene with different biological properties and tissue distribution pattern. FEBS Lett 495(1–2):56–60

Atamas SP (1997) Alternative splice variants of cytokines: making a list. Life Sci 61(12):1105–1112

Bae J, Leo CP, Hsu SY, Hsueh AJW (2000) Mcl-1S, a splicing variant of the antiapoptotic bcl-2 family member Mcl-1, encodes a proapoptotic protein possessing only the BH3 domain. J Biol Chem 275:25255–25261

Balvay L, Libri D, Gallego M, Fiszman MY (1992) Intronic sequence with both negative and positive effects on the regulation of alternative transcripts of the chicken beta tropomyosin transcripts. Nucleic Acids Res 20:3987–3992

Benedict MA, Hu Y, Inohara N, Nunez G (2000) Expression and functional analysis of Apaf-1 isoforms. Extra Wd-40 repeat is required for cytochrome c binding and regulated activation of procaspase-9. J Biol Chem 275(12):8461–8468

Berget SM (1995) Exon recognition in vertebrate splicing. J Biol Chem 2:2411–2414

Bertin J, DiStefano PS (2000) The PYRIN domain: a novel motif found in apoptosis and inflammation proteins. Cell Death Differ 7(12):1273–1274

Black DL (1995) Finding splice sites within a wilderness of RNA. RNA 1:763–771

Blanchette M, Chabot B(1999) Modulation of exon skipping by high affinity hnRNP A1 binding sites and by intron elements that repress splice site utilisation. EMBO J 18:1939–1952

Blencowe BJ (2000) Exonic splicing enhancers: mechanism of action, diversity and role in human genetic diseases. Trends Biochem Sci 25(3):106–110

Boise LH, González-García M, Postema CE, Ding L, Lindsten T, Turka LA, Mao X, Núñez G, Thompson CB (1993) bcl-x, a bcl-2-related gene that functions as a dominant regulator of apoptotic cell death. Cell 74:597–608

Boucher L, Ouzounis CA, Enright AJ, Blencowe BJ (2001) A genome-wide survey of RS domain proteins. RNA 7(12):1693–1701

Bourgeois CF, Popielarz M, Hildwein G, Stevenin J (1999) Mol Cell Biol 19:7347–7356

Brink R, Lodish HF (1998) Tumor necrosis factor receptor (TNFR)-associated factor 2A (TRAF2A), a TRAF2 splice variant with an extended RING finger domain that inhibits TNFR2-mediated NF-kappaB activation. J Biol Chem 273(7):4129–4134

Carlo T, Sterner DA, Berget SM (1996) RNA 2:342–353

Carstens RP, McKeehan WL, Garcia-Blanco MA (1998) An intronic sequence element mediates both activation and repression of rat fibroblast growth factor receptor 2 pre-mRNA splicing. Mol Cell Biol 18:2205–2217

Cascino I, Fiucci G, Papoff G, Ruberti G (1995) Three functional soluble forms of the human apoptosis-inducing Fas molecule are produced by alternative splicing. J Immunol 154: 2706–2713

Cascino I, Papoff G, Maria D, Testi R, Ruberti G (1996) Fas/Apo-1/CD95 receptor lacking the intracytoplasmic signaling domain protects tumor cells from Fas-mediated apoptosis. J Immunol 156:13

Chabot B (1996) Directing alternative splicing: cast and scenarios. Trends Genet 12(11):472–478

Chan RC, Black DL (1995) Conserved intron elements repress splicing of a neuron-specific c-src exon in vitro. Mol Cell Biol 15:6377–6385

Cheng J, Zhou T, Liu C, Shapiro JP, Brauer MJ, Kiefer MC, Barr PJ, Mountz JD (1994) Protection from Fas-mediated apoptosis by a soluble form of the Fas molecule. Science 263:1759

Chew SL, Baginsky L, Eperon IC (2000) An exonic splicing silencer in the testes DNA specific DNA ligase III beta exon. Nucleic Acids Res 28:402–410

Chou MY, Rooke N, Turck CW, Black DL (1999) hnRNP H is a component of a splicing enhancer complex that activates a c-src alternative exon in neuronal cells. Mol Cell Biol 19:69–77

Chinnaiyan AM, O'Rourke K, Lane BR, Dixit VM (1997) Interaction of CED-4 with CED-3 and CED-9: a molecular framework for cell death. Science 275:1122–1126

Clary DO, Reichardt LF (1994) An alternatively spliced form of the nerve growth factor receptor TrkA confers an enhanced response to neurotrophin 3. Proc Natl Acad Sci USA 91(23): 11133–11137

Conway EM, Pollefeyt S, Cornelissen J, DeBaere I, Steiner-Mosonyi M, Ong K, Baens M, Collen D, Schuh AC (2000) Three differentially expressed survivin cDNA variants encode proteins with distinct antiapoptotic functions. Blood 95(4):1435–1442

Cooper TA, Mattox W (1997) The regulation of splice-site selection, and its role in human disease. Am J Hum Genet 61(2):259–266

Cote J, Simard MJ, Chabot B (1999) An element in the 5′ common exon of the NCAM alternative splicing unit interacts with SR proteins and modulates 5′ splice site selection. Nucleic Acids Res 27(12):2529–2537

Cote J, Dupuis S, Jiang Z, Wu JY (2001a) Caspase-2 pre-mRNA alternative splicing: Identification of an intronic element containing a decoy 3′ acceptor site. Proc Natl Acad Sci USA 98(3): 938–943

Cote J, Dupuis S, Wu JY (2001b) Polypyrimidine track-binding protein binding downstream of caspase-2 alternative exon 9 represses its inclusion. J Biol Chem 276(11):8535–8543

Curtis D, Lehmann R, Zamore PD (1995) Translational regulation in development. Cell 81:171–178

Del Gatto F, Breathnach R(1995) Exon and intron sequences, respectively, repress and activate splicing of a fibroblast growth factor receptor 2 alternative exon. Mol Cell Biol 15:4825–4834

Donnini M, Lapucci A, Papucci L, Witort E, Tempestini A, Brewer G, Bevilacqua A, Nicolin A, Capaccioli S, Schiavone N (2001) Apoptosis is associated with modifications of bcl-2 mRNA AU-binding proteins. Biochem Biophys Res Commun 287(5):1063–1069

Dredge BK, Polydorides AD, Darnell RB (2001) The splice of life: alternative splicing and neurological disease. Nat Rev Neurosci 2(1):43–50

Du X, Poltorak A, Wei Y, Beutler B (2000) Three novel mammalian toll-like receptors: gene structure, expression, and evolution. Eur Cytokine Netw 11(3):362–371

Elrick LL, Humphrey MB, Cooper TA, Berget SM (1998) Mol Cell Biol 18:343–352

Fang W, Rivard JJ, Mueller DL, Behrens TW (1994) Cloning and molecular characterization of mouse bcl-x in B and T lymphocytes. J Immunol 153(10):4388–4398

Fernandes-Alnemri T, Litwack G, Alnemri ES (1995a) Mch2, a new member of the apoptotic Ced-3/Ice cysteine protease gene family. Cancer Res 55(13):2737–2742

Fernandes-Alnemri T, Takahashi A, Armstrong R, Krebs J, Fritz L, Tomaselli KJ, Wang L, Yu Z, Croce CM, Salveson G et al. (1995b) Mch3, a novel human apoptotic cysteine protease highly related to CPP32.Cancer Res 55(24):6045–6052

Fesik SW (2000) Insights into programmed cell death through structural biology. Cell 103(2):273–282

Förch P, Valcarcel J (2001) Molecular mechanisms of gene expression regulation by the apoptosis-promoting protein TIA-1. Apoptosis 6(6):463–468

Förch P, Puig O, Kedersha N, Martinez C, Granneman S, Seraphin B, Anderson P, Valcarcel J (2000) The apoptosis-promoting factor TIA-1 is a regulator of alternative pre-mRNA splicing. Mol Cell 6(5):1089–1098

Fu X-D (1995) The superfamily of arginine/serine rich splicing factors. RNA 1:663–680

Gallego ME, Gattoni R, Stevenin J, Marie J, Expert-Bezancon A (1997) The SR splicing factors ASF/SF2 and SC35 have antagonistic effects on intronic enhancer-dependent splicing of the beta tropomyosin alternative exon. EMBO J 16:1772–1784

Gamper C, Omene CO, van Eyndhoven WG, Glassman GD, Lederman S (2001) Expression and function of TRAF-3 splice-variant isoforms in human lymphoma cell lines. Hum Immunol 62(10):1167–1177

Gibson L, Holmgreen SP, Huang DCS, Bernard O, Copeland NG, Jenkins NA, Sutherland GR, Baker E, Adams JM, Cory S (1996) Bcl-w, a novel member of the bcl-2 family, promotes cell survival. Oncogene 13:665–675

Gleichmann M, Gillen C, Czardybon M, Bosse F, Greiner-Petter R, Auer J, Muller HW (2000) Cloning and characterization of SDF-1gamma, a novel SDF-1 chemokine transcript with developmentally regulated expression in the nervous system. Eur J Neurosci 12(6):1857–1866

Goldstrohm AC, Greenleaf AL, Garcia-Blanco MA (2001) Co-transcriptional splicing of pre-messenger RNAs: considerations for the mechanism of alternative splicing. Gene 277(1–2): 31–47

Gontarek RR, McNally MT, Beemon K (1993) Genes Dev 7:1926–1936

González-García M, Perez-Ballestero R, Ding LY Duan L, Boise LH, Thompson CB, Núñez G (1994) Bcl-xL is the major bcl-x mRNA form expressed during murine development and its product localizes to mitochondria. Development 120:3033–3042

Grabowski PJ, Black DL (2001) Alternative RNA splicing in the nervous system. Prog Neurobiol 65:289–308

Grillot DAM, González-García M, Ekhterae D, Duan L, Inohara N, Ohta S, Seldin MF, and Núñez G (1997) Genomic organization, promoter region analysis, and chromosome localization of the mouse bcl-x gene. J Immunol 158:4750–4757

Hastings ML, Krainer AR (2001) Pre-mRNA splicing in the new millennium. Curr Opin Cell Biol 13(3):302–309

Hardiman G, Rock FL, Balasubramanian S, Kastelein RA, Bazan JF (1996) Molecular characterization and modular analysis of human MyD88. Oncogene 13(11):2467–2475

Haut DD, Pintel DJ (1999) Inclusion of the NS2 specific exon in minute virus of mice mRNA is facilitated by an intronic splicing enhancer that affects definition of the downstream small intron. Virology 258:84–94

Heinrichs V, Ryner LC, Baker BS (1998) Regulation of sex specific selection of fruitless 5′ splice site by transformer and transformer 2. Mol Cell Biol 18:450–458

Hockenbery D, Nunez G, Milliman C, Schreiber RD, Korsmeyer SJ (1990) Bcl-2 is an inner mitochondrial membrane protein that blocks programmed cell death. Nature 348:334–336

Horiuchi T, Himeji D, Tsukamoto H, Harashima S, Hashimura C, Hayashi K (2000) Dominant expression of a novel splice variant of caspase-8 in human peripheral blood lymphocytes. Biochem Biophys Res Commun 272(3):877–881

Horvitz HR (1999) Genetic control of programmed cell death in the nematode *Caenorhabditis elegans*. Cancer Res 59(7 Suppl):1701s–1706s

Horvitz HR, Shaham S, Hengartner MO (1994) The genetics of programmed cell death in the nematode *caenorhabditis elegans*. Cold Spring Harb Symp Quat Biol 59:377–385

Huang Y, Shin NH, Sun Y, Wang KK (2001) Molecular cloning and characterization of a novel caspase-3 variant that attenuates apoptosis induced by proteasome inhibition. Biochem Biophys Res Commun 18283(4):762–769

Huh GS, Hynes RO (1994) Regulation of alternative pre-mRNA splicing by a novel repeated hexanucleotide element. Genes Dev 8:1561–1574

Humphrey MB, Bryan J, Cooper TA, Berget SM (1995) A 32-nucleotide exon-splicing enhancer regulates usage of competing 5′ splice sites in a differential internal exon. Mol Cell Biol 15:3979–3988

Ito A, Uehara T, Nomura Y (2000). Isolation of Ich-1 S (caspase-2 S)-binding protein that partially inhibits caspase activity. FEBS Lett 470:360–364

Iwami KI, Matsuguchi T, Masuda A, Kikuchi T, Musikacharoen T, Yoshikai Y (2000) Cutting edge: naturally occurring soluble form of mouse Toll-like receptor 4 inhibits lipopolysaccharide signaling. J Immunol 165(12):6682–6686

Jacobson MD, Weil M, Raff MC (1997) Programmed cell death in animal development. Cell 88:347–354

Jensen LE, Whitehead AS (2001) IRAK1b, a novel alternative splice variant of interleukin-1 receptor-associated kinase (IRAK), mediates interleukin-1 signaling and has prolonged stability. J Biol Chem 276(31):29037–29044

Jiang ZH, Wu JY (1999) Alternative splicing and programmed cell death. Proc Soc Exp Biol Med 220(2):64–72

Jiang ZH, Zhang WJ, Rao Y, Wu JY (1998) Regulation of Ich-1 pre-mRNA alternative splicing and apoptosis by mammalian splicing factors. Proc Natl Acad Sci USA 95(16):9155–9160

Jiang W, Kumar JM, Matters GL, Bond JS (2000) Structure of the mouse metalloprotease meprin beta gene (Mep1b): alternative splicing in cancer cells. Gene 248(1–2):77–87

Jin W, Huang ES, Bi W, Cote GJ (1999) Redundant intronic repressor function to inhibit fibroblast growth factor receptor one alpha exon recognition in glioblastoma cells. J Biol Chem 274:28035–28041

Kanopka A, Mühlemann O, Akūsjarvi G (1996) Inhibition by SR proteins of splicing of a regulated adenovirus pre-mRNA. Nature 381:535–538

Kanuka H, Sawamoto K, Inokara N, Matsuno K, Okano H, Mirra H (1999) Control of the cell death pathway by Dapaf-1, a *Drosophila* Apaf-1/CED 4-related caspase activation. Mol Cell 4:757–769

Keene JD (2001) Ribonucleoprotein infrastructure regulating the flow of genetic information between the genome and the proteome. Proc Natl Acad Sci USA 98(13):7018–7024

Kerr JF, Wyllie AH, Currie AR (1972) Apoptosis: a basic biological phenomenon with wide-ranging implications in tissue kinetics. Br J Cancer 26:239–257

Konarska MM (1998) Recognition of the 5′ splice site by the spliceosome. Acta Biochim Pol 45(4):869–881

Korsmeyer SJ, Gross A, Harada H, Zha J, Wang K, Yin XM, Wei M, Zinkel S (1999) Death and survival signals determine active/inactive conformations of pro-apoptotic BAX, BAD, and BID molecules. Cold Spring Harbor Symp Quant Biol 64:343–350

Korte A, Kochling J, Badiali L, Eckert C, Andreae J, Geilen W, Kebelmann-Betzing C, Taube T, Wu S, Henze G, Seeger K (2000) Expression analysis and characterization of alternatively spliced transcripts of human IL-7Ralpha chain encoding two truncated receptor proteins in relapsed childhood ALL. Cytokine 12(11):1597–1608

Krämer A (1996) The structure and function of proteins involved in mammalian pre-mRNA splicing. Annu Rev Biochem 65:367–409

Ladner MB, Martin GA, Noble JA, Nikoloff DM, Tal R (1987) Human CSF-1: gene structure and alternative splicing of mRNA precursors. EMBO J 6:2693–2698

Lavigueur A, La Branche H, Kornblihtt AR, Chabot B (1993) A splicing enhancer in the human fibronectin alternate ED1 exon interacts with SR proteins and stimulates U2 snRNP binding. Genes Dev 7:2405–2417

Liu X, Li P, Widlak P, Zou H, Luo X, Garrard WT, Wang X (1998) The 40-kDa subunit of DNA fragmentation factor induces DNA fragmentation and chromatin condensation during apoptosis. Proc Natl Acad Sci USA 95(15):8461–8466

Lou H, Yang Y, Cote GJ, Berget SM, Gagel RF (1995) An intron enhancer of a 5′ splice site sequence in the human calcitonin/ CGRP gene. Mol Cell Biol 15:7135–7142

Masuoka J, Shiraishi T, Ichinose M, Mineta T, Tabuchi K (2001) Expression of ICAD-l and ICAD-S in human brain tumor and its cleavage upon activation of apoptosis by anti-Fas antibody. Jpn J Cancer Res 92(7):806–812

McCullough AJ, Berget SM (1997) G triplets located throughout clss of small vertebrate introns enforce intron borders and regulate splice site selection. Mol Cell Biol 17:4562–4571

Milbrandt J, de Sauvage FJ, Fahrner TJ, Baloh RH, Leitner ML, Tansey MG, Lampe PA, Heuckeroth RO, Kotzbauer PT, Simburger KS, Golden JP, Davies JA, Vejsada R, Kato AC, Hynes M, Sherman D, Nishimura M, Wang LC, Vandlen R, Moffat B, Klein RD, Poulsen K, Gray C, Garces A, Johnson EM Jr et al. (1998) Persephin, a novel neurotrophic factor related to GDNF and neurturin. Neuron 20(2):245–253

Min H, Chan RC, Black DL (1995) The generally expressed hnRNP F is involved in a neural-specific pre-mRNA splicing event. Genes Dev 9:2659–2671

Min H, Turck CW, Nikolic JM, Black DL (1997) Genes Dev 11:1023–1036

Muro AF, Caputi M, Pariyarath R, Pagani, F, Buratti E, Baralle FE (1999) Regulation of fibronectin EDA exon alternative splicing; possible role of RNA secondary structure for enhancer display. Mol Cell Biol 19:2657–2671

Miyashita T, Shikama Y, Tadokoro K, Yamada M (2001) Molecular cloning and characterization of six novel isoforms of human Bim, a member of the proapoptotic Bcl-2 family FEBS Lett 509(1):135–141

Moore JM, Query CC, Sharp PA (1993) Splicing of precursor to messenger RNAs by the spliceosome. In: Gesteland RF, Atkins JF (eds) The RNA world. Cold Spring Harbor Laboratory Press, Cold Spring Harbor, NY, pp 303–358

Motoyama N, Wang F, Roth KA, Sawa H, Nakayama KI et al. (1995) Massive cell death of immature hematopoietic cells and neurons in Bcl-x deficient mice. Science 267:1506–1510

Manley JL, Tacke R (1996) SR proteins and splicing control. Genes Dev 10(13):1569–1579

Nakano K, Vousden KH (2001) PUMA, a novel proapoptotic gene, is induced by p53. Mol Cell 7(3):683–694

Nicholson DW, Thornberry NA (1997) Caspases: killer proteases. Trends Biochem Sci 22:299–306

Nijhawan D, Honarpour N, Wang X (2000) Apoptosis in neural development and disease. Annu Rev Neurosci 23:73–87

Nunez G, Benedict MA, Hu Y, Inohara N (1998) Caspases: the proteases of the apoptotic pathway. Oncogene 17(25):3237–3245

O'Connor L, Strasser A, O' Reilly LA, Hausmann G, Adams JM, Cory S, Huang DCS (1998) Bim: a novel member of the Bcl-2 family that promotes apoptosis. EMBO J 17:384–395

Oltvai ZN, Milliman CL, Korsmeyer SJ (1993) Bcl-2 heterodimerizes in vivo with a conserved homolog, Bax, that accelerates programmed cell death. Cell 74:609–619

Pan G, O'Rourke K, Chinnaiyan AM, Gentz R, Ebner R, Ni J, Dixit VM (1997) The receptor for the cytotoxic ligand TRAIL. Science 276:111–113

Papin S, Duquesnoy P, Cazeneuve C, Pantel J, Coppey-Moisan M, Dargemont C, Amselem S (2000) Alternative splicing at the MEFV locus involved in familial Mediterranean fever regulates translocation of the marenostrin/pyrin protein to the nucleus. Human Mol Genet 9(20): 3001–3009

Papoff G, Cascino I, Eramo A, Starace G, Lynch DH, Ruberti G (1996) An N-terminal domain shared by Fas/Apo-1 (CD95) soluble variants prevents cell death in vitro. J Immunol 156(12): 4622–4630

Philips AV, Cooper TA (2000) RNA processing and human disease. Cell Mol Life Sci 57(2):235–249

Pret AM, Balvay L, Fiszman MY (1999) Regulated splicing of an alternative exon of beta-tropomyosin pre-mRNA s in myogenic cells depends on the strength of pyrimidine-rich intronic enhancer elements. DNA Cell Biol 18:671–683

Puttaraju M, DiPasquale J, Baker CC, Mitchell LG, Garcia-Blanco MA (2001) Messenger RNA repair and restoration of protein function by spliceosome-mediated RNA trans-splicing. Mol Ther 4(2):105–114

Rasper DM, Vaillancourt JP, Hadano S, Houtzager VM, Seiden I, Keen SL, Tawa P, Xanthoudakis S, Nasir J, Martindale D, Koop BF, Peterson EP, Thornberry NA, Huang J, MacPherson DP, Black SC, Hornung F, Lenardo MJ, Hayden MR, Roy S, Nicholson DW (1998) Cell death attenuation by 'Usurpin', a mammalian DED-caspase homologue that precludes caspase-8 recruitment and activation by the CD-95 (Fas, APO-1) receptor complex. Cell Death Differ 5(4):271–288

Rathmell JC, Thompson CB (1999) The central effectors of cell death in the immune system. Annu Rev Immunol 17:781–828
Reed R (1996) Initial splice site recognition and base pairing during pre-mRNA splicing. Curr Opin Genet Dev 6:215–220
Robinson CJ, Stringer SE (2001) The splice variants of vascular endothelial growth factor (VEGF) and their receptors. J Cell Sci 114(Pt 5):853–865
Ruberti G, Cascino I, Papoff G, Eramo A (1996) Fas splicing variants and their effect on apoptosis. Adv Exp Med Biol 406:125–134
Ryan KJ, Cooper TA (1996) Mol Cell Biol 16:4014–4023
Samejima K, Earnshaw WC (2000) Differential localization of ICAD-L and ICAD-S in cells due to removal of a C-terminal NLS from ICAD-L by alternative splicing. Exp Cell Res 255(2):314–320
Schiavone N, Rosini P, Quattrone A, Donnini M, Lapucci A, Citti L, Bevilacqua A, Nicolin A, Capaccioli S (2000) A conserved AU-rich element in the 3′ untranslated region of bcl-2 mRNA is endowed with a destabilizing function that is involved in bcl-2 down-regulation during apoptosis. FASEB J 14(1):174–184
Screaton GR, Xu XN, Olsen AL, Cowper AE, Tan R, McMichael AJ, Bell JI (1997a) LARD: a new lymphoid-specific death domain containing receptor regulated by alternative pre-mRNA splicing. Proc Natl Acad Sci USA 94:4615–4619
Screaton GR, Mongkolsapaya J, Xu XN, Cowper AE, McMichael AJ, Bell JI (1997b) TRICK2, a new alternatively spliced receptor that transduces the cytotoxic signal from TRAIL. Curr Biol 7:693–696
Shaham S, Horvitz HR (1996) An alternatively spliced *C. elegans* ced-4 RNA encodes a novel cell death inhibitor. Cell 86:201–208
Shav-Tal Y, Cohen M, Lapter S, Dye B, Patton JG, Vandekerckhove J, Zipori D (2001) Nuclear relocalization of the pre-mRNA splicing factor PSF during apoptosis involves hyperphosphorylation, masking of antigenic epitopes, and changes in protein interactions. Mol Biol Cell 12(8):2328–2340
Shaw G, Kamen R (1986) A conserved AU sequence from the 3′ untranslated region of GM-CSF mRNA mediates selection mRNA degradation. Cell 46:659–667
Shiraiwa N, Inohara N, Okaka S, Yuzaki M, Shoji S, Ohta S (1996) An additional form of rat Bcl-x, Bcl-xbeta, generated by an unspliced RNA, promotes apoptosis in promyeloid cells. J Biol Chem 271:13258
Siegmund D, Mauri D, Peters N, Juo P, Thome M, Reichwein M, Blenis J, Scheurich P, Tschopp J, Wajant H (2001) Fas-associated death domain protein (FADD) and caspase-8 mediate up-regulation of c-Fos by Fas ligand and tumor necrosis factor-related apoptosis-inducing ligand (TRAIL) via a FLICE inhibitory protein (FLIP)-regulated pathway. J Biol Chem 276(35):32585–32590
Sirand-Pugnet P, Durosay P, Brody E, Marie J (1995) An intronic (A/U)GGG repeat enhances the splicing of an alternative intron of the chicken beta-tropomyosin pre-mRNA. Nucleic Acids Res 23:3501–3507
Smith CW, Valcarcel J (2000) Alternative pre-mRNA splicing: the logic of combinatorial control. Trends Biochem Sci 25(8):381–388
Smith CW, Patton JG, Nadal-Ginard B (1989) Alternative splicing in the control of gene expression. Annu Rev Genet 23:527–577
Spector MS, Desnoyers S, Hoeppner DJ, Hengartner MO (1997) Interaction between the *C. elegans* cell-death regulators CED-9 and CED-4. Nature 385:653–656
Srinivasula SM, Ahmad M, Guo Y, Zhan Y, Lazebnik Y, Fernandes-Alnemri T (1999) Identification of an endogenous dominant-negative short isoform of caspase-9 that can regulate apoptosis. Cancer Res 59(5):999–1002
Staley JP, Guthrie C (1998) Mechanical devices of the spliceosome: motors, clocks, springs, and things. Cell 92(3):315–326
Sterner DA, Carlo T, Berget SM (1996) Architectural limits on split genes. Proc Natl Acad Sci USA 93:15081–15085

Sun YF, Yu LY, Saarma M, Timmusk T, Arumae U (2001) Neuron-specific Bcl-2 homology 3 domain-only splice variant of Bak is anti-apoptotic in neurons, but pro-apoptotic in non-neuronal cells. J Biol Chem 276(19):16240–16247

Tacke R, Manley JL (1999) Determinants of SR protein specificity. Curr Opin Cell Biol 11:358–362

Talapatra S, Thompson CB (2001) Growth factor signaling in cell survival: implications for cancer treatment. J Pharmacol Exp Ther 298(3):873–878

Tanaka S, Saito K, Reed JC (1993) Structure-function analysis of the Bcl-2 oncoprotein, addition of a heterologous transmembrane domain portions of the Bcl-2b protein restores function as a regulator of cell survival. J Biol Chem 268:10920–10926

Thiede B, Dimmler C, Siejak F, Rudel T (2001) Predominant identification of RNA-binding proteins in Fas-induced apoptosis by proteome analysis. J Biol Chem 276:26044–26050

Thompson CB (1995) Apoptosis in the pathogenesis and treatment of disease. Science 267:1456–1462

Tian Q, Streuli M, Saito H, Schlossman SF, Anderson P (1991) A polyadenylate binding protein localized to the granules of cytolytic lymphocytes induces DNA fragmentation in target cells. Cell 67(3):629–639

Tsujimoto Y, Croce C (1986) Analysis of the structure, transcripts, and protein products of bcl-2, the gene involved in human follicular lymphoma. Proc Natl Acad Sci USA 83:5214–5218

Utz PJ, Hottelet M, Venrooij WJV, Anderson P (1998) Association of phosphorylated serine/arginine (SR) splicing factors with the U1-small ribonucleoprotein (snRNP) autoantigen complex accompanies apoptotic cell death. J Exp Med 187:547–560

Valcarcel J, Green MR. (1996) The SR protein family: pleiotropic functions in pre-mRNA splicing. Trends Biochem Sci 21(8):296–301

Walke DW, Morgan JI (2000) A comparison of the expression and properties of Apaf-1 and Apaf-1L. Brain Res 886(1–2):73–81

Wang J, Manley JL (1997) Regulation of pre-mRNA splicing in Metazoa. Curr Opin Genet Dev 7(2):205–211

Wang L, Miura M, Bergeron L, Zhu H, Yuan J (1994) Ich-1, an Ice/ced-3-related gene, encodes both positive and negative regulators of programmed cell death. Cell 78:739–750

Wang S, Miura M, Jung YK, Zhu H, Li E, Yuan T (1998) Murine caspase-11, an ICE-interacting protease, is essential for the activation of ICE. Cell 92:501–509

Watakabe A, Tanaka K, Shimura Y (1993) The role of exon sequences in splice site selection. Genes Dev 7:407–418

Wei N, Lin CQ, Modafferi EF, Gomes WA, Black DL (1997) A unique intronic splicing enhancer controls the inclusion of the agrin Y exon. RNA 3:1275–88

Whitman S, Wang X, Shalaby R, Shtivelman E (2000) Alternatively spliced products CC3 and TC3 have opposing effects on apoptosis. Mol Cell Biol 20(2):583–593

Wu D, Wallen HD, Nunez G (1997) Interaction and regulation of subcellular localization of CED-4 by CED-9. Science 275:1126–1129

Wu W, Wallen HD, Inohara N, Nunez G (1997) Interaction and regulation of the *Caenorhabditis elegans* death protease CED-3 by CED-4 and CED-9. J Biol Chem 272:21449–21454

Xu R, Teng J, Cooper TA (1993) A unique intronic splicing enhancer controls the inclusion of the agrin Y exon. Mol Cell Biol 13:3660–3674

Yang XF, Weber GF, Cantor H (1997) A novel Bcl-x isoform connected to the T cell receptor regulates apoptosis in T cells. Immunity 7:629–639

Youn BS, Zhang SM, Broxmeyer HE, Cooper S, Antol K, Fraser M Jr, Kwon BS (1998) Characterization of CKbeta8 and CKbeta8-1: two alternatively spliced forms of human beta-chemokine, chemoattractants for neutrophils, monocytes, and lymphocytes, and potent agonists at CC chemokine receptor 1. Blood 91(9):3118–3126

Yuan J, Shaham S, Ledoux S, Ellis HM, Horvitz HR (1993) The *C. elegans* cell death gene ced-3 encodes a protein similar to mammalian interleukin-1 beta-converting enzyme. Cell 75:641–652

Zhang L, Ashiya M, Sherman TG, Grabowski PJ (1996) Essential nucleotides direct neuron-specific splicing of gamma 2 pre-mRNA. RNA 2:682–698

Zhao Q, Schoborg RV, Pintel DJ (1994) Alternative splicing of pre-mRNAs encoding the non-structural proteins of minute virus of mice is facilitated by sequences within the downstream exon. J Virol 68:2849–2859
Zou H, Henzel WJ, Liu XS, Lutschg A, Wang XD (1997) Apaf-1, a human protein homologous to *C. elegans* CED-4, participates in cytochrome c-dependent activation of caspase-3. Cell 90:405–413

Alternative Pre-mRNA Splicing and Neuronal Function

D.L. Black[1] and P.J. Grabowski[2]

The protein output of a gene is often regulated by splicing the primary RNA transcript into multiple mRNAs that differ in their coding exon sequences. These alternative splicing patterns are found in all kinds of genes and tissues. However, in the nervous system, proteins involved in two processes show particularly high levels of molecular diversity created by alternative splicing. These are proteins that determine the formation of neuronal connections during development and proteins that mediate cell excitation. Although some systems of splicing are highly complex, work on simpler model systems has started to identify the molecular components that determine these splicing switches. This review describes how alternative splicing is central to the control of neuronal function, and what is currently known about its mechanisms of regulation. How errors in splicing might contribute to diseases of the nervous system is also discussed.

1 Introduction

The recent genomic description of ~35,000 human genes and the earlier *Drosophila* genome sequence with ~13,600 genes generated much discussion of how a complex organism can be described by so few genes (Adams et al. 2000; Claverie 2001; Consortium 2001; Venter et al. 2001). It is clear that the protein complexity of an organism far outstrips the number of transcription units (Black 2000; Graveley 2001). The most common means of producing multiple proteins from one gene is through the alternative splicing of the gene's pre-mRNA. The processing of a primary gene transcript can be altered in the inclusion of exons, or the position of individual splice sites or polyadenylation sites to produce a variety of transcripts that differ in their encoded polypeptides (Fig. 1). Such mRNA sequence alterations make crucial changes in protein activity that are precisely regulated by the cellular environment. Alternative splicing is particularly common in the mammalian nervous system. Proteins

[1] Howard Hughes Medical Institute, University of California, Los Angeles, MRL 5-748, 675 Charles E. Young Dr. South, Los Angeles, California 90095, USA
[2] Howard Hughes Medical Institute and Department of Biological Sciences, University of Pittsburgh, 4249 Fifth Avenue, Pittsburgh, Pennsylvania 15260, USA

Progress in Molecular and Subcellular Biology, Vol. 31
Philippe Jeanteur (Ed.)

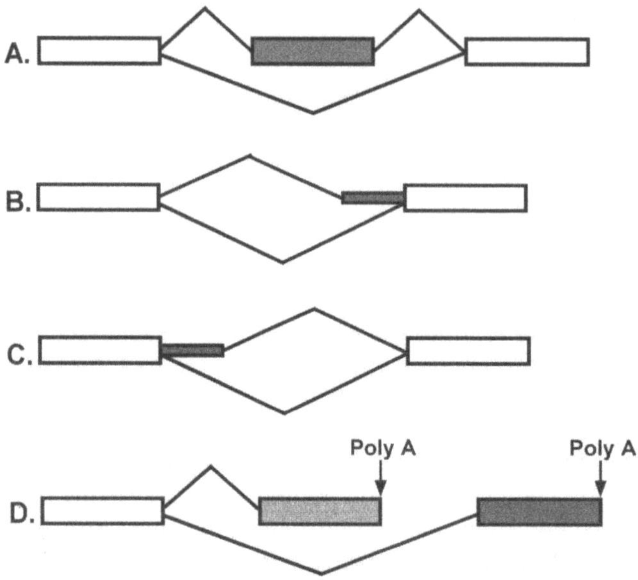

Fig. 1A–D. Four common patterns of alternative splicing. *White boxes* represent constitutive exons present in every mRNA product. *Gray boxes* represent optional sequences that are included or excluded from the mRNA depending on the conditions. Patterns of alternative splicing include: **A** cassette exon; **B** alternative 3′ splice sites; **C** alternative 5′ splice sites; **D** alternative 3′ splice sites combined with alternative polyadenylation sites. These patterns can occur singly or in combination. Other patterns of alternative splicing are not shown

involved in all aspects of neuronal development and function are diversified in this manner. From hundreds of examples, we focus on a few systems where either the regulatory mechanisms or the effects of altered splicing on protein function are better understood.

2 Cell–Cell Interactions and Neuronal Differentiation

The most complex patterns of splicing so far described are in molecules important for the differentiation of neurons and the formation of their intricate connections. The Down's syndrome cell adhesion molecule (DSCAM) and neurexin transcripts show a spectacular variety of splicing patterns numbering in the thousands and have been reviewed elsewhere (Missler and Sudhof 1998; Missler et al. 1998; Black 2000; Grabowski and Black 2001; Graveley 2001). DSCAM plays a role in axon guidance in the developing *Drosophila* nervous system (Schmucker et al. 2000). Neurexin may act later during synaptic differentiation (Scheiffele et al. 2000). N-cadherins are another example of a diverse family of related proteins that are thought to be involved in the formation of neuronal connections (Wu and Maniatis 1999, 2000). How the splicing variation changes the activity of these proteins is not well understood, although an Eph receptor provides an interesting precedent for different spliced forms of a protein mediating cell attraction or repulsion (Holmberg et al. 2000). The role of splicing in altering cell adhesion and synaptic development has been most thoroughly studied in the agrin and neural cell adhesion molecule (NCAM) transcripts.

Agrin is an extracellular matrix protein, identified as an essential mediator of synapse formation at the neuromuscular junction (Bowe and Fallon 1995; Hoch 1999). The clustering of muscle acetylcholine receptors at the neuromuscular synapse is induced by agrin secreted by the neuronal growth cone. Although muscle cells also make agrin, only the neuronal protein has clustering activity. Neuronal agrin mRNA contains several exons absent from muscle-expressed agrin (the Y and Z exons). The Y exon confers heparin binding on the protein, whereas the Z exons (exons 32 and 33) increase the affinity of agrin for α-dystroglycan and confer clustering activity on the protein (Campanelli et al. 1996; Gesemann et al. 1996; Hopf and Hoch 1996; O'Toole et al. 1996). Mice with the Z exons selectively knocked out of the agrin gene are defective in neuromuscular synaptogenesis. In contrast, synaptogenesis is normal when the Y exon is selectively removed (Burgess et al. 1999). Thus, the agrin splicing phenotype of neurons and muscle cells controls synaptic differentiation at the neuromuscular junction.

Neural cell adhesion molecule (NCAM) is thought to be involved both in maintaining cell–cell interactions in the adult nervous system and in stimulating axonal outgrowth during development (Walsh and Doherty 1997). There are numerous spliced isoforms of NCAM (Santoni et al. 1989). Two groups of proteins, NCAM 140 and NCAM 180, differ in the inclusion of a large exon (exon 18) encoding a 266-amino acid intracellular domain. NCAM exon 18 is included in neuronal NCAM, but skipped in muscle and this inclusion is controlled by a number of *cis*-acting RNA regulatory sequences (Tacke and Goridis 1991; Cote and Chabot 1997; Cote et al. 1999). NCAM can stimulate neurite outgrowth in neurons plated over cells expressing particular NCAM isoforms (Walsh and Doherty 1997). This activity is regulated by alternative splicing. NCAM 180, containing the exon 18 domain, is found at sites of cell–cell contact in the adult, and interacts with the cytoskeleton. However, NCAM 180 is poor at stimulating neurite outgrowth compared to isoforms with a small intracellular domain (such as NCAM 140). This outgrowth-promoting activity is also inhibited by another of the many regulated NCAM exons, a short 30-nucleotide exon named variable alternatively spliced exon (VASE). VASE modifies the fourth extracellular immunoglobulin domain of the protein, and may alter a direct contact between the developing axon and the cellular substrate. Thus, NCAM activity is modulated both inside and outside the cell by different alternative exons.

In addition to affecting neuronal interactions during development, the splicing phenotype of a cell also contributes to its identity as a mature neuron. Transcripts from the gene for the transcription factor neuron restrictive silencer factor (NRSF/REST) are spliced to make two different proteins. In non-neuronal cells, the NRSF/REST transcripts encode a transcriptional repressor of neuron-specific genes (Chong 1995; Schoenherr and Anderson 1995). In neurons, an alternative splicing event produces a truncated NRSF/REST protein that apparently releases the NRSF/REST repression (Palm et al. 1998; Shimojo et al. 1999).

The first description of neuronal alternative splicing was in the gene encoding the calcitonin/calcitonin gene related peptide (CGRP) neuropeptides, whose transcripts are spliced to produce the calcitonin mRNA in thyroid and the CGRP mRNA in neurons (Rosenfeld et al. 1983). This system has proven to be an exemplar of numerous alternatively spliced neuropeptide transcripts (Buck et al. 1987; Angers and DesGroseillers 1998; Seong et al. 1999). In *Limnaea*, *Aplysia*, *C. elegans* and other invertebrates, splicing of the FMRFa neuropeptide transcript determines the function of neurons controlling heartbeat, egg laying, copulation and other activities (Burke et al. 1992; Rosoff et al. 1992; Benjamin and Burke 1994; Santama et al. 1995; van Golen et al. 1995). Alternative splicing of the FMRFa transcript switches expression between families of peptides that have different and sometimes antagonistic effects. The production of these splice variants is precisely controlled such that individual neurons within a particular ganglion can express completely different sets of peptides.

3 Neurotransmission: Splicing Control of Neurotransmitter Receptors and Ion Channels

Nearly all the proteins involved in neurotransmission exhibit subtle, but important, functional variation brought about by alternative splicing. The electrical properties of a neuron or other excitable cell are carefully tuned to coordinate many different ion channels and receptors, and to match the properties of cells within a circuit. To meet these needs, the proteins involved in these processes exhibit great diversity. Functional diversity is generated by a variety of mechanisms including extensive alternative splicing. Changes in the splicing of an ion channel can regulate its voltage and ligand dependence, its activation and inactivation kinetics, and its coupling to intracellular signaling pathways. Some of these properties are set during the differentiation of a neuron, while other properties can be altered in a mature neuron in response to particular stimulus.

3.1 AMPA Receptors

Alpha-amino-3-hydroxy-5-methyl-4-isoxalone propionic acid (AMPA)-type glutamate receptors are the major excitatory neurotransmitter receptor in the central nervous system. They are encoded by a multigene family GluR1, GluR2, GluR3, GluR4, GluRA and GluRD. Each of these subtypes is generated in two spliced isoforms containing either the flip or the flop exon. In electrophysiological measurements of recombinant channels, Flop isoforms of GluR2 exhibited faster desensitization kinetics compared to the Flip isoforms (Koike et al. 2000). Single cell studies have correlated AMPA receptor function to splicing pattern in a variety of neuronal cell types. In nonpyramidal neurons of the cortex where AMPA receptors exhibit faster desensitization kinetics, flop isoforms predominate (Lambolez et al. 1996). Conversely, slower GluR desensitization is seen in pyramidal cells, where the flip isoforms are more abundant. The flip and flop variants of AMPA receptors also vary in their sensitivity to the allosteric regulator cyclothiazide (Partin et al. 1995). AMPA receptors were one of the first examples of electrophysiology being regulated by alternative splicing. This has become a very common theme in many different ion channels.

3.2
NMDA R1 Receptor

N-methyl-D-aspartate (NMDA) receptors are another major class of glutamate-gated ion channels important in synaptic plasticity and neuronal development. There are multiple spliced forms of the NMDA R1 receptor (Zukin 1995). The best understood splicing derived changes in NMDA receptor activity result from the regulated inclusion of exons 5 and 21. Exon 5 encodes what is called the N1 peptide cassette within the extracellular domain of the protein. Exon 5 inclusion has complex effects on the physiology of the receptor. The N1 cassette increases the current amplitude and alters receptor potentiation by polyamines and zinc (Zukin 1995; Traynelis et al. 1998; Rumbaugh et al. 2000). Exon 21 encodes what is called the C1 cassette within the intracellular domain of the protein. Exon 22 contains two different 3′ splice sites whose use determines the inclusion of the C2 cassette. C1 and C2 mediate targeting of the receptor to the plasma membrane, and allow the association of the receptor with neurofilament L. C1 also contains a phosphorylation site and a binding site for calcium/calmodulin (Zukin 1995; Hisatsune et al. 1997; Ehlers et al. 1998). There have been several interesting studies of the role of the C1 and C2 cassettes in targeting the receptor to the plasma membrane. C1 contains an ER retention signal that blocks export of the R1 receptor alone to the plasma. If the R1 receptor is assembled with the R2 subunit in the ER, this masks the retention signal allowing expression on the cell surface. Either the exclusion of the C1 cassette by splicing or the inclusion of C2 will allow R1 receptor transport to the synapse without R2 expression. There are also a number of interesting reports of the splicing of these NMDA exons being altered by cell excitation and other stimuli (Musshoff et al. 2000; Rafiki et al. 1998; Vallano et al. 1999; Vallano et al. 1996; Vezzani et al. 1995; Xie and Black 2001). Exons 5 and 21 show divergent patterns of inclusion within the brain. Exon 5 is more strongly included in hind brain structures such as the cerebellum where Exon 21 is low. Conversely, Exon 21 is high in the cortex and Exon 5 is low. The splicing regulator neuroblastoma apoptosis-related RNA-binding protein (NAPOR)1 is implicated in maintaining this pattern by repressing Exon 5 and activating Exon 21 (Zhang et al., submitted).

3.3 Dopamine Receptors

Dopamine receptors are G protein-coupled receptors of diverse subtypes that mediate particular physiological responses to the neurotransmitter dopamine. The D2 and D3 dopamine receptors have multiple spliced isoforms that vary according to cell type (Dal Toso, 1989; Giros et al. 1989; Grandy et al. 1989). Two isoforms of the D2 receptor, D2L and D2S, differentially affect adenylyl cyclase activity and hence downstream signaling (Guiramand et al. 1995). The D2L cassette exon is within the third cytoplasmic loop of the receptor and is thought to mediate the interaction of the receptor with the G protein alpha subunit. Other G protein-coupled receptors show splicing variation that modulates downstream signaling. Notably the pituitary adenylyl cyclase-activating polypeptide (PACAP) receptor is produced in several forms that vary in their ability to stimulate adenylyl cyclase and phospholipase C (Journot et al. 1994).

3.4 GABA Receptors

Synaptic inhibition is often mediated by the neurotransmitter gamma-amino butyric acid (GABA) that binds the $GABA_A$ receptor. The $GABA_A$ receptor is a ligand-gated chloride channel assembled from different combinations of subunits. Alternative splicing of the $\gamma 2$ subunit controls the binding of benzodiazepine agonists to the receptor (Wafford et al. 1993). The $\gamma 2L$ and $\gamma 2S$ forms differ in the inclusion of a short eight amino acid exon within an intracellular loop domain of the protein. $\gamma 2L$ knockout mice, which express only the $\gamma 2S$ form, have increased affinity for benzodiazepines, and show increased sensitivity in behavioral responses to these agonists compared to parent strains (Quinlan et al. 2000). The $\gamma 2L$ exon is thus thought to make the receptor less sensitive to benzodiazepine. This eight amino acid segment also contains a protein kinase C phosphorylation site that is believed to modulate the channel response to ethanol (Whiting et al. 1990; Krishek et al. 1994). The $\gamma 2$ exon has been extensively studied biochemically and is the best understood ion channel exon from the standpoint of its mechanisms of regulation(Zhang et al. 1996, 1999; Ashiya and Grabowski 1997).

4
Voltage-Dependent Potassium and Sodium Channels

Voltage-dependent potassium and sodium channels mediate action potential propagation and determine many aspects of the electrical activity in neurons (Hille 1992; Coetzee et al. 1999). These proteins are encoded by large families of related genes. Alternative splicing within these transcripts generates still more functional diversity. The best understood of these proteins is the Shaker Potassium Channel. *Drosophila* Shaker is differentially processed at both the N and C termini through the use of alternative promoters, splice sites and poly-A sites (Schwarz et al. 1988; Sutcliffe and Milner 1988). Shaker channels in other organisms show additional sites of alternative splicing (Kim et al. 1997). Since Shaker protein exists as a tetramer, the combination of variants produces a wide variety of functionally distinct channels. In *Drosophila* Shaker, the variable processing leads to distinct inactivation kinetics for the channel. Fast inactivation kinetics are altered by the choice of amino terminal sequences. The different carboxyl termini create differences in the rate of recovery from inactivation (Iverson et al. 1988; Timpe et al. 1988; Hoshi et al. 1990, 1991; Iverson and Rudy 1990).

That differential splicing is responsible for the different currents observed by electrophysiology was shown for the C-terminal variants using an elegant reporter system (Mottes and Iverson 1995; Iverson et al. 1997). The alternatively spliced segments of Shaker were fused to LacZ, such that β-galactosidase (β-Gal) would be expressed only if certain splicing patterns were used. These reporter genes were expressed in transgenic flies and the flies were stained for β-Gal. It was found that the choice of C-terminal splicing pattern, as reported by tissue staining, correlated with the electrophysiology for each tissue. Similar splicing reporters have been used to good effect in other systems (Adams et al. 1997). Mammalian potassium channels also show functional variation derived from alternative splicing (Liu and Kaczmarek 1998).

Similar to potassium channels, sodium channels show extensive alternative splicing in both vertebrates and *Drosophila*. An extreme example is the channel encoded by the *Drosophila* Paralytic (Para) gene whose transcripts encode over a thousand different polypeptides through alternative splicing and many more through RNA editing events (Thackeray and Ganetzky 1994, 1995; Graveley 2001). Genetic studies have uncovered modifiers of Para splicing including one fascinating gene, mle^{napte}, encoding an RNA Helicase, whose mutation has drastic effects on several Para exons(Reenan et al. 2000). For Shaker, Para and other potassium and sodium channels, electrophysiology is intricately adjusted through the differential use of numerous spliced isoforms. These genetic studies are promising approaches both for uncovering the regulators of this splicing and to understand the roles of individual isoforms.

4.1
Calcium Channels

Calcium channels are multi-subunit voltage-gated channels that play diverse cellular roles regulating calcium-dependent genes, controlling firing patterns, and affecting neurotransmitter release at the synapse. The N- and P/Q-type channels function, in part, to couple cellular excitation to neurotransmitter release. The α_1 subunit of the channel forms the ion pore and is encoded in multiple genes. Two extracellular loops of the α_{1B} subunit, found in N-type channels, show variation due to alternative splicing. In one loop, insertion of a small cassette exon encoding just two amino acids, ET (glutamic acid, threonine), confers slower activation kinetics on the channel (Lin et al. 1999). The inclusion of the ET exon is tissue-specific. In the central nervous system, most of the α_{1B} subunits lack the ET exon, whereas the predominant form in sympathetic and sensory ganglia contains the ET segment (Lin et al. 1997). P- and Q-type channels apparently both contain the α_{1A} subunit, but differ in their physiology due to cell type-specific changes in the splicing of the α_{1A} transcript (Beam 1999; Bourinet et al. 1999). The inclusion of a single valine in the domain I-II linker of the Q channel gives it slower inactivation kinetics than the P channel. The inclusion of the dipeptide NP in domain 4 of α_{1A} makes the Q channel less sensitive to ω-agatoxin IVA than the P channel. Changes in splicing also alter the regulation of the channel by G proteins. The calcium channel illustrates how very small peptide cassettes, sometimes encoding a single amino acid, can determine important changes in protein activity.

4.2
Calcium and Voltage-Activated Potassium Channels

The Ca^{2+} and voltage-activated K^+ channel (also called BK or *slo* channel) provides an example of how the controlled production of subtly different spliced forms of a protein is important for modulating the function of a particular cell type (Vergara et al. 1998; Black 1998; Jones et al. 1999a). Potassium gating by BK channels requires both a voltage change across the membrane and the presence of intracellular calcium. These channels contribute to shaping the repolarization phase of action potentials in many cell types (Lingle et al. 1996; Coetzee et al. 1999). More than 500 different BK transcripts are potentially generated by the combinatorial inclusion of numerous cassette exons. These different exon cassettes specify changes in the channel physiology by altering its calcium or voltage dependence, its conductance, or other properties. Changes in BK channel kinetics control, in part, the tuning of hair cells in the inner ear to transduce particular sound frequencies (Black 1998; Fettiplace 1999). Several groups have shown that individual hair cells along the tonotopic gradient of the Cochlea express different repertoires of BK channel splice variants (Navaratnam et al. 1997; Rosenblatt et al. 1997; Jones et al. 1999b). This splicing allows hair cells tuned for low or high frequency responses to have the channels appropriate for these frequencies. BK channels have also been extensively studied in adrenal chromaffin cells. In these cells, specific changes in excitation induced by stress hormones were shown to derive from BK channel alternative splicing (Xie and McCobb 1998). The hormonally activated inclusion of the stress axis regulated exon (STREX) exon increases the sensitivity of the channel to calcium and leads to a higher rate of action potential spiking in the chromaffin cells (Solaro et al. 1995). It seems likely that, similar to hair cells, the tuning of action potential kinetics in neurons will also involve the controlled expression of particular splice variants for multiple ion channel proteins.

5 Mechanisms of Alternative Splicing Control

The molecular mechanisms that control changes in splice site choice are only beginning to emerge (Lopez 1998; Smith and Valcarcel 2000). We will discuss a few model systems directly related to the nervous system. Additional model systems are discussed in other chapters of this volume. Many transcripts undergo changes in splicing during neuronal development or are spliced differently in different types of neurons. In addition to these changes with development or cell type, some splicing patterns are regulated in mature cells in response to particular stimuli, and may vary from cell to cell even within a defined population of neurons (Thomas et al. 1993; Wang and Grabowski 1996). Thus, the splicing apparatus must be able to respond to myriad signals in controlling production of the thousands of alternatively spliced transcripts.

Each intron in a pre-mRNA is excised by a particle called the spliceosome that must assemble onto the splice sites, pair them across the intron, and finally catalyze the intron excision and exon ligation. Splice sites adhere to defined consensus sequences: the 5′ splice site at the 5′ end of the intron and the 3′ splice site with its associated branch point and polypyrimidine tract at the 3′ end of the intron (Burge et al. 1999). The spliceosome is made up of five small nuclear ribonucleoproteins (snRNPs), U1, U2, U4, U5, and U6 and a number of accessory proteins. Spliceosome assembly is a stepwise process requiring the sequential binding and disruption of multiple RNA and protein contacts (Staley and Guthrie 1998). Alternative splicing patterns result from changes in the assembly of spliceosomes. These changes are thought to be primarily regulated during the early steps of spliceosome assembly, such as the recognition of the 5′ splice site by the U1 snRNP and the 3′ splice site by the U2 auxiliary factor U2AF. Splicing patterns can be altered in a variety of ways (Lopez 1998). It is clear that the transcription reaction itself, as well as the secondary structure of the pre-mRNA, can both affect splice site choice. However, here we will discuss the best understood systems of neuronal splicing regulation where the hnRNP structure of the mRNA precursor seems to be the primary determinant of the change in splicing.

RNA sequence elements in the pre-mRNA can direct changes in splicing pattern by activating or repressing nearby spliceosome assembly (Fig. 2). Regulatory sequences that activate splicing at particular sites are called exonic or intronic splicing enhancers, depending on where they are located. There are also splicing repressor elements, or silencers, that inhibit splicing. The tissue-specific splicing of a transcript is often under a combination of positive and negative control (Zhang et al. 1996; Modafferi and Black 1999; Smith and Valcarcel 2000). These regulatory elements are bound by sequence-specific RNA binding proteins that induce changes in spliceosome assembly. Regulatory elements and proteins have been identified in a number of systems. However, there is limited understanding of how they actually work.

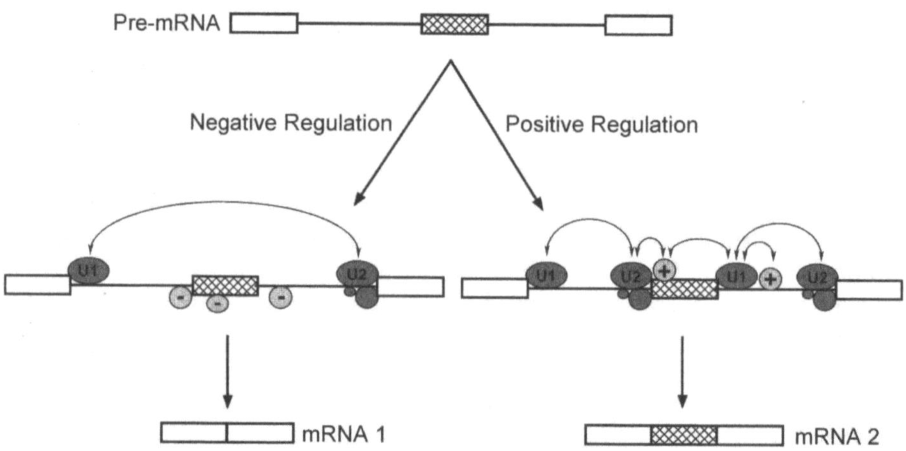

Fig. 2. Positive and negative regulation of splicing. *Above* The introns and constitutive (*white*) or regulated (*hatched*) exons of a pre-mRNA are defined by where the spliceosome assembles. During early spliceosome assembly, the U1 snRNP binds the 5′ splice site of an intron and the U2 snRNP, with the auxiliary proteins U2AF and SF1, binds to the branch point/3′ splice site region. The binding of these initial components and the assembly of the early spliceosome complexes is thought to define the intron to be excised. The spliceosome can assemble on a single large intron to form a two exon mRNA (exon skipping or exclusion, *bottom left*). Alternatively, spliceosomes can assemble on each of two smaller introns, to include a new exon in the mRNA (exon inclusion, *bottom right*) Negative regulatory proteins, such as sxl, PTB or hnRNP A1 (*light gray, left pathway*), can block spliceosome assembly at particular sites by preventing U1 or U2 binding or by preventing later snRNP assembly. Positive regulatory proteins, such as SR proteins or TIA-1 (*light gray, right pathway*), enhance spliceosome assembly at sites that are otherwise recognized poorly by the general splicing factors. Systems of alternative splicing are usually controlled by a combination of positive and negative regulatory proteins. Although not shown here, splicing patterns can also be affected by factors such as RNA secondary structure and transcription rate. (Adapted from Black 2000)

5.1 Splicing Repression

We will focus on two mRNAs exhibiting neural-specific splicing patterns that have been analyzed biochemically. In most neurons, transcripts for the tyrosine kinase c-src include an exon (N1) that is skipped in nonneuronal tissue. In contrast, a regulated exon in the $GABA_A$ receptor $\gamma 2$ is spliced differentially between different neuronal subsets.

Regulated exons often have features, such as short exon length and weak splice sites, that prevent their efficient recognition by general splicing factors and thus contribute to the general repression of these exons. Reducing the length of a normally constitutive exon below ~50 nucleotides can cause exon skipping (Dominski and Kole 1991). This is believed to result from the loss of exon bridging interactions that promote splice site recognition in a process called exon definition (Berget 1995). Many regulated exons including the c-src N1 exon (18 nucleotides) and the $GABA_A$ receptor $\gamma 2$ exon (24 nucleotides) are well below this length. When the N1 exon is lengthened to ~100 nucleotides it becomes constitutively spliced (Black 1991). The splice sites in the $GABA_A$ receptor $\gamma 2$ exon are also suboptimal, and mutations that strengthen the 5′ splice site result in efficient splicing in nonneural cells (Zhang et al. 1996). Similar effects of exon length and splice site strength have been seen in many other regulated exons (Smith and Valcarcel 2000).

Another common feature of c-src and $GABA_A$ splicing, as well as most other characterized systems of tissue-specific splicing in mammals, is the involvement of the polypyrimidine tract binding protein (PTB; also called hnRNP I; Valcarcel and Gebauer 1997; Wagner and Garcia-Blanco 2001). PTB contains four RRM domains (Varani and Nagai 1998; Conte et al. 2000), and binds to pyrimidine-rich RNA elements that act as splicing repressor or silencer elements (Valcarcel and Gebauer 1997; Wagner and Garcia-Blanco 2001). Neuronally regulated exons of c-src, $GABA_A$ receptor $\gamma 2$, clathrin light chain B (CLCB) and NMDA R1 exon 5 all contain PTB binding repressor elements within the polypyrimidine tract of their 3′ splice sites (Chan and Black 1995; Ashiya and Grabowski 1997; Zhang et al. 1999). Regulated exons in nonneural systems also can have PTB binding sites in their polypyrimidine tracts (Mulligan et al. 1992; Valcarcel and Gebauer 1997; Gooding et al. 1998; Wagner and Garcia-Blanco 2001). The region surrounding a regulated exon often contains additional PTB binding sites. Mutating these binding sites or depleting PTB from an in vitro splicing system generally activates splicing of a repressed exon. PTB thus seems to be required for splicing repression.

Different arrangements of PTB binding sites has led to two models for PTB-mediated splicing repression (Fig. 3). For the $GABA_A$ receptor $\gamma 2$ exon, PTB binds to an extended polypyrimidine tract starting about 100 nucleotides upstream of the 3′ splice site and extending through the exon. Since the branch site is embedded in this polypyrimidine tract, it is thought to be the target of negative regulation by PTB.

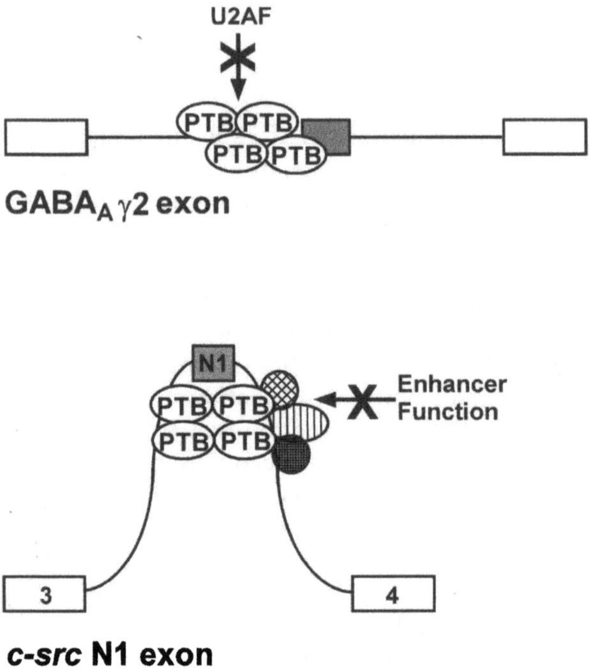

Fig. 3. Models for PTB-mediated splicing repression. In the GABA$_A$ γ2 transcript, multiple PTB proteins are thought to bind tightly to the 3′ splice site region and block spliceosome assembly on the alternative exon. For the c-src N1 exon, PTB binding sites on both sides of the exon are required for splicing repression. These are thought to block the function of an intronic splicing enhancer that surrounds the downstream PTB sites, as well as block the upstream 3′ splice site. In both cases, PTB binding must be released in neuronal cells to allow splicing to proceed

Repression of the c-src N1 exon is dependent upon distributed PTB binding sites on both sides of the N1 exon (Chan and Black 1995, 1997; Chou et al. 2000). Mutating the binding sites downstream of the exon reduces PTB binding to the upstream 3′ splice site and activates splicing in vitro. The PTB molecules binding to these separated sites are thought to interact and sequester the N1 exon in a loop. However, the contacts that bridge this loop are not clear. Interestingly, the downstream binding sites for PTB are located within an intronic splicing enhancer (see below). One possibility is that the arrangement of PTB sites is needed to prevent splicing activation by this enhancer. Although they have not yet been shown to bind PTB, similar pyrimidine-rich repressor elements control the splicing of neuronally regulated exons in β amyloid pre-mRNA (Shibata et al. 1996; Bergsdorf et al. 2000).

Several muscle-specific exons are also repressed by PTB in a similar manner to src N1 and GABA$_A$ γ2 (Valcarcel and Gebauer 1997; Smith and Valcarcel 2000). These two patterns of repression by PTB are also reminiscent of models for splicing repression by the *Drosophila* sxl protein (Lopez 1998). In repressing splicing of the tra transcript, sxl appears primarily to block a 3′ splice site, whereas for an exon in the *sxl* transcript itself, sxl protein binds on both sides of the exon to repress its splicing.

An important question for exons repressed by PTB is how the tissue specificity of splicing is generated. The exons in the $GABA_A$ $\gamma 2$ receptor and c-src are repressed by PTB in nonneural cells. There are several skeletal muscle-specific exons that are repressed by PTB in nonmuscle tissues. One common feature in these systems is that PTB is thought to maintain repression of a highly tissue-specific exon outside one particular tissue. Thus, the tissue specificity of splicing apparently derives from the controlled loss of PTB repression.

PTB is widely expressed, although there are tissue-specific differences in its expression. In most cell lines (HeLa, for example), PTB is expressed as several spliced isoforms. These isoforms have different activity in some assays and their ratio changes in different cell lines (Christopher Smith, pers. comm.; Wagner et al. 1999). In nuclear extracts from rat brain (Ashiya and Grabowski 1997) and a neural cell line (Chan and Black 1997), these isoforms are reduced, and a PTB-like protein is present, termed neural- or brain-enriched PTB (nPTB or brPTB; Ashiya and Grabowski 1997; Chan and Black 1997; Markovtsov et al. 2000; Polydorides et al. 2000). nPTB shows reduced splicing repression activity on the N1 exon (Markovtsov et al. 2000). Interestingly, a switch in PTB and nPTB expression occurs during rat cerebellar development that correlates with the loss of splicing repression in several transcripts (Zhang et al. 1999).

Although changes in PTB expression likely contribute to the release of splicing repression, they are not the only mechanism. Some cell lines with abundant PTB, and little nPTB, can splice the c-src N1 exon (Markovtsov et al. 2000). There must be additional factors that affect splicing repression by PTB. In a neural extract where PTB does not repress c-src N1 splicing, PTB is present and binds to the N1 3′ splice site in the absence of ATP (Chou et al. 2000). However, this PTB binding is lost when ATP is added to initiate the splicing reaction. In HeLa extracts, where N1 is repressed, PTB remains bound even in the presence of ATP. This ATP dependence of PTB dissociation may imply an active removal process where the protein is specifically detached from particular sites to allow splicing to proceed (Chou et al. 2000).

Different splicing silencer elements have been found in other systems. The hnRNP A1 protein was identified as a splicing repressor that binds to elements within the repressed exon of several transcripts (Chabot et al. 1997; Caputi et al. 1999; Tange et al. 2001). Similar results are seen with the hnRNP H protein repressing several exons (Chen et al. 1999; Jacquenet et al. 2001). SnRNPs are also implicated in the negative control of splicing through exonic repressor elements (Siebel et al. 1994; Kan and Green 1999). Although these systems do not show neuronal regulation, these factors are also likely to affect exons in neuronal transcripts.

5.2 Splicing Enhancement

Splicing patterns are also positively regulated through splicing enhancers. Exonic enhancers are diverse in sequence, but the best characterized are purine-rich elements with GAR (R = purine) triplet repeats. GAR elements and other exonic enhancers are bound by proteins of the SR family (Lopez 1998). These proteins contain one or more RRM-type RNA binding domains joined to a carboxy terminal SR domain containing repeated serine-arginine dipeptides. SR proteins are needed for constitutive splicing, and also play important roles in alternative splicing (for reviews, see Manley and Tacke 1996; Graveley 2000). The best-studied enhancer is in the *Drosophila* doublesex transcript, where the specific regulatory proteins tra and tra-2 cooperate with SR proteins to assemble an enhancer complex on the regulated exon. The tethering of the SR domain to the exon stimulates spliceosome assembly on the upstream 3′ splice site (Graveley and Maniatis 1998). This induced assembly is thought to be due to interactions between the SR domain of the SR protein and similar domains on the U2AF heterodimer. Phosphorylation of the SR domain also regulates its biochemical properties (Colwill et al. 1996; Du et al. 1998; Xiao and Manley 1998; Prasad et al. 1999; Yeakley et al. 1999; Yun and Fu 2000). Through the interactions of the effector (SR) domain with spliceosomal components, SR proteins are thought to be mediators of a range of splicing choices (Graveley 2000).

Individual enhancers require specific members of the SR family and thus, tissue-specific variation in SR protein expression can affect enhancer activity (Hanamura et al. 1998). For example, levels of ASF/SF2, an SR protein that promotes activation of proximal 5′ splice sites, vary considerably relative to the protein hnRNP A1, which promotes activation of distal 5′ splice sites (Mayeda et al. 1993; Caceres et al. 1994). These proteins have mostly been studied in nonneuronal transcripts. However, SR proteins as well as the mammalian homologue of Tra-2 are present in neurons, and many neuronal splicing events will involve exonic enhancers controlled by these proteins (Beil et al. 1997).

In addition to exonic enhancers, intronic splicing enhancers have been identified downstream of numerous regulated exons (Black 1992; Huh and Hynes 1994; Sirand-Pugnet et al. 1995; Carlo et al. 1996; Ryan and Cooper 1996; Modafferi and Black 1997; Del Gatto-Konczak et al. 2000; Wei et al. 1997). Intronic enhancers are not known to bind SR proteins, and may function differently from exonic enhancers. Several small sequence elements have been shown to have intronic enhancer activity, including a G-rich element found in the c-src, cardiac troponin T, α-globin, and β-tropomyosin transcripts (Sirand-Pugnet et al. 1995; Carlo et al. 1996; Modafferi and Black 1997; McCullough and Berget 2000), and a UGCAUG element in the c-src, fibronectin, nonmuscle myosin II, and calcitonin/CGRP transcripts (Black 1992; Huh and Hynes 1994; Kawamoto 1996; Hedjran et al. 1997; Modafferi and Black 1997). A bioinformatic analysis of the intron sequences downstream from a group of neuronally regulated exons found that UGCAUG was the most common hexamer in these intron regions indicating its likely involvement in further splicing events (Brudno et al. 2001).

The complex intronic enhancer downstream of the c-src N1 exon is required for N1 splicing in vivo, and is relatively nontissue-specific, activating splicing in many cell types (Modafferi and Black 1997, 1999). The tissue specificity of N1 splicing stems from the interaction of the enhancer with the system of PTB-mediated repression (see above). The most conserved core of the src N1 enhancer (called the DCS) contains a GGGGG element (G5) and a UGCAUG element flanking a PTB/nPTB binding site. In splicing active neural extracts, the DCS assembles a complex of proteins including hnRNPs F and H bound to G5, nPTB bound to CUCUCU, and the KH-type splicing regulatory protein (KSRP) bound to the UGCAUG (Min et al. 1995, 1997; Chou et al. 1999; Markovtsov et al. 2000). The roles for these proteins in splicing enhancement are not yet clear. Antibody inhibition experiments imply a positive role for hnRNPs F and H and KSRP in N1 splicing. However, their interaction with PTB may indicate a function in repression or derepression rather than true splicing enhancement. Reports of G-rich and UGCAUG elements activating splicing in vitro is encouraging for identifying the proteins that mediate their enhancement activity (Carlo et al. 2000, 1996; Guo and Kawamoto 2000).

Other intronic enhancer proteins were also recently identified. A study of a regulated intron in the *Drosophila* msl2 transcript identified the protein TIA-1 as binding to an intronic enhancer in HeLa extract and promoting U1 snRNP binding to the upstream 5′ splice site (Forch et al. 2000). This work is important since this is the only intronic enhancer binding protein shown to affect spliceosome assembly. TIA-1 is also involved in FGF receptor 2 alternative splicing (Del Gatto-Konczak et al. 2000). Interestingly, TIA-1 is a mammalian protein involved in apoptosis and has homologues in *Drosophila* (Rox-1) and Yeast (NAM8).

Another significant group of intronic enhancer binding factors are the CELF proteins (CUG binding proteins and ETR3-like factors; Ladd et al. 2001). Several members of this family, including CUG binding protein (CUG-BP), enhance cardiac troponin T exon 5 inclusion when overexpressed in vivo (Philips et al. 1998). The developmental regulation of these proteins is proposed to mediate control of muscle-specific splicing events. CUG binding protein also binds to the intronic repressor region of the neural pre-mRNAs CLCB and NMDA R1, although its functional effect is not known (Zhang et al. 1999). A neuronal member of this family, NAPOR1, was recently shown to have strong effects on the splicing of NMDA R1 exons 5 and 21. Interestingly, NAPOR1 represses exon 5; but enhances exon 21 splicing. Thus, these proteins are likely to play important roles in diverse neuronal splicing events(Zhang et al., submitted).

A puzzling question for all the analyzed mammalian systems of splicing is how the tissue specificity is generated (Smith and Valcarcel 2000). The characterized muscle and neuron-specific exons all show multiple features of positive and negative control. For example, the src N1 exon is generally repressed by its short length, generally activated by multiple factors affecting both the intronic enhancer and an exonic enhancer, and finally; specifically repressed in nonneural cells by PTB. Thus, the control of N1 is the product of a highly combinatorial system of inputs, but the key to its tissue specificity seems to be the loss of PTB repression in neural cells (Modafferi and Black 1999). Most other systems of splicing regulation show similar combinations of relatively nontissue-specific factors affecting the level of each splicing pattern (Smith and Valcarcel 2000).

5.3 Neuronal RNA Binding Proteins

Although splicing patterns are at least partly controlled by combinations of relatively unspecific factors, there are some RNA binding proteins that show highly neural-specific expression. The *Drosophila* embryonic lethal abnormal visual system (*elav*) gene encodes a neuron-specific RNA binding protein implicated in the regulation of alternative splicing. When *elav* is genetically depleted from *Drosophila* photoreceptor neurons, a decrease is observed in a spliced variant of neuroglian (Koushika et al. 1996). Conversely, this neural form of neuroglian is generated ectopically when *elav* is expressed in nonneural cells. Similar effects on the splicing of *Drosophila erect wing* transcripts are observed later in development (Koushika et al. 1999). ELAV was recently shown to directly bind the neurologin transcript (Lisbin et al. 2001). It will be very interesting to determine how it interacts with other regulators and the spliceosome itself.

In mammals there are numerous *elav* family members, some of which are specifically expressed in neurons (Chung et al. 1996; Gao 1996). These proteins are most strongly implicated in the control of mRNA stability through their recognition of 3′ untranslated regions (Antic and Keene 1997, 1999; Chung et al. 1997; Fan and Steitz 1998; Peng et al. 1998; Ford et al. 1999). However, given that some *elav* proteins are cytoplasmic while others predominantly nuclear, it seems possible that they will serve a variety of roles in mRNA metabolism.

Two other highly neuron-specific RNA binding proteins are Nova-1 and Nova-2. These are closely related proteins of the hnRNP K homology type (KH type), first identified as autoantigens in patients with paraneoplastic neurological disease (Darnell 1996). The Nova proteins are enriched in the cell nucleus of neurons, and like other hnRNP proteins may be involved in multiple processes. However, Nova-1 clearly effects alternative splicing. Nova-1 interacts with nPTB (brPTB) in the yeast two hybrid assay, and recognition elements for both nPTB and Nova-1 are located adjacent to a regulated exon in the glycine receptor pre-mRNA (GlyRα2; Buchanovich 1997; Jensen et al. 2000b;

Polydorides et al. 2000). Nova-1 knockout mice exhibit postnatal lethality due to the loss of spinal and brainstem neurons. This is accompanied by changes in the splicing of both the GlyRα2 exon and the regulated $GABA_A$ receptor γ2 exon (Jensen et al. 2000a). The partial effect of the knockout suggests that Nova-1 is one of several factors that affect these exons, as might be predicted from the multiple factors implicated in other systems. This genetic knockout approach offers an important route for dissecting these highly combinatorial systems of splicing control (Wang et al. 1996; Wang et al. 2001).

In a more refined genetic system, the mec-8 gene product of *C. elegans* was defined as a splicing regulator (Lundquist and Herman 1994; Lundquist et al. 1996). Mec-8 encodes an RNA binding protein with two RRM domains. Mutations in Mec-8 reduce the abundance of several spliced isoforms of UNC52, a proteoglycan important for body wall muscle function. Mec-8 mutations are pleiotropic and also affect the function of mechanosensory and chemosensory neurons. It will be very interesting to determine the role of Mec-8 in regulating splicing in these cells.

5.4 Splicing and Cell Excitation

How signal transduction pathways affect splicing patterns is an area of great interest, but little data. The level of many spliced transcripts is altered by particular extracellular stimuli (Wang et al. 1991; Collett and Steele 1993; Shifrin and Neel 1993; Chalfant et al. 1995; Zacharias and Strehler 1996; Smith et al. 1997; Rodger et al. 1998;; Berke et al. 2001). In cases where the change in transcript occurs rapidly, a change in splicing must be distinguished from a change in the decay rates of the different spliced forms. In two systems (CD44 and CD45), specific RNA elements within the regulated exon are needed for the splicing to respond to protein kinase C and Ras signaling. (Konig et al. 1998; Lynch and Weiss 2000). The activation of CD44 exon V5 upon T-cell stimulation requires the MEK (mitogen-activated protein kinase K/ERK kinase) and ERK (extracellular signal-regulated kinase) kinases (Weg-Remers et al. 2001). However, the complete path from an initial stimulus to the splicing apparatus is not clear. Significantly, it was recently shown that the protein hnRNP A1 is controlled in its nucleo/cytoplasmic localization by the MKK-p38 signaling pathway (van der Houven van Oordt et al. 2000). Since hnRNP A1 affects the splicing of multiple transcripts, this control of splicing factor localization should prove to be important in determining splicing patterns. Although the direct modification of splicing factors or the induction of splicing regulator genes are also likely mechanisms.

The splicing of NMDA receptors and of BK potassium channels is known to be altered by cell activity (Daoud et al. 1999; Vallano et al. 1999). It was recently shown that depolarization of GH3 pituitary cells and signaling by CaM Kinase IV represses splicing of both the BK channel STREX exon and NMDA exons 5 and 21 (Xie and Black 2001). Particular sequence elements surrounding the STREX exon respond to the CaMK IV signal. These depolarization-induced changes in the BK channel and NMDA R1 proteins will alter the subsequent excitatory properties of the stimulated cells. Thus, it will be interesting to look for effects of CaMK IV signaling on many other ion channel and neurotransmitter receptor transcripts.

The role of signal transduction pathways in controlling splicing is particularly interesting in neuronal systems, where learning and memory, as well as cellular homeostasis, depend on activity-dependent changes in cell excitation. The mechanisms that control these changes are known to include alterations in transcription and translation (Finkbeiner and Greenberg 1998; Lisman and Fallon 1999). However, nearly all the receptor and channel molecules that determine cell electrical excitation are also regulated by alternative splicing. Thus, altering the splicing of these transcripts in response to electrical activity provides an additional mechanism for inducing long term changes in excitability.

5.5 Alternative Splicing and Neurological Disease

The importance of alternative splicing to gene expression in the nervous system is highlighted by several forms of neurological disease caused by splicing misregulation or other errors of RNA metabolism. There are many examples of mutations in splice sites or splicing enhancer or silencer elements that disrupt constitutive splicing patterns (Krawczak et al. 1992; Nakai and Sakamoto 1994; O'Neill et al. 1998; Liu et al. 2001). There are also disease pathologies where the regulation of splicing is known or suspected to be in error (Cooper 1997; Philips and Cooper 2000). Several systems provide examples of how seemingly subtle changes in a ratio of spliced isoforms can have disastrous consequences. Mutations in the microtubule associated protein Tau cause the disorder frontotemporal dementia with parkinsonism linked to chromosome 17 (Wilhelmsen 1999). Spinal muscular atrophy (SMA) is a common congenital disorder resulting from mutations in the survival of motor neurons (SMN) gene (Lefebvre et al. 1998; Mattaj 1998; Matera 1999; Sendtner 2001). Both of these systems involve changes in exon inclusion and have been discussed extensively elsewhere (Lefebvre et al. 1998; Wilhelmsen 1999; Grabowski and Black 2001; Sendtner 2001).

Myotonic dystrophy (MD) provides an interesting picture of how a mutation in a gene unrelated to splicing might lead to a dominant and generalized effect on splicing regulation. MD is an inherited neuromuscular disease exhibiting a complex pathology in both skeletal and smooth muscle and in other systems (Timchenko 1999). MD is caused by a CTG triplet expansion mutation in the 3′ UTR of the myotonic dystrophy protein kinase (DMPK) protein kinase gene. The CTG expansion reduces the expression of DMPK and other genes in adjacent chromatin, and this may lead to portions of the complex disease phenotype (Alwazzan et al. 1999). However, the disease shows autosomal dominant inheritance and this is thought to result from the presence of the expanded CUG repeat in the expressed DMPK RNA. In MD cells, the mutant DMPK RNA becomes aggregated in cell nuclei (Taneja et al. 1995). DMPK is alternatively spliced and one set of spliced mRNAs lacking the CUG repeats is transported to the cytoplasm normally (Groenen et al. 2000; Tiscornia and Mahadevan 2000). Thus, the CUG expansion will change the ratio of expressed DMPK isoforms and this could, in part, explain the dominant phenotype. However, transgenic mouse experiments argue for a dominant effect from the CUG repeat RNA alone (Mankodi et al. 2000). When the CUG repeat is expressed in the muscle of transgenic mice from a transcript unrelated to DMPK, the mice develop a disease very similar to MD. This is strong evidence for an "RNA gain of function" basis for the disease. The nuclear RNA accumulation is thought to sequester RNA binding proteins, including CUG-BP, hnRNP C, PTB, U2AF, PTB-associated splicing factor (PSF), and Muscleblind, thus changing their effective concentration in the cell (Timchenko et al. 1996; Miller et al. 2000; Tiscornia and Mahadevan 2000). CUG-BP is a ubiquitous hnRNP protein identified by its binding to CUG repeat RNA, although its preferred RNA binding site may be a somewhat different sequence. CUG-BP is normally distributed between nucleus and cytoplasm, but in DM cells a hypophosphorylated form of the protein accumulates in the nucleus (Roberts et al. 1997). CUG-BP is known to affect splicing of exon 5 in cardiac troponin T (cTNT), and in DM muscle cells, the inclusion of cTNT exon 5 is increased (Philips et al. 1998). CUG-BP also affects the translation initiation of C/EBPB mRNA and this process is also altered in DM cells (Timchenko et al. 1999). However, it is not clear how the activity of CUG-BP might be altered by the DM mutation. Troponin exon 5 is thought to be positively regulated by CUG-BP and so the increase in exon 5 splicing is not easily explained by the sequestration of the protein by the CUG repeats. Instead, the protein may be altered in its modification or localization. Moreover, other proteins binding to CUG repeats such as Muscleblind could be altered in activity (Miller et al. 2000). Nevertheless, the genetic dominant nature of the disease is nicely explained by an RNA gain of function, something that is likely to be seen again in other triplet expansion diseases, many of which show neurological manifestation (Imbert et al. 1996; Pulst et al. 1996; Sanpei et al. 1996).

The currently known diseases that are caused by errors in splicing regulation are likely to be the first of many. Tau and SMN provide demonstrations of how changes in spliced isoform ratios can have highly deleterious consequences in specific cell types. MD points to how triplet expansion mutations can have indirect effects on splicing. There are a number of studies associating changes in alternative splicing with various psychiatric disorders (Huntsman et al. 1998; Le Corre et al. 2000; Vawter et al. 2000; Weiland et al. 2000), although these changes may indirectly result from some other dysfunction. We clearly need to know more about how the vast network of splicing regulatory machinery contributes to function both in individual cells and at the level of integrating larger systems.

6 Conclusions

We discuss here only a tiny portion of the huge number of important neuronal and glial cell proteins that exhibit multiple splice variants. It is clear that alternative splicing plays a central role in controlling neuronal development and function. However, the most interesting questions remain unanswered. How does protein diversity from alternative splicing contribute to the diversity neuronal function? For BK channels in the cochlea, it seems clear that the multiple splice variants are needed to tune the frequency response of the hair cell. However, for most proteins, it is not known why it is important to have a particular ratio of protein isoforms in a particular cell type. More genetic data, such as seen with the agrin knockouts are needed to answer these questions.

Our understanding of the mechanisms of splice site choice is also in its infancy. Very little is known about how regulatory factors actually interact with a spliceosome. How is specificity determined by combinations of common factors and what is the contribution of the tissue-specific proteins? Progress on these fronts will require new biochemical assays that allow examination of single or small groups of proteins for their effects on spliceosomal assembly.

How do signaling pathways affect the splicing apparatus and how are inducible changes in splicing mediated? Are splicing regulatory protein genes induced during the primary response to a stimulus? How does the phosphorylation or methylation of splicing factors alter their localization, RNA binding or interactions with other factors? What is the ensemble of splicing changes occurring after a stimulus and how long do they last? Answers to these questions will be important to understanding the mechanisms controlling a neuron's ability to adapt to excitation and its remarkable plastic properties.

Acknowledgement. We thank Kim Simmonds and Anna Callahan for their help in preparing the manuscript and our many colleagues for discussions and help in providing references. The authors are supported by the Howard Hughes Medical Institute and a grant from the National Institutes of Health to Douglas L. Black (R01 GM 49662).

References

Adams MD, Tarng RS, Rio DC (1997) The alternative splicing factor PSI regulates P-element third intron splicing in vivo. Genes Dev 11:129–138

Adams MD, Celniker SE, Holt RA, Evans CA, Gocayne JD, Amanatides PG, Scherer SE, Li PW, Hoskins RA, Galle RF et al. (2000) The genome sequence of *Drosophila melanogaster*. Science 287:2185–2195

Alwazzan M, Newman E, Hamshere MG, Brook JD (1999) Myotonic dystrophy is associated with a reduced level of RNA from the DMWD allele adjacent to the expanded repeat. Hum Mol Genet 8:1491–1497

Angers A, DesGroseillers L (1998) Alternative splicing and genomic organization of the L5–67 gene of *Aplysia californica*. Gene 208:271–277

Antic D, Keene JD (1997) Embryonic lethal abnormal visual RNA-binding proteins involved in growth, differentiation, and posttranscriptional gene expression. Am J Hum Genet 61:273–278

Antic D, Lu N, Keene JD (1999) ELAV tumor antigen, Hel-N1, increases translation of neurofilament M mRNA and induces formation of neurites in human teratocarcinoma cells. Genes Dev 13:449–461

Ashiya M, Grabowski PJ (1997) A neuron-specific splicing switch mediated by an array of pre-mRNA repressor sites: evidence of a regulatory role for the polypyrimidine tract binding protein and a brain-specific PTB counterpart. RNA 3:996–1015

Beam K (1999) Calcium channel splicing: mind your Ps and Qs, Nat Neurosci 2:393–394

Beil B, Screaton G, Stamm S (1997) Molecular cloning of htra2-beta-1 and htra2-beta-2, two human homologs of tra-2 generated by alternative splicing. DNA Cell Biol 16:679–690

Benjamin PR, Burke JF (1994) Alternative mRNA splicing of the FMRFamide gene and its role in neuropeptidergic signalling in a defined neural network. Bioessays 16:335–342

Berget SM (1995) Exon recognition in vertebrate splicing. J Biol Chem 270:2411–2414

Bergsdorf C, Paliga K, Kreger S, Masters CL, Bayreuther K (2000) Identification of cis-elements regulating exon 15 splicing of the amyloid precursor protein pre-mRNA. J Biol Chem 275:2046–2056

Berke JD, Sgambato V, Zhu P, Lavoie B, Vincent M, Krause M, Hyman SE (2001) Dopamine and glutamate induce distinct striatal splice forms of ania-6, an RNA polymerase II-associated cyclin. Neuron 32:277–287

Black DL (1991) Does steric interference between splice sites block the splicing of a short c-src neuron-specific exon in nonneuronal cells? Genes Dev 5:389–402

Black DL (1992) Activation of c-src neuron-specific splicing by an unusual RNA element in vivo and in vitro. Cell 69:795–807

Black DL (1998) Splicing in the inner ear: a familiar tune, but what are the instruments? Neuron 20:165–168

Black DL (2000) Protein diversity from alternative splicing: a challenge for bioinformatics and post-genome biology. Cell 103:367–370

Bourinet E, Soong TW, Sutton K, Slaymaker S, Mathews E, Monteil A, Zamponi GW, Nargeot J, Snutch TP (1999) Splicing of alpha 1 A subunit gene generates phenotypic variants of P- and Q-type calcium channels. Nat Neurosci 2:407–415

Bowe MA, Fallon JR (1995) The role of agrin in synapse formation, Annu Rev Neurosci 18:443–462

Brudno M, Gelfand MS, Spengler S, Zorn M, Dubchak I, Conboy JG (2001) Computational analysis of candidate intron regulatory elements for tissue-specific alternative pre-mRNA splicing. Nucleic Acids Res 29:2338–2348

Buchanovich RJaDRB (1997) The neuronal RNA binding protein Nova-1 recognizes specific RNA targets in vitro and in vivo. Mol Cell Biol 17:3194–3201

Buck LB, Bigelow JM, Axel R (1987) Alternative splicing in individual Aplysia neurons generates neuropeptide diversity. Cell 51:127–133

Burge CB, Tuschl T, Sharp, PA (1999) Splicing of precursors to mRNAs by the spliceosomes. In: Gesteland RF, Cech TR, Atkins JF (eds) The RNA world, 2nd edn. Cold Spring Harbor Laboratory Press, New York, pp 525–560

Burgess RW, Nguyen QT, Son YJ, Lichtman JW, Sanes JR (1999) Alternatively spliced isoforms of nerve and muscle derived agrin: their roles at the neuromuscular junction. Neuron 23:33–44

Burke JF, Bright KE, Kellett E, Benjamin PR, Saunders SE (1992) Alternative mRNA splicing in the nervous system. Prog Brain Res 92:115–125

Caceres JF, Stamm S, Helfman DM, Krainer AR (1994) Regulation of alternative splicing in vivo by overexpression of antagonistic splicing factors. Science 265:1706–1709

Campanelli JT, Gayer GG, Scheller RH (1996) Alternative RNA splicing that determines agrin activity regulates binding to heparin and alpha-dystroglycan. Development 122:1663–1672

Caputi M, Mayeda A, Krainer AR, Zahler AM (1999) hnRNP A/B proteins are required for inhibition of HIV-1 pre-mRNA splicing. EMBO J 18:4060–4067

Carlo T, Sterner DA, Berget SM (1996) An intron splicing enhancer containing a G-rich repeat facilitates inclusion of a vertebrate micro-exon. RNA 2:342–353

Carlo T, Sierra R, Berget SM (2000) A 5′ splice site-proximal enhancer binds SF1 and activates exon bridging of a microexon. Mol Cell Biol 20:3988–3995

Chabot B, Blanchette M, Lapierre I, La Branche H (1997) An intron element modulating 5′ splice site selection in the hnRNP A1 pre-mRNA interacts with hnRNP A1. Mol Cell Biol 17: 1776–1786

Chalfant CE, Mischak H, Watson JE, Winkler BC, Goodnight J, Farese RV, Cooper DR (1995) Regulation of alternative splicing of protein kinase C beta by insulin. J Biol Chem 270:13326–13332

Chan RC, Black DL (1995) Conserved intron elements repress splicing of a neuron-specific c-src exon in vitro. Mol Cell Biol 15:6377–6385 [published erratum appears in Mol Cell Biol 1997;17(5):2970]

Chan RC, Black DL (1997) The polypyrimidine tract binding protein binds upstream of neural cell-specific c-src exon N1 to repress the splicing of the intron downstream. Mol Cell Biol 17:4667–4676

Chen CD, Kobayashi R, Helfman DM (1999) Binding of hnRNP H to an exonic splicing silencer is involved in the regulation of alternative splicing of the rat beta-tropomyosin gene. Genes Dev 13:593–606

Chong JA ea (1995) REST: a mammalian silencer protein that restricts sodium channel gene expression to neurons. Cell 80:949–957

Chou MY, Rooke N, Turck CW, Black DL (1999) hnRNP H is a component of a splicing enhancer complex that activates a c-src alternative exon in neuronal cells. Mol Cell Biol 19:69–77

Chou MY, Underwood JG, Nikolic J, Luu MH, Black DL (2000) Multisite RNA binding and release of polypyrimidine tract binding protein during the regulation of c-src neural-specific splicing. Mol Cell 5:949–957

Chung SL, Jiang S, Cheng S, Furneaux H (1996) Purification and properties of HuD, a neuronal RNA-binding protein. J Biol Chem 271:11518–11524

Chung S, Eckrich M, Perrone-Bizzozero N, Kohn D, Furneaux H (1997) The Elav-like proteins bind to a conserved regulatory element in the 3′ untranslated region of GAP-43 mRNA. J Biol Chem 272:6595–6598

Claverie JM (2001) What if there are only 30,000 human genes? Science 291:1255–7

Coetzee WA, Amarillo Y, Chiu J, Chow A, Lau D, McCormack T, Moreno H, Nadal MS, Ozaita A, Pountney D et al. (1999) Molecular diversity of K^+ channels. Ann New York Acad Sci 868:233–285

Collett JW, Steele RE (1993) Alternative splicing of a neural-specific Src mRNA (Src+) is a rapid and protein synthesis-independent response to neural induction in *Xenopus laevis*. Dev Biol 158:487–495

Colwill K, Pawson T, Andrews B, Prasad J, Manley JL, Bell JC, Duncan PI (1996) The Clk/Sty protein kinase phosphorylates SR splicing factors and regulates their intranuclear distribution. EMBO J 15:265–275

Consortium IHGS (2001) Initial sequencing and analysis of the human genome. Nature 409:860–921

Conte MR, Grune T, Ghuman J, Kelly G, Ladas A, Matthews S, Curry S (2000) Structure of tandem RNA recognition motifs from polypyrimidine tract binding protein reveals novel features of the RRM fold. EMBO J 19:3132–3141

Cooper TAMW (1997) The regulation of splice site selection, and its role in human disease. Am J Hum Genet 61:259–266

Cote J, Chabot B (1997) Natural base-pairing interactions between 5′ splice site and branch site sequences affect mammalian 5′ splice site selection. RNA 3:1248–1261

Cote J, Simard MJ, Chabot B (1999) An element in the 5′ common exon of the NCAM alternative splicing unit interacts with SR proteins and modulates 5′ splice site selection. Nucleic Acids Res 27:2529–2537

Dal Toso R, Sommer B, Ewert M, Herb A, Pritchett DB, Bach A, Shivers BD, Seeburg PH (1989) The dopamine D2 receptor: two molecular forms generated by alternative splicing. EMBO J 8:4025–4034

Daoud R, Da Penha Berzaghi M, Siedler F, Hubener M, Stamm S (1999) Activity-dependent regulation of alternative splicing patterns in the rat brain. Eur J Neurosci 11:788–802

Darnell RB (1996) Onconeural antigens and the paraneoplastic neurological disorders: at the intersection of cancer, immunity and the brain. Proc Natl Acad Sci USA 93:4529–4536

Del Gatto-Konczak F, Bourgeois CF, Le Guiner C, Kister L, Gesnel MC, Stevenin J, Breathnach R (2000) The RNA-binding protein TIA-1 is a novel mammalian splicing regulator acting through intron sequences adjacent to a 5′ splice site. Mol Cell Biol 20:6287–6299

Dominski Z, Kole R (1991) Selection of splice sites in pre-mRNAs with short internal exons. Mol Cell Biol 11:6075–6083

Du C, McGuffin ME, Dauwalder B, Rabinow L, Mattox W (1998) Protein phosphorylation plays an essential role in the regulation of alternative splicing and sex determination in *Drosophila*. Mol Cell 2:741–750

Ehlers MD, Fung ET, O'Brien RJ, Huganir RL (1998) Splice variant-specific interaction of the NMDA receptor subunit NR1 with neuronal intermediate filaments. J Neurosci 18:720–730

Fan XC, Steitz JA (1998) Overexpression of HuR, a nuclear-cytoplasmic shuttling protein, increases the in vivo stability of ARE containing mRNAs. EMBO J 17:3448–3460

Fettiplace RFP (1999) Mechanisms of hair cell tuning. Annu Rev Physiol 61:809–83

Finkbeiner S, Greenberg ME (1998) Ca^{2+} channel-regulated neuronal gene expression. J Neurobiol 37:171–189

Forch P, Puig O, Kedersha N, Martinez C, Granneman S, Seraphin B, Anderson P, Valcarcel J (2000) The apoptosis-promoting factor TIA-1 is a regulator of alternative pre- mRNA splicing. Mol Cell 6:1089–1098

Ford LP, Watson J, Keene JD, Wilusz J (1999) ELAV proteins stabilize deadenylated intermediates in a novel in vitro mRNA deadenylation/degradation system. Genes Dev 13:188–201

Gao FB, Keene JD (1996) Hel-N1/Hel-N2 proteins are bound to poly(A)+ mRNA in granular RNP structures and are implicated in neuronal differentiation. J Cell Sci 109:579–589

Gesemann M, Cavalli V, Denzer AJ, Brancaccio A, Schumacher B, Ruegg MA (1996) Alternative splicing of agrin alters its binding to heparin, dystroglycan, and the putative agrin receptor. Neuron 16:755–767

Giros BSP, Martres MP, Riou JF, Emorine LJ, Schwartz JC (1989) Alternative splicing directs the expression of two D2 dopamine receptor isoforms. Nature 342:923–926

Gooding C, Roberts GC, Smith CW (1998) Role of an inhibitory pyrimidine element and polypyrimidine tract binding protein in repression of a regulated alpha-tropomyosin exon. RNA 4:85–100

Grabowski PJ, Black DL (2001) Alternative RNA splicing in the nervous system. Prog Neurobiol 65:289–308

Grandy DKMM, Makam H, Stofko RE, Alfano M, Frothingham L, Fischer JBB-HK, Bunzow JR, Server AC et al. (1989) Cloning of the cDNA and gene for a human D2 dopamine receptor. Proc Natl Acad Sci USA 86:9762–9766

Graveley BR (2000) Sorting out the complexity of SR protein functions. RNA 6:1197–1211

Graveley BR (2001) Alternative splicing: increasing diversity in the proteomic world. Trends Genet 17:100–107

Graveley BR, Maniatis T (1998) Arginine/serine-rich domains of SR proteins can function as activators of pre-mRNA splicing. Mol Cell 1:765–771

Groenen PJ, Wansink DG, Coerwinkel M, van den Broek W, Jansen G, Wieringa B (2000) Constitutive and regulated modes of splicing produce six major myotonic dystrophy protein kinase (DMPK) isoforms with distinct properties. Hum Mol Genet 9:605–616

Guiramand J, Montmayeur JP, Ceraline J, Bhatia M, Borrelli E (1995) Alternative splicing of the dopamine D2 receptor directs specificity of coupling to G-proteins. J Biol Chem 270:7354–7358
Guo N, Kawamoto S (2000) An intronic downstream enhancer promotes 3′ splice site usage of a neural cell-specific exon. J Biol Chem 275:33641–33649
Hanamura A, Caceres JF, Mayeda A, Franza BR Jr, Krainer AR (1998) Regulated tissue-specific expression of antagonistic pre-mRNA splicing factors. RNA 4:430–444
Hedjran F, Yeakley JM, Huh GS, Hynes RO, Rosenfeld MG (1997) Control of alternative pre-mRNA splicing by distributed pentameric repeats. Proc Natl Acad Sci USA 94:12343–12347
Hille B (1992) Ionic channels of excitable membranes, 2nd edn. Sinauer Associates, Sunderland, MA
Hisatsune C, Umemori H, Inoue T, Michikawa T, Kohda K, Mikoshiba K, Yamamoto T (1997) Phosphorylation-dependent regulation of *N*-methyl-D-aspartate receptors by calmodulin. J Biol Chem 272:20805–20810
Hoch W (1999) Formation of the neuromuscular junction. Agrin and its unusual receptors. Eur J Biochem 265:1–10
Holmberg J, Clarke DL, Frisen J (2000) Regulation of repulsion versus adhesion by different splice forms of an Eph receptor. Nature 408:203–206
Hopf C, Hoch W (1996) Agrin binding to alpha-dystroglycan. Domains of agrin necessary to induce acetylcholine receptor clustering are overlapping but not identical to the alpha-dystroglycan-binding region. J Biol Chem 271:5231–5236
Hoshi T, Zagotta WN, Aldrich RW (1990) Biophysical and molecular mechanisms of Shaker potassium channel inactivation. Science 250:533–538
Hoshi T, Zagotta WN, Aldrich RW (1991) Two types of inactivation in Shaker K^+ channels: effects of alterations in the carboxy-terminal region. Neuron 7:547–556
Huh GS, Hynes RO (1994) Regulation of alternative pre-mRNA splicing by a novel repeated hexanucleotide element. Genes Dev 8:1561–1574
Huntsman NMTB, Potkin SG, Bunney WE Jr, Jones EG (1998) Altered ratios of alternatively spliced long and short gamma2 subunit mRNAs of the gamma-amino butyrate type A receptor in prefrontal cortex of schizophrenics. Proc Natl Acad Sci USA 95:15066–15071
Imbert G, Saudou F, Yvert G, Devys D, Trottier Y, Garnier JM, Weber C, Mandel JL, Cancel G, Abbas N et al. (1996) Cloning of the gene for spinocerebellar ataxia 2 reveals a locus with high sensitivity to expanded CAG/glutamine repeats. Nat Genet 14:285–291
Iverson LE, Rudy B (1990) The role of the divergent amino and carboxyl domains on the inactivation properties of potassium channels derived from the Shaker gene of *Drosophila.* J Neurosci 10:2903–2916
Iverson LE, Tanouye MA, Lester HA, Davidson N, Rudy B (1988) A-type potassium channels expressed from Shaker locus cDNA. Proc Natl Acad Sci USA 85:5723–5727
Iverson LE, Mottes JR, Yeager SA, Germeraad SE (1997) Tissue-specific alternative splicing of Shaker potassium channel transcripts results from distinct modes of regulating 3′ splice choice. J Neurobiol 32:457–468
Jacquenet S, Mereau A, Bilodeau PS, Damier L, Stoltzfus CM, Branlant C (2001) A second exon splicing silencer within human immunodeficiency virus type 1 tat exon 2 represses splicing of Tat mRNA and binds protein hnRNP H. J Biol Chem 276:40464–40475
Jensen KB, Dredge BK, Stefani G, Zhong R, Buckanovich RJ, Okano HJ, Yang YY, Darnell RB (2000a) Nova-1 regulates neuron-specific alternative splicing and is essential for neuronal viability. Neuron 25:359–371
Jensen KB, Musunuru K, Lewis HA, Burley SK, Darnell RB (2000b) The tetranucleotide UCAY directs the specific recognition of RNA by the Nova K-homology 3 domain. Proc Natl Acad Sci USA 97:5740–5745
Jones EM, Gray-Keller M, Art JJ, Fettiplace R (1999a) The functional role of alternative splicing of Ca(2+)-activated K+ channels in auditory hair cells, Ann NY Acad Sci 868:379–385
Jones EMC, Gray-Keller M, Fettiplace R (1999b) The role of Ca2+-activated K+ channel spliced variants in the tonotopic organization of the turtle cochlea. J Physiol 518:653–665

Journot L, Spengler D, Pantaloni C, Dumuis A, Sebben M, Bockaert J (1994) The PACAP receptor: generation by alternative splicing of functional diversity among G protein-coupled receptors in nerve cells. Semin Cell Biol 5:263–272

Kan JL, Green MR (1999) Pre-mRNA splicing of IgM exons M1 and M2 is directed by a juxtaposed splicing enhancer and inhibitor. Genes Dev 13:462–471

Kawamoto S (1996) Neuron-specific alternative splicing of nonmuscle myosin II heavy chain-B pre-mRNA requires a cis-acting intron sequence. J Biol Chem 271:17613–17616

Kim M, Baro DJ, Lanning CC, Doshi M, Farnham J, Moskowitz HS, Peck JH, Olivera BM, Harris-Warrick RM (1997) Alternative splicing in the pore-forming region of shaker potassium channels. J Neurosci 17:8213–8224

Koike M, Tsukada S, Tsuzuki K, Kijima H, Ozawa S (2000) Regulation of kinetic properties of GluR2 AMPA receptor channels by alternative splicing. J Neurosci 20:2166–2174

Konig H, Ponta H, Herrlich P(1998) Coupling of signal transduction to alternative pre-mRNA splicing by a composite splice regulator. EMBO J 17:2904–2913

Koushika SP, Lisbin MJ, White K (1996) ELAV, a *Drosophila* neuron-specific protein, mediates the generation of an alternatively spliced neural protein isoform. Curr Biol 6:1634–1641

Koushika SP, Soller M, DeSimone SM, Daub DM, White K (1999) Differential and inefficient splicing of a broadly expressed *Drosophila* erect wing transcript results in tissue-specific enrichment of the vital EWG protein isoform. Mol Cell Biol 19:3998–4007

Krawczak M, Reiss J, Cooper DN (1992) The mutational spectrum of single base-pair substitutions in mRNA splice junctions of human genes: causes and consequences, Hum Genet 90:41–54

Krishek BJ, Xie X, Blackstone C, Huganir RL, Moss SJ, Smart TG (1994) Regulation of GABAA receptor function by protein kinase C phosphorylation. Neuron 12:1081–1095

Ladd AN, Charlet N, Cooper TA (2001) The CELF family of RNA binding proteins is implicated in cell-specific and developmentally regulated alternative splicing. Mol Cell Biol 21:1285–1296

Lambolez B, Ropert N, Perrais D, Rossier J, Hestrin S (1996) Correlation between kinetics and RNA splicing of alpha-amino-3-hydroxy-5-methylisoxazole-4-propionic acid receptors in neocortical neurons. Proc Natl Acad Sci USA 93:1797–1802

Le Corre S, Harper CG, Lopez P, Ward P, Catts S (2000) Increased levels of expression of an NMDARI splice variant in the superior temporal gyrus in schizophrenia. Neuroreport 11:983–986

Lefebvre S, Burglen L, Frezal J, Munnich A, Melki J (1998) The role of the SMN gene in proximal spinal muscular atrophy. Hum Mol Genet 7:1531–1536

Lin Z, Haus S, Edgerton J, Lipscombe D (1997) Identification of functionally distinct isoforms of the N-type Ca^{2+} channel in rat sympathetic ganglia and brain. Neuron 18:153–166

Lin Z, Lin Y, Schorge S, Pan JQ, Beierlein M, Lipscombe D (1999) Alternative splicing of a short cassette exon in alpha1B generates functionally distinct N-type calcium channels in central and peripheral neurons. J Neurosci 19:5322–5331

Lingle CJ, Solaro CR, Prakriya M, Ding JP (1996) Calcium-activated potassium channels in adrenal chromaffin cells. Ion Channels 4:261–301

Lisbin MJ, Qiu J, White K (2001) The neuron-specific RNA-binding protein ELAV regulates neuroglian alternative splicing in neurons and binds directly to its pre-mRNA. Genes Dev 15: 2546–2561

Lisman JE, Fallon JR (1999) What maintains memories? Science 283:339–340

Liu HX, Cartegni L, Zhang MQ, Krainer AR (2001) A mechanism for exon skipping caused by nonsense or missense mutations in BRCA1 and other genes. Nat Genet 27:55–58

Liu S-QJ, Kaczmarek LK (1998) The expression of two splice variants of the Kv3.1 potassium channel gene is regulated by different signaling pathways. J Neurosci 18:2881–2890

Lopez AJ (1998) Alternative splicing of pre-mRNA: developmental consequences and mechanisms of regulation. Annu Rev Genet 32:279–305

Lundquist EA, Herman RK (1994) The mec-8 gene of *Caenorhabditis elegans* affects muscle and sensory neuron function and interacts with three other genes: unc-52, smu-1 and smu-2. Genetics 138:83–101

Lundquist EA, Herman RK, Rogalski TM, Mullen GP, Moerman DG, Shaw JE (1996) The mec-8 gene of *C. elegans* encodes a protein with two RNA recognition motifs and regulates alternative splicing of unc-52 transcripts. Development 122:1601–1610

Lynch KW, Weiss A (2000) A model system for activation-induced alternative splicing of CD45 pre-mRNA in T cells implicates protein kinase C and Ras. Mol Cell Biol 20:70–80

Mankodi A, Logigian E, Callahan L, McClain C, White R, Henderson D, Krym M, Thornton CA (2000) Myotonic dystrophy in transgenic mice expressing an expanded CUG repeat. Science 289:1769–1773

Manley JL, Tacke R (1996) SR proteins and splicing control. Genes Dev 10:1569–1579

Markovtsov V, Nikolic JM, Goldman JA, Turck CW, Chou MY, Black DL (2000) Cooperative assembly of an hnRNP complex induced by a tissue-specific homolog of polypyrimidine tract binding protein. Mol Cell Biol 20:7463–7479

Matera AG (1999) RNA splicing: more clues from spinal muscular atrophy. Curr Biol 9:R140–R142

Mattaj IW (1998) Ribonucleoprotein assembly: clues from spinal muscular atrophy. Curr Biol 8:93–95

Mayeda A, Helfman DM, Krainer AR (1993) Modulation of exon skipping and inclusion by heterogeneous nuclear ribonucleoprotein A1 and pre-mRNA splicing factor SF2/ASF. Mol Cell Biol 13:2993–3001 [published erratum appears in Mol Cell Biol 1993;13(7):4458]

McCullough AJ, Berget SM (2000) An intronic splicing enhancer binds U1 snRNPs to enhance splicing and select 5′ splice sites. Mol Cell Biol 20:9225–9235

Miller JW, Urbinati CR, Teng-Umnuay P, Stenberg MG, Byrne BJ, Thornton CA, Swanson MS (2000) Recruitment of human muscleblind proteins to (CUG)(n) expansions associated with myotonic dystrophy. EMBO J 19:4439–4448

Min H, Chan RC, Black DL (1995) The generally expressed hnRNP F is involved in a neural-specific pre-mRNA splicing event. Genes Dev 9:2659–2671

Min H, Turck CW, Nikolic JM, Black DL (1997) A new regulatory protein, KSRP, mediates exon inclusion through an intronic splicing enhancer. Genes Dev 11:1023–1036

Missler M, Sudhof TC (1998) Neurexins: three genes and 1001 products. Trends Genet 14:20–26

Missler M, Fernandez-Chacon R, Sudhof TC (1998) The making of neurexins. J Neurochem 71:1339–1347

Modafferi EF, Black DL (1997) A complex intronic splicing enhancer from the c-src pre-mRNA activates inclusion of a heterologous exon. Mol Cell Biol 17:6537–6545

Modafferi EF, Black DL (1999) Combinatorial control of a neuron-specific exon. RNA 5:687–706

Mottes JR, Iverson LE (1995) Tissue-specific alternative splicing of hybrid Shaker/lacZ genes correlates with kinetic differences in Shaker K+ currents in vivo. Neuron 14:613–623

Mulligan GJ, Guo W, Wormsley S, Helfman DM (1992) Polypyrimidine tract binding protein interacts with sequences involved in alternative splicing of beta-tropomyosin pre-mRNA. J Biol Chem 267:25480–25487

Musshoff U, Schunke U, Kohling R, Speckmann EJ (2000) Alternative splicing of the NMDAR1 glutamate receptor subunit in human temporal lobe epilepsy. Mol Brain Res 76:377–384

Nakai K, Sakamoto H (1994) Construction of a novel database containing aberrant splicing mutations of mammalian genes. Gene 141:171–177

Navaratnam DS, Bell TJ, Tu TD, Cohen EL, Oberholtzer JC (1997) Differential distribution of Ca2+-activated K+ channel splice variants among hair cells along the tonotopic axis of the chick cochlea. Neuron 19:1077–1085

O'Neill JP, Rogan PK, Cariello N, Nicklas JA (1998) Mutations that alter RNA splicing of the human HPRT gene: a review of the spectrum. Mutat Res 411:179–214

O'Toole JJ, Deyst KA, Bowe MA, Nastuk MA, McKechnie BA, Fallon JR (1996) Alternative splicing of agrin regulates its binding to heparin alpha-dystroglycan, and the cell surface. Proc Natl Acad Sci USA 93:7369–7374

Palm K, Belluardo N, Metsis M, Timmusk T (1998) Neuronal expression of zinc finger transcription factor REST/NRSF/XBR gene. J Neurosci 18:1280–1296

Partin KM, Bowie D, Mayer ML (1995) Structural determinants of allosteric regulation in alternatively spliced AMPA receptors. Neuron 14:833–843

Peng SS, Chen CY, Xu N, Shyu AB (1998) RNA stabilization by the AU-rich element binding protein, HuR, an ELAV protein. EMBO J 17:3461–3470

Philips AV, Cooper TA (2000) RNA processing and human disease. Cell Mol Life Sci 57:235–249

Philips AV, Timchenko LT, Cooper TA (1998) Disruption of splicing regulated by a CUG-binding protein in myotonic dystrophy. Science 280:737–741

Polydorides AD, Okano HJ, Yang YY, Stefani G, Darnell RB (2000) A brain-enriched polypyrimidine tract-binding protein antagonizes the ability of Nova to regulate neuron-specific alternative splicing. Proc Natl Acad Sci USA 97:6350–6355

Prasad J, Colwill K, Pawson T, Manley JL (1999) The protein kinase Clk/Sty directly modulates SR protein activity: both hyper- and hypophosphorylation inhibit splicing. Mol Cell Biol 19: 6991–7000

Pulst SM, Nechiporuk A, Nechiporuk T, Gispert S, Chen XN, Lopes-Cendes I, Pearlman S, Starkman S, Orozco-Diaz G, Lunkes A et al. (1996) Moderate expansion of a normally biallelic trinucleotide repeat in spinocerebellar ataxia type 2. Nat Genet 14:269–276

Quinlan JJ, Firestone LL, Homanics GE (2000) Mice lacking the long splice variant of the gamma 2 subunit of the GABA (A) receptor are more sensitive to benzodiazepines, Pharmacol Biochem Behav 66:371–374

Rafiki A, Ben-Ari Y, Khrestchatisky M, Represa A (1998) Long-lasting enhanced expression in the rat hippocampus of NMDAR1 splice variants in a kainate model of epilepsy. Eur J Neurosci 10:497–507

Reenan RA, Hanrahan CJ, Barry G (2000) The mle(napts) RNA helicase mutation in *Drosophila* results in a splicing catastrophe of the para Na+ channel transcript in a region of RNA editing. Neuron 25:139–149

Roberts R, Timchenko NA, Miller JW, Reddy S, Caskey CT, Swanson MS, Timchenko LT (1997) Altered phosphorylation and intracellular distribution of a (CUG)n triplet repeat RNA-binding protein in patients with myotonic dystrophy and in myotonin protein kinase knockout mice. Proc Natl Acad Sci USA 94:13221–13226

Rodger J, Davis S, Laroche S, Mallet J, Hicks A (1998) Induction of long-term potentiation in vivo regulates alternate splicing to alter syntaxin 3 isoform expression in rat dentate gyrus. J Neurochem 71:666–675

Rosenblatt K, Sun PZP, Heller S, Hudspeth AJ (1997) Distribution of Ca2+-activated K+ channel isoforms along the tonotopic gradient of the chicken's cochlea. Neuron 19:1061–1075

Rosenfeld MG, Mermod JJ, Amara SG, Swanson LW, Sawchenko PE, Rivier J, Vale WW, Evans RM (1983) Production of a novel neuropeptide encoded by the calcitonin gene via tissue-specific RNA processing. Nature 304:129–135

Rosoff ML, Burglin TR, Li C (1992) Alternatively spliced transcripts of the flp-1 gene encode distinct FMRFamide-like peptides in *Caenorhabditis elegans.* J Neurosci 12:2356–2361

Rumbaugh G, Prybylowski K, Wang JF, Vicini S (2000) Exon 5 and spermine regulate deactivation of NMDA receptor subtypes. J Neurophys 83:1300–1306

Ryan KJ, Cooper TA (1996) Muscle-specific splicing enhancers regulate inclusion of the cardiac troponin T alternative exon in embryonic skeletal muscle. Mol Cell Biol 16:4014–4023

Sanpei K, Takano H, Igarashi S, Sato T, Oyake M, Sasaki H, Wakisaka A, Tashiro K, Ishida Y, Ikeuchi T et al. (1996) Identification of the spinocerebellar ataxia type 2 gene using a direct identification of repeat expansion and cloning technique, DIRECT. Nat Genet 14:277–284

Santama N, Benjamin PR, Burke JF (1995) Alternative RNA splicing generates diversity of neuropeptide expression in the brain of the snail *Lymnaea*: in situ analysis of mutually exclusive transcripts of the FMRFamide gene. Eur J Neurosci 7:65–76

Santoni MJ, Barthels D, Vopper G, Boned A, Goridis C, Wille W (1989) Differential exon usage involving an unusual splicing mechanism generates at least eight types of NCAM cDNA in mouse brain. EMBO J 8:385–392

Scheiffele P, Fan J, Choih J, Fetter R, Serafini T (2000) Neuroligin expressed in nonneuronal cells triggers presynaptic development in contacting axons. Cell 101:657–669

Schmucker D, Clemens JC, Shu H, Worby CA, Xiao J, Muda M, Dixon JE, Zipursky SL (2000) *Drosophila* Dscam is an axon guidance receptor exhibiting extraordinary molecular diversity. Cell 101:671–684

Schoenherr CJ, Anderson DJ (1995) The neuron-restrictive silencer factor (NRSF): a coordinate repressor of multiple neuron-specific genes. Science 267:1360–1363

Schwarz TL, Tempel BL, Papazian DM, Jan YN, Jan LY (1988) Multiple potassium-channel components are produced by alternative splicing at the Shaker locus in *Drosophila*. Nature 331:137–142 [published erratum appears in Nature 1988 332(6166):740]

Sendtner M (2001) Molecular mechanisms in spinal muscular atrophy: models and perspectives. Curr Opin Neurol 14:629–634

Seong JY, Park S, Kim K (1999) Enhanced splicing of the first intron from the gonadotropin-releasing hormone (GnRH) primary transcript is a prerequisite for mature GnRH messenger RNA: presence of GnRH neuron-specific splicing factors. Mol Endocrinol 13:1882–1895

Shibata A, Hattori M, Suda H, Sakaki Y (1996) Identification of cis-acting elements involved in an alternative splicing of the amyloid precursor protein (APP) gene. Gene 175:203–208

Shifrin VI, Neel BG (1993) Growth factor-inducible alternative splicing of nontransmembrane phosphotyrosine phosphatase PTP-1B pre-mRNA. J Biol Chem 268:25376–25384

Shimojo M, Paquette AJ, Anderson DJ, Hersh LB (1999) Protein kinase A regulates cholinergic gene expression in PC12 cells: REST4 silences the silencing activity of neuron-restrictive silencer factor/REST. Mol Cell Biol 19:6788–6795

Siebel CW, Kanaar R, Rio DC (1994) Regulation of tissue-specific P-element pre-mRNA splicing requires the RNA-binding protein PSI. Genes Dev 8:1713–1725

Sirand-Pugnet P, Durosay P, Brody E, Marie J (1995) An intronic (A/U)GGG repeat enhances the splicing of an alternative intron of the chicken beta-tropomyosin pre-mRNA. Nucleic Acids Res 23:3501–3507

Smith CW, Valcarcel J (2000) Alternative pre-mRNA splicing: the logic of combinatorial control. Trends Biochem Sci 25:381–388

Smith MA, Fanger GR, O'Connor LT, Bridle P, Maue RA (1997) Selective regulation of agrin mRNA induction and alternative splicing in PC12 cells by Ras-dependent actions of nerve growth factor. J Biol Chem 272:15675–15681

Solaro CR, Nelson C, Wei A, Salkoff L, Lingle CJ (1995) Cytoplasmic Mg-2+ modulates Ca-2+-dependent activation of MSLO by binding to a low affinity site on the channel core. Biophysical J 68:A30

Staley JP, Guthrie C (1998) Mechanical devices of the spliceosome: motors, clocks, springs, and things. Cell 92:315–326

Sutcliffe JG, Milner RJ (1988) Alternative mRNA splicing: the Shaker gene. Trends Genet 4:297–299

Tacke R, Goridis C (1991) Alternative splicing in the neural cell adhesion molecule pre-mRNA: regulation of exon 18 skipping depends on the 5′-splice site. Genes Dev 5:1416–1429

Taneja KL, McCurrach M, Schalling M, Housman D, Singer RH (1995) Foci of trinucleotide repeat transcripts in nuclei of myotonic dystrophy cells and tissues. J Cell Biol 128:995–1002

Tange TO, Damgaard CK, Guth S, Valcarcel J, Kjems J (2001) The hnRNP A1 protein regulates HIV-1 tat splicing via a novel intron silencer element. EMBO J 20:5748–5758

Thackeray JR, Ganetzky B (1994) Developmentally regulated alternative splicing generates a complex array of *Drosophila* para sodium channel isoforms. J Neurosci 14:2569–2578

Thackeray JR, Ganetzky B (1995) Conserved alternative splicing patterns and splicing signals in the *Drosophila* sodium channel gene para. Genetics 141:203–214

Thomas WS, O'Dowd DK, Smith MA (1993) Developmental expression and alternative splicing of chick agrin RNA. Dev Biol 158:523–535

Timchenko LT (1999) Myotonic dystrophy: the role of RNA CUG triplet repeats, Am J Hum Genet 64:360–364

Timchenko LT, Miller JW, Timchenko NA, DeVore DR, Datar KV, Lin L, Roberts R, Caskey CT, Swanson MS (1996) Identification of a (CUG)n triplet repeat RNA-binding protein and its expression in myotonic dystrophy. Nucleic Acids Res 24:4407–4414

Timchenko NA, Welm AL, Lu X, Timchenko LT (1999) CUG repeat binding protein (CUGBP1) interacts with the 5′ region of C/EBPbeta mRNA and regulates translation of C/EBPbeta isoforms. Nucleic Acids Res 27:4517–4525

Timpe LC, Schwarz TL, Tempel BL, Papazian DM, Jan YN, Jan LY (1988) Expression of functional potassium channels from Shaker cDNA in *Xenopus* oocytes. Nature 331:143–145

Tiscornia G, Mahadevan MS (2000) Myotonic dystrophy: the role of the CUG triplet repeats in splicing of a novel DMPK exon and altered cytoplasmic DMPK mRNA isoform ratios. Mol Cell 5:959–967

Traynelis SF, Burgess MF, Zheng F, Lyuboslavsky P, Powers JL (1998) Control of voltage-independent zinc inhibition of NMDA receptors by the NR1 subunit. J Neurosci 18:6163–6175

Valcarcel J, Gebauer F (1997) Post-transcriptional regulation: the dawn of PTB. Curr Biol 7:R705–R708

Vallano ML, Lambolez B, Audinat E, Rossier J (1996) Neuronal activity differentially regulates NMDA receptor subunit expression in cerebellar granule cells. J Neurosci 16:631–639

Vallano ML, Beaman-Hall CM, Benmansour S (1999) Ca2+ and pH modulate alternative splicing of exon 5 in NMDA receptor subunit 1. Neuroreport 10:3659–3664

van der Houven van Oordt W, Diaz-Meco MT, Lozano J, Krainer AR, Moscat J, Caceres JF (2000) The MKK(3/6)-p38-signaling cascade alters the subcellular distribution of hnRNP A1 and modulates alternative splicing regulation. J Cell Biol 149:307–316

van Golen FA, Li KW, de Lange RP, Jespersen S, Geraerts WP (1995) Mutually exclusive neuronal expression of peptides encoded by the FMRFa gene underlies a differential control of copulation in *Lymnaea*. J Biol Chem 270:28487–28493

Varani G, Nagai K (1998) RNA recognition by RNP proteins during RNA processing. Annu Rev Biophys Biomol Struct 27:407–445

Vawter MP, Frye MA, Hemperly JJ, VanderPutten DM, Usen N, Doherty P, Saffell JL, Issa F, Post RM, Wyatt RJ, Freed WJ (2000) Elevated concentration of N-CAM VASE isoforms in schizophrenia. J Psychiatr Res 34:25–34

Venter JC, Adams MD, Myers EW, Li PW, Mural RJ, Sutton GG, Smith HO, Yandell MD, Evans CA, Holt RA (2001) The sequence of the human genome. Science 291:1304–1351

Vergara C, Latorre R, Marrion NV, Adelman JP (1998) Calcium-activated potassium channels. Curr Opin Neurobiol 8:321–329

Vezzani A, Speciale C, Della Vedova F, Tamburin M, Benatti L (1995) Alternative splicing at the C-terminal but not at the N-terminal domain of the NMDA receptor NR1 is altered in the kindled hippocampus. Eur J Neurosci 7:2513–2517

Wafford KA, Bain CJ, Whiting PJ, Kemp JA (1993) Functional comparison of the role of gamma subunits in recombinant human gamma-aminobutyric acidA/benzodiazepine receptors. Mol Pharmacol 44:437–442

Wagner EJ, Garcia-Blanco MA (2001) Polypyrimidine tract binding protein antagonizes exon definition. Mol Cell Biol 21:3281–3288

Wagner EJ, Carstens RP, Garcia-Blanco MA (1999) A novel isoform ratio switch of the polypyrimidine tract binding protein. Electrophoresis 20:1082–1086

Walsh FS, Doherty P(1997) Neural cell adhesion molecules of the immunoglobulin superfamily: role in axon growth and guidance. Annu Rev Cell Devel Biol 13:425–456

Wang A, Cohen DS, Palmer E, Sheppard D (1991) Polarized regulation of fibronectin secretion and alternative splicing by transforming growth factor. J Biol Chem 266:15598–15601

Wang HY, Xu X, Ding JH, Bermingham JR Jr, Fu XD (2001) SC35 plays a role in T cell development and alternative splicing of CD45. Mol Cell 7:331–342

Wang J, Takagaki Y, Manley JL (1996) Targeted disruption of an essential vertebrate gene: ASF/SF2 is required for cell viability. Genes Dev 10:2588–2599

Wang Z, Grabowski PJ (1996) Cell- and stage-specific splicing events resolved in specialized neurons of the rat cerebellum. RNA 2:1241–1253

Weg-Remers S, Ponta H, Herrlich P, Konig H (2001) Regulation of alternative pre-mRNA splicing by the ERK MAP-kinase pathway. EMBO J 20:4194–4203

Wei N, Lin CQ, Modafferi EF, Gomes WA, Black DL (1997) A unique intronic splicing enhancer controls the inclusion of the agrin Y exon. RNA 3:1275–1288

Weiland S, Bertrand D, Leonard S (2000) Neuronal nicotinic acetylcholine receptors: from the gene to the disease. Behav Brain Res 113:43–56

Whiting P, McKernan RM, Iversen LL (1990) Another mechanism for creating diversity in gamma-aminobutyrate type A receptors: RNA splicing directs expression of two forms of gamma 2 phosphorylation site. Proc Natl Acad Sci USA 87:9966–9970

Wilhelmsen KC (1999) The tangled biology of tau. Proc Natl Acad Sci USA 96:7120–7121

Wu Q, Maniatis T (1999) A striking organization of a large family of human neural cadherin-like cell adhesion genes. Cell 97:779–790

Wu Q, Maniatis T (2000) Large exons encoding multiple ectodomains are a characteristic feature of protocadherin genes. Proc Natl Acad Sci USA 97:3124–3129

Xiao SH, Manley JL (1998) Phosphorylation-dephosphorylation differentially affects activities of splicing factor ASF/SF2. EMBO J 17:6359–6367

Xie J, Black DL (2001) A CaMK IV responsive RNA element mediates depolarization-induced alternative splicing of ion channels. Nature 410:936–939

Xie J, McCobb DP (1998) Control of alternative splicing of potassium channels by stress hormones. Science 280:443–446

Yeakley JM, Tronchere H, Olesen J, Dyck JA, Wang HY, Fu XD (1999) Phosphorylation regulates in vivo interaction and molecular targeting of serine/arginine-rich pre-mRNA splicing factors. J Cell Biol 145:447–455

Yun CY, Fu XD (2000) Conserved SR protein kinase functions in nuclear import and its action is counteracted by arginine methylation in *Saccharomyces cerevisiae.* J Cell Biol 150:707–718

Zacharias DA, Strehler EE (1996) Change in plasma membrane Ca2(+)-ATPase splice-variant expression in response to a rise in intracellular Ca2+. Curr Biol 6:1642–1652

Zhang L, Ashiya M, Sherman TG, Grabowski PJ (1996) Essential nucleotides direct neuron-specific splicing of gamma 2 pre- mRNA. RNA 2:682–698

Zhang L, Liu W, Grabowski PJ (1999) Coordinate repression of a trio of neuron-specific splicing events by the splicing regulator PTB. RNA 5:117–130

Zukin R (1995) Alternatively spliced isoforms of the NMDAR1 receptor subunit. Trends Neurosci 7:306–313

Modulation of Alternative Splicing by Antisense Oligonucleotides

P. Sazani and R. Kole[1]

1 Introduction

Sequence-specific base pairing is the principle mechanism by which antisense oligonucleotides function as therapeutics and sequence-specific research tools. The same basic mechanism of action is used for regulation of expression of certain genes by living cells. Antisense genes were found in *E. coli* (Simons and Kleckner 1988) and more recently in *C. elegans* (Ambros 2000) and in human cells (Hastings et al. 2000). Inhibition of Rous sarcoma virus replication by a synthetic 13-mer antisense DNA molecule (Zamecnik and Stephenson, 1978) provides one of the first examples of the effects of antisense oligonucleotides. For antisense genes and oligonucleotides the assumed or determined function was to decrease the expression of their sense target RNAs. Numerous subsequent papers, involving downregulation of gene expression, reported that DNA oligonucleotides were downregulating gene expression by base pairing to mRNA and eliciting destruction of the RNA in the RNA–DNA duplex by ribonuclease H (RNase H). Modified oligonucleotides that do not elicit RNase H activity or triplex-forming oligonucleotides that interact with DNA targets were also extensively used for downregulation of gene expression. The antisense field was extensively reviewed in a recent volume (Crooke 2001).

A growing body of research shows that RNase H inactive oligonucleotides, not only downregulate, but also alter or restore the expression of a given gene. This goal is accomplished by modification of alternative splicing patterns of pre-mRNA or by correction of an aberrant, disease causing splicing event (Kole and Sazani 2001). This review will discuss the ability of antisense oligonucleotides to: (1) manipulate alternative splicing pathways in terms of clinical applications and (2) probe the interactions of splicing factors with splice sites and other sequence elements involved in splicing. While the consensus has been that cytoplasmic mRNA is the sole target of oligonucleotides, the fact that these compounds modify splicing of nuclear pre-mRNA challenges this view. Thus, the intracellular site of action of antisense oligonucleotides will also be discussed.

[1] Lineberger Comprehensive Cancer Center and Department of Pharmacology, University of North Carolina, CB #7295, Chapel Hill, North Carolina 27599-7295, USA

Progress in Molecular and Subcellular Biology, Vol. 31
Philippe Jeanteur (Ed.)

2
Manipulation of Splicing by Antisense Oligonucleotides

Inhibition of splicing by antisense oligonucleotides has been accomplished by targeting oligonucleotides to small nuclear RNAs (snRNAs), which participate in spliceosome formation and are thus essential for splicing, and to splice sites and adjacent sequences (reviewed in Kole et al. 1991). The sections below review more recent work that resulted in antisense driven restoration or modification of splicing of genes with clinical importance.

2.1
RNA Repair in Genetic Disorders

2.1.1
β-Globin

β-Thalassemia is an inherited blood disease caused by mutations in the β-globin gene that compromise or totally prevent production of β-globin subunits of hemoglobin. Deficits in functional hemoglobin lead to anemia and compensatory expansion of the bone marrow. Furthermore, defects in β-globin production result in an excess of unpaired α-globin subunits, causing their precipitation into inclusion bodies, lysis of red blood cells and iron overload. The latter can be fatal if untreated. β-thalassemia is caused by close to 200 mutations in the β-globin gene and is most common in a large region of the world east of the Mediterranean, stretching through the Middle East to Southeast Asia, and in Africa. Mutations that damage splicing of β-globin pre-mRNA are responsible for a majority of β-thalassemia cases (Schwartz and Benz 1995).

Splicing mutations that are clinically quite common affect splicing of both β-globin introns. In intron 1 (IVS1) the mutations weaken the 5′ splice site at nucleotide 5 and 6 (IVS1–5, –6) or create a new 3′ splice site at nucleotide 110 of the intron (IVS1–110). Mutations in intron 2, IVS2–654, –705 and –745, create aberrant 5′ splice sites and activate a cryptic 3′ splice site within the intron. As a consequence of IVS1–5 and –6, three cryptic 5′ splice sites are activated upstream and downstream from the correct one. As a result of IVS1–110 and intron 2 mutations, portions of the introns are included into the spliced mRNA. The included fragments generate stop codons that truncate the gene product and abrogate β-globin and therefore, hemoglobin expression in thalassemic patients. A mutation in exon 1 of β-globin creates a hemoglobin variant (β^E) and also activates a 5′ cryptic splice site, leading to truncation of spliced RNA and decreased β^E production (Schwartz and Benz 1995).

It has been demonstrated in splicing HeLa extracts that 18-mer 2′-*O*-methyl-oligoribonucleoside phosphorothioates directed at the IVS1 branch point partially restored correct splicing of IVS1–110 mutant pre-mRNA.

Targeting the aberrant splice sites in IVS2–654 and –705 mutants was even more effective in restoring correct β-globin splicing. The oligonucleotides successfully competed with splicing factors for access to the aberrant splice sites formed at nucleotides 652 or 705 of the intron, forcing re-selection of the correct and still functional splice sites by the spliceosome (Dominski and Kole 1993).

Correction of aberrant splicing was also accomplished in HeLa cells that stably expressed IVS2–654, –705 and –745 β-globin genes. The cells were treated with 18-mer 2′-*O*-methyl-, and morpholino oligomers directed at the 5′ (ON-654, ON-705, and ON-745) or 3′ (ON-3′cr) aberrant splice sites (Fig. 1A). Correction was sequence-specific and dose-dependent, with EC_{50} values ranging from 50 nM to 1 μM depending on the chemistry of the oligomer and the delivery method used. Interestingly, ON-3′cr was significantly less effective against the IVS2–654 mutation in cell culture than in vitro in splicing extracts (see below; Sierakowska et al. 1996, 1999; Schmajuk et al. 1999).

Ex vivo correction of β-globin pre-mRNA and upregulation of protein was accomplished in erythropoietic progenitor cells from patients with IVS2–654 and –745 thalassemia (Lacerra et al. 2000) using morpholino oligonucleotides. These compounds were delivered via temporary mechanical disruption of cell membranes by passing the cells through a syringe needle in the presence of the oligomer. Though the doses required to elicit the effect were high (up to 45 μM), sequence specificity and dose and time dependence were seen. Recent results showed significant correction of β-globin splicing in thalassemic erythroid cells cultured with the morpholino oligonucleotide in the absence of any delivery method. Similar results were obtained with morpholino oligonucleotides targeted to IVS1–5 and β^E aberrant splice sites (Suwanmanee and Kole, unpubl.). A key obstacle to further advancement of this treatment in vivo is efficient delivery of the oligonucleotides to the site of β-globin production in the bone marrow. Experiments addressing this problem are carried out in the mouse model of IVS2–654 thalassemia (Lewis et al. 1998).

2.1.2 *Cystic Fibrosis Transmembrane Conductance Regulator*

Over 800 different mutations in the cystic fibrosis transmembrane conductance regulator (CFTR) gene are known to cause cystic fibrosis (CF; Tsui 1992). The most common mutation, Δ-508, deletes a codon in-frame and is thought to interfere with proper trafficking of the protein to the membrane. A less common but prevalent mutation in intron 19, 3849+10 kb C to T, creates an aberrant 5′ splice site and activates a 3′ cryptic splice site within the same intron (see Fig. 1A), much like the thalassemic mutations (Highsmith et al. 1994). Approximately 2% of all CF patients have this mutation; it is more common among the Ashkenazi Jewish community (Abeliovich et al. 1992). 18-mer 2′-O-methyl oligonucleotides directed at the aberrant 5′ and 3′ splice

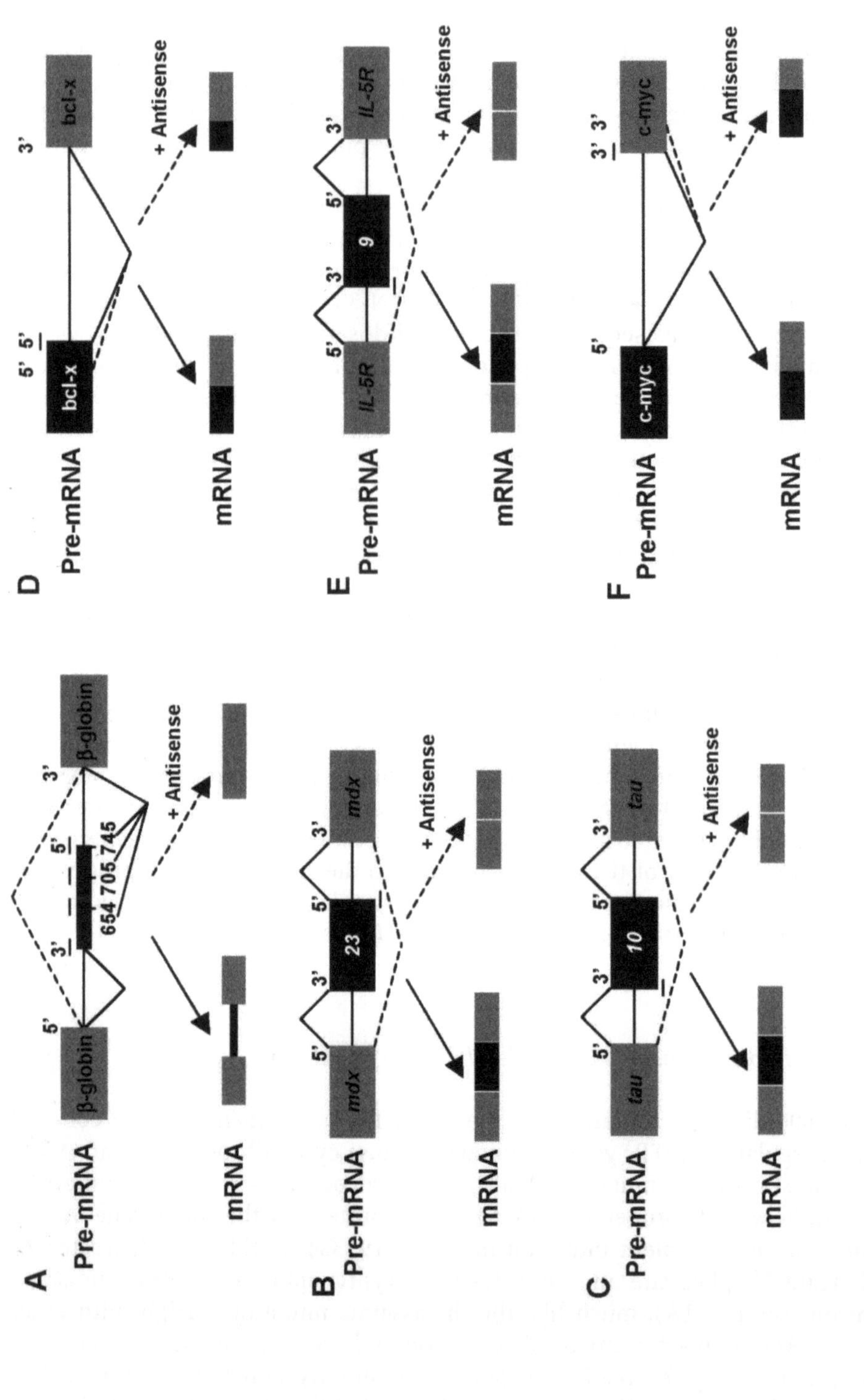
A
Pre-mRNA
mRNA
β-globin
β-globin
654 705 745
+ Antisense
B
Pre-mRNA
mRNA
mdx
23
mdx
+ Antisense
C
Pre-mRNA
mRNA
tau
10
tau
+ Antisense
D
Pre-mRNA
mRNA
bcl-x
bcl-x
+ Antisense
E
Pre-mRNA
mRNA
IL-5R
9
IL-5R
+ Antisense
F
Pre-mRNA
mRNA
c-myc
c-myc
+ Antisense

sites in the 3849+10 kb C to T intron 19 pre-mRNA restored correct splicing and protein expression of CFTR in mouse NIH 3T3, CFT1 or C127 cells (Friedman et al. 1999). Simultaneous treatment of cells with oligonucleotides directed at both the 5′ and 3′ sites had an additive effect. Surprisingly, this effect was not seen in the thalassemic β-globin models (Sierakowska and Kole, unpubl.).

2.1.3
Dystrophin

Duchenne muscular dystrophy (DMD) is caused by mutations in the dystrophin gene that prevent the complete translation of dystrophin protein (Koenig et al. 1989). A milder form of the disease, Becker muscular dystrophy (BMD), arises due to internal deletions that preserve the reading frame and therefore the N- and C-terminal domains of dystrophin. Since internal deletions are less deleterious, it was hypothesized that antisense oligonucleotides could be used to cause skipping of the exons containing the stop codons and convert the DMD phenotype to a milder BMD-like form. The dystrophin gene is huge and the full-length human dystrophin mRNA is 14 kb long, making gene replacement by viral gene delivery methods difficult. Antisense therapy provides a more feasible, albeit less permanent, alternative and therefore attracted a lot of attention.

In the *mdx* mouse model of DMD a mutation in exon 23 creates a premature stop codon and leads to symptoms of dystrophy in mice (Bulfield et al. 1984). Wilton et al. showed that 2′-*O*-methyl oligonucleotides could be used to cause skipping of exon 23. The antisense compound directed at the 5′ splice site of exon 23 (Fig. 1B) was delivered by cationic lipid transfection to C2C12 cell lines and cultured primary mouse muscle expressing *mdx*. Reverse transcription-PCR (RT-PCR) showed a dose-dependent increase in the levels of *mdx* mRNA lacking exon 23, a deletion of 213 nucleotides. In contrast, oligonucleotides directed at the 3′ splice site had no effect (Wilton et al. 1999). In the work of others, effects on the 3′ splice site oligonucleotide were noted (Dunckley et al. 1998). More recently, Mann et al. showed that the

Fig. 1A–F. Modulation of alternative splicing by antisense oligonucleotides. **A** An aberrant exon (*black box*) in the intron 2 of β-globin intron (IVS2) is formed by point mutation at positions 654, 705 or 745 that create a 5′ splice site and activate a cryptic 3′ splice site. Similar splicing patterns are also seen in CFTR pre-mRNA. Antisense oligonucleotides targeted to either site cause a dose-dependent shift in pre-mRNA splicing from the aberrant (*solid arrow*) to correct (*dotted arrow*) pathway. **B** Antisense oligonucleotides targeted to the 5′ splice site of *mdx* exon 23 or **C** the 3′ splice site of exon 10 in *tau* pre-mRNA cause exclusion of those exons. **D** The selection of a distal alternative 5′ splice sites is forced by antisense oligonucleotides targeted at the proximal 5′ splice site in bcl-x pre-mRNA. **E** Targeting of exon 9 of IL-5R with oligonucleotides causes exon skipping. **F** Blocking the proper 3′ splice site of c-myc causes the activation of a cryptic 3′ splice site further downstream within exon 2

oligonucleotide directed at the same 5′ splice site was also effective locally when injected into mouse muscle in the presence of a cationic lipid. Western blotting and immunohistochemistry showed that the oligonucleotide caused an increase in muscle fibers containing the dystrophin product lacking exon 23 (Mann et al. 2001). A similar approach resulted in deletion of exon 46 in antisense treated myotube cultures from two DMD patients carrying an exon 45 deletion. Van Deutekom et al. (2001) found that a 15% level of exon skipping was sufficient to restore significant amounts of properly localized dystrophin in at least 75% of myotubes.

2.1.4
Tau

Tau is a neuron-specific gene coding for a protein involved in microtubule assembly and stability (Goedert et al. 1989). There are six splice variants expressed in the brain, all of which normally skip exon 10 because of a stem loop structure that sequesters the 5′ splice site from the spliceosome. Mutations that disrupt the stem and expose the splice site allow exon 10 to be included in the spliced mRNA. This RNA is translated into a gravely dysfunctional protein involved in the disorder termed frontotemporal dementia and parkinsonism linked to chromosome 17 (FTDP-17; Hutton 2001).

Kalbfuss et al. (2001) took advantage of a PC12 cell line that endogenously expresses the FTDP-17 mutant form of *tau*. Under control conditions, exon 10 was included in the majority of mRNAs. Treatment of the cells with an 18-mer 2′-*O*-methyl-oligonucleotide directed at the 5′ splice site of exon 10 in the disrupted region of the stem loop (Fig. 1C) was able to restore skipping of exon 10 in a dose-dependent fashion. Furthermore, restoration of the functional *tau* protein was demonstrated by the ablation of the offending splice variant and by favorable changes in cell morphology. Interestingly, a control oligonucleotide directed at the 3′ splice site of exon 10 had no effect on splicing of *tau* pre-mRNA (Kalbfuss et al. 2001).

2.2
Modification of Splicing in Cancer

2.2.1
Bcl-x

Bcl-x is a member of the bcl-2 family of apoptotic genes. It is alternatively spliced into two splice variants, bcl-xS (short) and bcl-xL (long), that have opposing pro- and anti-apoptotic functions, respectively (Boise et al. 1993). Overexpression of bcl-xL occurs in several cancers, including prostate and lung carcinoma (Krajewska et al. 1996). The splice variants result from the use of two alternative 5′ splice sites and a common 3′ splice site. Therefore, preventing access of the splicing factors to the proximal 5′ splice site by blocking it with an antisense oligonucleotide should increase the ratio of Bcl-xS to -xL, favor the pro-apoptotic form and inhibit growth of cancer cells (Fig. 1D).

Two reports show that shifting the balance of the two isoforms of bcl-x indeed had the desired effect (Taylor et al. 1999; Mercatante et al. 2001). Taylor et al. showed that a 2′-*O*-methoxyethyl-oligonucleoside-phosphorothioate (2′-*O*-MOE oligonucleotide) antisense to the bcl-xL 5′ splice site shifted splicing towards the bcl-xS pathway in human A549 lung carcinoma cells (Taylor et al. 1999). Doses to a maximum of 200 nM of the oligonucleotide delivered by cationic lipid transfection shifted splicing in a dose-dependent and sequence-specific manner. The newly formed bcl-xS transcript was fully functional and efficiently exported from the nucleus. In these cells, induction of bcl-xS alone was insufficient to induce apoptosis (Taylor et al. 1999). However, antisense treatment potentiated the effects of classical chemotherapeutic agents such as-*cis*-platinum. In contrast, Mercatante et al. showed that in PC3 prostate cancer cell line treatment with 2′-*O*-methyl oligonucleotides directed towards the bcl-xL 5′ splice site alone was sufficient to induce significant increases in apoptosis. High levels of apoptosis were seen at 80 nM oligonucleotide, with effects lasting up to 48 h after cationic lipid-mediated delivery. It was also found that shifting the balance between the anti- and pro-apoptotic forms of bcl-x was more effective than simply downregulating the bcl-x mRNA. In MCF7 breast cancer cells, the antisense oligonucleotide was unable to induce apoptosis by itself, but did sensitize the cells to apoptotic stimuli such as thapsigargin (Mercatante et al. 2001).

2.2.2
IL-5R

The cytokine interleukin-5 and its receptor (IL-5R) are critical regulators of hematopoiesis and mediators in certain inflammatory diseases such as allergies and asthma (Imamura et al. 1994; Adachi and Alam 1998). The IL-5R gene generates 2 pre-mRNA splice variants that either include or exclude exon 9. The former includes the transmembrane domain of the receptor and forms functional membrane-bound protein, while the latter is a soluble, nonfunctional "decoy" receptor that binds IL-5, but does not mediate a response (Kotsimbos and Hamid 1997). Shifting the balance of splicing to favor the decoy form of IL-5R was hypothesized to be favorable in the treatment of the above conditions.

2′-*O*-MOE-oligonucleotides complementary to specific regions of exon 9 were delivered to B-cell lymphoma cells by electroporation. (Karras et al. 2000). A 20-mer targeted to the 5′ end of exon 9, including the intron/exon junction (Fig. 1E), led to a dramatic decrease in the level of transcripts containing exon

9, with a concomitant increase in those that lacked exon 9. As expected, the results were sequence-specific, with 3 mismatches in the oligonucleotide decreasing its effects by 60%. Dose-dependent effects were seen at concentrations up to 12 μM, with confirmation by Western blot that the insoluble IL-5R protein was downregulated. Interestingly, if the above oligonucleotide contained 7 deoxynucleotides at its center and therefore formed a duplex in which RNase H cleaved the target RNA, then the mRNA levels of both the soluble and insoluble forms of IL-5R were downregulated. This was a surprising result since the mRNA of the soluble form is completely devoid of the target sequence (exon 9). As will be discussed in more detail below, these data strongly argue that the site of action of antisense oligonucleotides is predominantly nuclear.

2.2.3
C-myc

C-myc is a powerful oncogene that is mutated in a number of cancers. While no known alternative splicing forms have been reported, antisense oligonucleotides were able to disrupt normal processing of the c-myc pre-mRNA and effectively downregulate the gene (Giles et al. 1999). Morpholino 28-mers spanning from the initiation AUG codon in exon 2 to the upstream intron/exon border (Fig. 1F) were delivered to KYO1 cells expressing c-myc by streptolysin-O reversible permeabilization. The pre-mRNA was consequently spliced such that a large segment of exon 2 (including the initiation codon) was omitted from the mRNA. Translation then initiated in-frame at a downstream AUG codon, effectively truncating the c-myc gene product. This protein was detectable by c-myc antibodies, and was not found in cells treated with sequence control oligomers or those directed at just the AUG codon and not the intron/exon junction. These results indicate that the morpholino oligomer most likely bound the pre-mRNA in the nucleus and affected RNA processing before translation took place.

3
Oligonucleotides as Probes for Splicing Mechanisms

Alternative splicing is emerging as one of the key processes responsible for the size and diversity of mammalian and human proteomes. It is estimated that in humans 35–85% of genes are alternatively spliced (Krawczak et al. 1992; Consortium 2001). Moreover, several single genes can encode a vast array of protein isoforms. In the most extreme examples, the *slowpoke* calcium ion channel (Brenner et al. 2000) and the *Drosophila* gene *dscam* (Schmucker et al. 2000) can potentially be spliced into at least 500 and 38,000 isoforms, respectively.

The process of pre-mRNA splicing has been intensely studied and basic mechanisms have been elucidated (Krainer 1997; Burge et al. 1999; Hastings and Krainer 2001). Although the chemical transesterification steps appear to be driven by RNA (Valadkhan and Manley 2001), additional protein and ribonucleoprotein factors are essential for formation of the spliceosomes and recognition of the splicing elements involved in splicing. Which sequences, in addition to the 3′ and 5′ splice sites, the branch point and the polypyrimidine tracts are involved, in particular in alternative splicing, is still not completely understood. This is exemplified by the fact that prediction of legitimate splice sites *in silico* still generates substantial false positives/negatives (Thanaraj and Robinson 2000; Kan et al. 2001; Pertea et al. 2001). The key to understanding how alternative splicing is accomplished and regulated seems to lie in the relationship between basic splicing elements and intron and exon splicing enhancers (ISE and ESE). In view of the complexity of the latter, (Liu et al. 2000, 2001) the difficulties are not surprising.

The sections below show that oligonucleotides can be used to identify and characterize pre-mRNA sequences involved in splicing and, indirectly, to measure the strength of the interactions between splice sites or other sequence elements and splicing factors.

3.1 Identification of Splicing Elements

The oligonucleotides that block splice sites can replace mutational analysis for identification of splicing elements. These oligonucleotides show that selection of splice sites occurs via competition between the splicing elements for splicing factors. When access to a splice site is blocked, the splicing machinery selects another one, which under normal circumstances might have been used less effectively, or not at all. The same reasoning applies to other sequence elements involved in splicing.

In the thalassemic mutation IVS1–110, a G to A mutation at position 110 β-globin intron 1 creates a new 3′ splice site at position 109, 21 nucleotides upstream from the correct 3′ splice site (Spritz et al. 1981) resulting in extension of exon 2 and lack of β-globin protein translation. Interestingly, targeting the aberrant splice site or sequences downstream with a 14-mer 2′-*O*-methyloligoribonucleoside phosphorothioate did not correct splicing of IVS1–110 β-globin pre-mRNA, but actually inhibited aberrant splicing. However, oligonucleotides targeted to the branch point and up to 22 nucleotides upstream from this element significantly corrected splicing. These results indicated that the interactions of the splicing factors (e.g., U2 or U2AF) with pre-mRNA extended past the branch point. Further analysis indicated that U2 must have reselected a previously identified cryptic branch point (Zhuang and Weiner 1989), which restored spliceosome formation at the normal 3′ splice site, restoring correct splicing (Dominski and Kole 1994).

Probing in vitro with a series of oligonucleotides spanning the regions adjacent to aberrant splice sites in the IVS2–654 and –705 thalassemic mutants showed that blocking the sequences downstream from the 3′ cryptic and upstream from the 5′ aberrant splice sites prevented their use by the splicing machinery and restored correct splicing (Dominski and Kole 1994). Correction was also observed in vivo and in cell culture models with an oligonucleotide targeted to nucleotide 623 (ON-623) of β-globin intron 2, i.e., 31 and 82 nucleotides upstream from the aberrant 5′ splice sites created by IVS2–654 and IVS2–705 mutations, respectively (unpublished data). The ability of ON-623 to correct aberrant splicing suggested the presence of an ESE. This theory is consistent with mutagenesis experiments that showed that a four-nucleotide deletion within this region corrected splicing of IVS2–654 (Gemignani et al. 2001). The likely interpretation of these results is that the oligonucleotides and the mutation prevented binding of SR-protein(s) to the ESE that helps to recruit the nearby 5′ splice site into the splicing complex that forms across the aberrant exon-like sequence (Hertel et al. 1997; Blencowe 2000; Hastings and Krainer 2001).

3.2 Splice Site Accessibility

The HeLa cell lines that express three thalassemic β-globin mutations IVS2–654, –705 and –745 (Sierakowska et al. 1996) offer an attractive model for in vivo studies of alternative splicing and splice site selection with antisense oligonucleotides. The mutations generate aberrant 5′ splice sites, each of which activates the same cryptic 3′ splice site at nucleotide 579 of β-globin intron 2. The pre-mRNAs are identical, except for a single nucleotide at the site of mutation; likewise, the spliced mRNAs differ only by the length of the fragment of the intron included in the RNA after splicing. The IVS2–654 and IVS2–705 pre-mRNA are spliced exclusively via the aberrant splicing pathway, while IVS2–745 is partially spliced correctly.

Previous work from this laboratory showed that treatment of the cells with 18-mer oligonucleotides targeted to the aberrant 5′ splice sites (ON-654, ON-705 or ON-745) led to significant restoration of correct splicing of β-globin mRNA in all three mutants. However, for the IVS2–654 mutation, maximal correction was achieved at approximately 200 nM oligonucleotide with an EC_{50} of about 115 nM. For the –705 and –745 mutations, efficient correction was detectable at much lower concentrations with EC_{50} values of 5 and 1 nM, respectively (Sierakowska et al. 1999). Although the oligonucleotides were targeted to different sequences within the intron, their affinity to targets, as predicted by Tm values, was essentially the same. Thus, the differences in effective oligonucleotide concentrations must have reflected the accessibility of the splice sites to the oligonucleotides. This accessibility could be determined by the structure of the RNA and/or its interactions with the splicing factors, such as U1, that compete with oligonucleotides for the splice site sequences.

The differences in the accessibility of the splice sites were also confirmed by treatment of the three cell lines with an antisense 17-mer oligonucleotide (ON-3′cr) targeted to the common 3′ cryptic splice site (Sierakowska et al. 1999). In this instance, the rank order of susceptibility to correction (654 << 705 < 745) was maintained. However, the differences in the EC_{50} values of the same ON-3′cr used in the three different contexts were particularly dramatic. Splicing of the IVS2–654 pre-mRNA was minimally affected, with ON-3′cr EC_{50} of 1500 nM; EC_{50} values for –705 and –745 mutations were 30 and 2 nM, respectively. Thus the accessibility, as reflected in EC_{50} values, of the same splice site to the same oligonucleotide changed by 750-fold depending on the 5′ splice site downstream. This is further reinforced by the observations that changing the IVS2–654 and –705 splice sites to consensus 5′ splice sites (IVS2–654con, and –705con) led to additional decreases in accessibility of the 3′ splice site to ON-3′cr. In IVS2–654con pre-mRNA this splice site was completely unresponsive to ON-3′cr, while that of IVS2–705con was reduced fourfold. These observations are consistent with the exon definition mechanism, which postulates that there is interaction between the 3′ and 5′ splice sites flanking the exon. The interactions are mediated through a series of RNA-protein and protein-protein interactions across the exon (Berget 1995). The fact that ON-623, targeted to the sequence between the splice sites, also restores correct splicing provides further evidence for this model and for the usefulness of the oligonucleotides as tools for the study of pre-mRNA-spliceosome interactions.

It is notable that IVS2–654 and IVS2–705 pre-mRNAs seem to be spliced exclusively via the aberrant pathway. Yet their susceptibility to oligonucleotide interference (23-fold difference at the 5′ splice site, 50-fold at the 3′) points to significant differences in the pre-mRNAs interactions with splicing factors. The oligonucleotides can quantify these effects or even those of relatively minor changes, such as increased match of IVS2–654 and –705 splice sites to consensus sequence. Even more subtle effects are detectable by antisense oligonucleotides.

The IVS2–654 intron was inserted into enhanced green fluorescent protein coding sequence (EGFP) and the resultant EGFP-654 construct was stably expressed in HeLa cells (Kang et al. 1998; Sazani et al. 2001). Similarly to the IVS2–654 β-globin construct, background levels of correct splicing were minimal to nil and efficient correction of splicing was seen when the aberrant 5′ splice was targeted by antisense oligonucleotides. Unlike in the β-globin context, however, ON-3′cr and ON-654 both efficiently corrected splicing to a similar extent and at similar concentrations (Sazani and Kole, unpubl. data). Likewise, a transiently expressed modified U7 antisense RNA that includes an antisense sequence targeted to the same region as ON-3′cr (Gorman et al. 1998) efficiently corrected splicing of EGFP-654 (Sazani and Kole unpubl.). The same construct exhibited only trace effects against β-globin IVS2–654 pre-mRNA. Furthermore, in the EGFP context, the IVS2–705con intron was spliced exclusively via the aberrant pathway, yet it was correctable with approximately fourfold lower concentration of ON-3′cr than IVS2–705con in the β-globin context. In other words, the EGFP exons flanking the IVS2–654 or IVS2–705con introns decreased the strength of the interactions between the aberrant 3′ splice site and the splicing factors, allowing the oligonucleotide to compete more successfully for binding to the pre-mRNA molecule than in the context of β-globin.

Further work will be needed to identify the nature of the splicing factors that were inhibited from binding to their targets. Nevertheless, oligonucleotides detect and quantify the strength of their interactions between the pre-mRNA and the splicing factors at the splice sites and the ESE regions. They also measure the long distance effects of exons on the interactions within the intron between pre-mRNA and splicing factors.

4 Splicing Assay for Antisense Activity of Oligonucleotides

As shown above, antisense oligonucleotides can be used to assess the accessibility of splice sites and susceptibility of alternative splicing to manipulation. Conversely, alternative splicing can easily be exploited as an assay for the antisense activity of oligonucleotides. This approach was developed into quantitative, high throughput assays for characterization of antisense activity and its dependence on the chemistry and methods of delivery of oligonucleotides (Kang et al. 1998; Sazani et al. 2001).

4.1 Synthetic Antisense Oligonucleotides

Several requirements must be met in the design of a modified oligonucleotide used to shift splicing: (1) it must form duplexes with RNA resistant to RNase H (Furdon et al. 1989), which would otherwise destroy the pre-mRNA in the duplex before it could be spliced and ultimately translated into protein (Dominski and Kole 1993), (2) it must be able to reach its site of action within the cell (Kole and Sazani 2001), and (3) it must have high enough affinity to the target sequence to effectively compete with splicing factors for binding. Over the past decade, a large number of modified oligonucleotides have emerged (Micklefield 2001); those applicable to modification of alternative splicing are discussed below.

4.1.1 2′-Modifications of Ribose

Oligoribonucleotides or those with modified 2′ hydroxyl groups render the RNA in oligonucleotide-RNA duplex resistant to RNase H. 2′-*O*-methyl- (Fig. 2A) and more recently 2′-*O*-methoxyethyl- (2′-*O*-MOE; Fig. 2B) derivatives

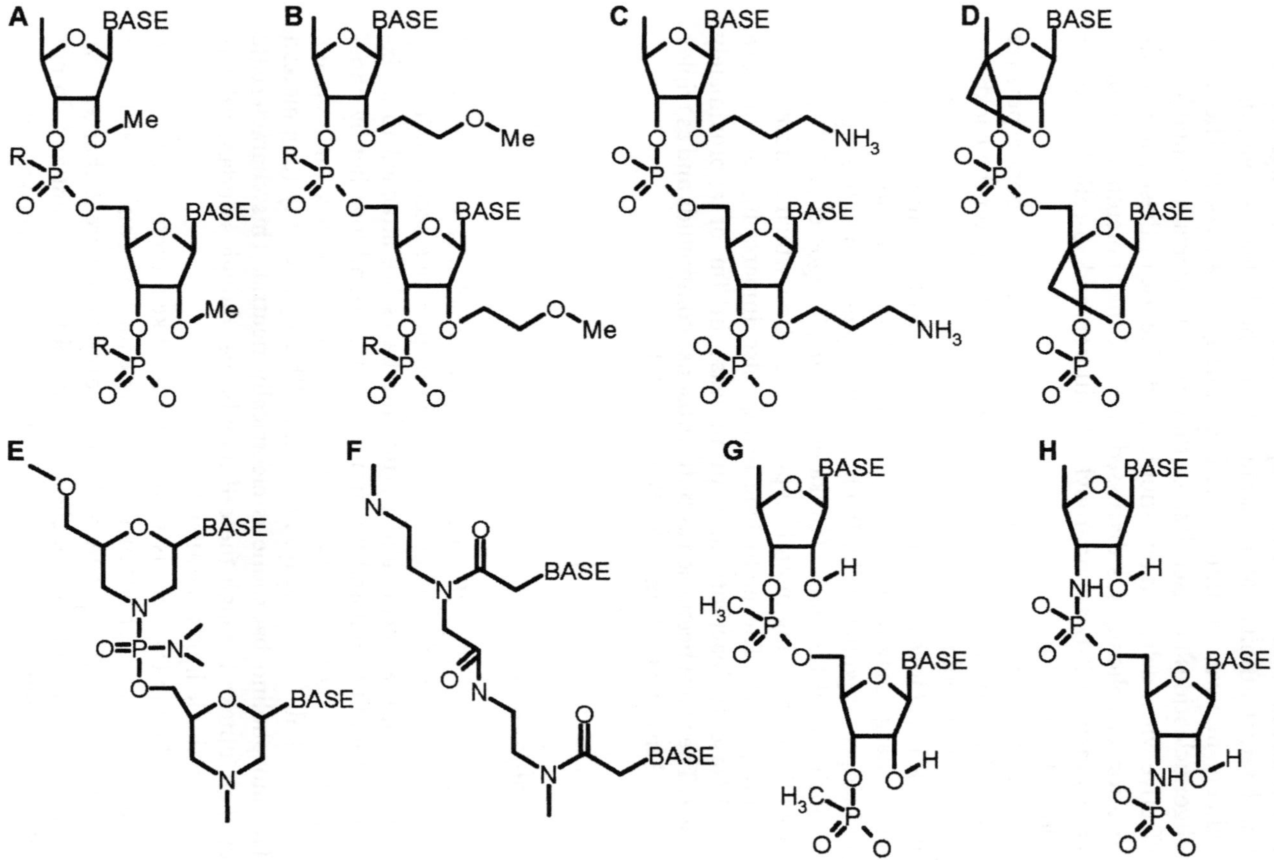

Fig. 2A–H. Synthetic oligonucleotides. **A** 2′-*O*-methyl- **B** 2′-*O*-methoxyethyl- **C** 2′-*O*-aminopropyl-oligoribonucleotides. **D** 2′-*O*, 4′-C-Methylene bridged nucleic acid (BNA). **E** Morpholino oligomers. **F** Peptide nucleic acid (PNA). **G** Methylphosphonate-. **H** Phosphoramidate- oligonucleotides. R = O, S in phosphodiester or phosphorothioate internucleotide bonds

were found to form RNase H-resistant duplexes and also exhibit high nuclease stability and target affinity (Manoharan 1999). The latter compounds also showed effects in vivo, in animal studies (Zhang et al. 2000). The 2′-*O*-methyl-oligonucleotides are useful, since they are commercially available and highly effective when delivered by cationic lipid. The use of these compounds in modification of splicing was discussed above. Other potentially useful 2′ modifications include 2′-*O*-aminopropyl substitution and so-called locked or bridged oligonucleotides.

The zwitterionic 2′-*O*-aminopropyl oligomer (Fig. 2C) harbors a positive charge at the 2′ position in addition to the negative charge on the backbone (Griffey et al. 1996; Teplova et al. 1999). It has increased Tm (+1 °C/base) and higher nuclease resistance than unmodified oligonucleotides, and the presence of a positive charge may improve its uptake and biodistribution properties. Locked or bridged oligomers (Fig. 2D) have a methylene group linking the 2′-*O*- and 4′ positions of the ribose (Obika et al. 2001; Torigoe et al. 2001). This bridge locks the ribose ring in an N-type conformation, which is more favorable for binding RNA. The methylene bridge also imparts higher nuclease resistance, RNase H inactivity and an increase in Tm of approximately +4–5 °C/base. They have been used as both antisense compounds and as triplex forming strand invaders (Orum and Wengel 2001).

4.1.2
Backbone Modifications

Morpholino oligomers (Fig. 2E) and peptide nucleic acids (PNA) were found useful in modification of splicing as discussed in previous sections. In morpholino oligomers, the purine and pyrimidine bases are attached to a six-membered morpholine ring and the monomers are bound together by phosphordiamidate linkages. They have high nuclease resistance, high duplex stability (Stein et al. 1997) and do not induce RNase H activity (Summerton 1999). The morpholino backbone is electrically neutral. This characteristic improves their uptake through the cell membrane, though standard cellular delivery with cationic lipids is difficult.

The PNA backbone (Fig. 2F) is an oligoglycine-like structure with nucleic acid base residues spaced at distances appropriate for base pairing (Nielsen et al. 1991; Hanvey et al. 1992). PNA molecules carry no inherent charge, have notably high target affinity, are nuclease and peptidase-resistant and do not activate RNase H. Their ability to cross cellular membranes has been reported as rather limited. As shown below, this obstacle can be overcome by further modifications (Sazani et al. 2001).

Methylphosphonate oligomers (Fig. 2G) have one of their nonbridging oxygen atoms replaced by a methyl group (Miller 1998). This eliminates the negative backbone charge present on normal oligonucleotides, and imparts nuclease resistance and RNase H inactivity to the compound (Furdon et al. 1989; Akhtar et al. 1991). However, this modification decreases the thermal stability of duplexes, and the cellular uptake of the oligonucleotide was not as good as anticipated. However, chimeric molecules incorporating methylphosphonate and 2′-*O*-methyl monomers had antiproliferative effects against HSV virus in Vero cells (Kean et al. 1995; Miller et al. 2000) and may potentially be useful for antiviral treatment.

Replacement of the oxygen atom linking the 3′ hydroxyl of the deoxyribose to the phosphorous with a nitrogen atom generates oligonucleotides with N3′→P5′ phosphoramidate (NP) internucleotide linkage (Fig. 2H; Letsinger and Mungall 1970). The NP oligonucleotides exhibit high affinity for target RNAs (+2.6 °C/base; Gryaznov et al. 1995), high nuclease stability and resistance of RNA/NP duplexes to RNase H (DeDionisio and Gryaznov 1995; Heidenreich et al. 1997). Recently, they have been found active in triplex formation (Faria et al. 2000) and in the inhibition of translation in cells (Faria et al. 2001) and in vivo (Wang et al. 1999), but were not tested in modification of splicing.

4.1.3
Base Modifications

Of the multitude of possible base modifications (Herdewijn 2000), few were amenable to antisense studies. All good candidates increase the thermal stability of duplexes, maintain sequence specificity and, for the most part, maintain the RNase H activity of the oligomer to which it is attached. The best studied of these modifications is the C-5 propynyl cytosine analog, which increased antisense activity of oligonucleotides carrying this modification at multiple positions in the oligodeoxynucleotide with a phosphorothioate backbone (Moulds et al. 1995; Raviprakash et al. 1995; Flanagan et al. 1996). Other cytosine analogs include the heterocyclic phenoxazine and G-clamp molecules (Lin et al. 1995; Lin and Matteucci 1998). They both pair with guanosine with high affinity; the latter forms an extra hydrogen bond at the hoogstein face, enhancing antisense activity in the context of a phosphorothioate oligodeoxynucleotide (Flanagan et al. 1999a,b). Finally, modifications to the 7 position of adenosine have been shown to increase thermal stability of the duplex by up to 4 °C/base (Balow et al. 1998), with only a slight increase in antisense activity seen. However, the combination of these substitutions with RNase H inhibiting backbones may generate oligonucleotides useful in modification of splicing.

4.2
Splicing Assay for Antisense Activity of Synthetic Oligonucleotides

As mentioned above, the IVS2–654 β-globin intron was inserted into the EGFP coding sequence. HeLa cells expressing the EGFP-654 construct were used as a splicing assay to compare the activity of the ON-654 oligonucleotide with four different backbones (Sazani et al. 2001): 2′-*O*-methyl-, 2′-*O*-MOE-, morpholino and PNA. The oligonucleotides were scrape loaded into HeLa cells and generation of EGFP was quantified by FACS analysis. Scrape loading temporarily damages the cell membrane, allowing the oligonucleotides to diffuse into the cells (Partridge et al. 1996). This strategy allows us to directly compare the intracellular activity of the oligonucleotides, thus bypassing the issue of the uptake through the cell membrane. Under these conditions 2′-*O*-MOE, morpholino and PNA derivatives were equally effective (Sazani et al. 2001); the effects of 2′-*O*-methyl-oligonucleotides were lower, likely because of their lower resistance to nucleases (Manoharan 1999).

Free uptake of the four oligonucleotides showed large differences in their antisense activity. Interestingly, the negatively charged 2′-*O*-methyl- and 2′-*O*-MOE-oligonucleotides were virtually inactive at concentrations of up to 10 μM. In contrast, the neutral morpholino and PNA-1K (with single positively charged lysine at the 3′ end) oligomers generated high levels of EGFP at a concentration as low as 1 μM. To determine if the positive charges contributed to antisense effects, the PNA conjugated to four lysines (PNA-4K) was tested. This oligomer had an EC_{50} 2.5-fold lower than that of PNA-1K (Sazani et al. 2001).

A similar assay based on a luciferase readout from a luciferase-IVS2–705 construct was also generated (Kang et al. 1998) and found useful in comparison of the efficacy of various cationic lipid and nonlipid oligonucleotide delivery adjuvants (Kang et al. 1999). The advantage of both splicing based assays was that the activity of oligonucleotides led to upregulation of easily quantifiable EGFP or luciferase signals, allowing for rapid acquisition of a large number of data points with increased sensitivity.

5 Site of Action of Antisense Oligonucleotides

5.1 The Nucleus vs. Cytoplasm

In most of the reports in the antisense field, the oligonucleotides are designed to target cytoplasmic mRNA. However, a number of results point to the nucleus as the predominant site of action for antisense oligonucleotides. The earliest data came from the development of cationic lipid carriers enhancing efficacy of antisense oligonucleotides (Bennett et al. 1992). It was found that the Lipofectin cationic lipid carrier dramatically increased antisense activity and concomitantly nuclear localization of an 18-mer phosphorothioate 2′-deoxyoligonucleotide targeted to the AUG codon of ICAM-1 mRNA in HUVECs. Later studies with double-labeled oligonucleotide/lipid complexes showed that the complex enters the cell by first associating with the cell surface and then localizing in punctate, endosomal, structures within the cytoplasm.

At later times, the lipid remains associated with the endosomes while the oligonucleotide escapes and freely translocates to and accumulates in the nucleus (Zelphati and Szoka 1996; Marcusson et al. 1998). While these data suggest that the oligonucleotide is at least transiently cytoplasmic, functional studies revealed that antisense effects of phosphorothioate oligomers directed towards PKC-α mRNA were seen only after significant nuclear accumulation (Marcusson et al. 1998).

The most obvious and strongest argument for the nuclear action of oligonucleotides is the fact that they are able to shift splicing, a strictly nuclear process (Kole and Sazani 2001). In particular, reports by several groups show that even though cytoplasmic mRNAs were the intended target, pre-mRNA splicing was actually affected. For example, targeting an alternatively spliced exon with an RNase -H-inactive antisense oligomer simultaneously decreased the level of mRNA that contained that exon and increased the mRNA in which that exon was not included (Karras et al. 2000). Since the oligomer could not bind to the mRNA lacking the exonic sequence, it must have bound to the pre-mRNA in the nucleus, thus promoting skipping of the alternatively spliced exon. Similarly, Condon and Bennett found that an RNase H-active oligonucleotide promoted retention of the 13th intron of E-selectin when targeted to the 3′ untranslated region (Condon and Bennett 1996). This result could only be obtained if the oligonucleotide affected the pre-mRNA before it was spliced. As mentioned above, targeting of the AUG codon of c-myc mRNA by a 28-mer morpholino oligomer led to activation of a cryptic splice site (Giles et al. 1999).

Additional studies from this laboratory confirmed the nucleus as a site of action of RNase H-inactive oligonucleotides. A morpholino oligomer was used to correct aberrant splicing of the β^E mutation that activates a cryptic 5′ splice site within exon 1 of β-globin. Since the oligonucleotide targets a coding sequence, there was a possibility that it may inhibit translation even if correct splicing is restored. However, treatment of model HeLa cell lines led to the expected increase in the level of the correctly spliced mRNA and in translation of β-globin protein. Thus, the oligonucleotide shifted splicing in the nucleus, but did not remain with the spliced mRNA to interfere with its export to the cytoplasm or its translation (Suwanmanee and Kole, unpubl.).

The results discussed above strongly argue that oligonucleotides act predominantly, if not exclusively, in the nucleus. However, other studies were consistent with the cytoplasmic action of these compounds. For example, morpholino oligomers targeted directly at the AUG or closely downstream, efficiently blocked translation of a hepatitis B virus/luciferase construct in vitro and in cell culture. Targeting the 5′ untranslated region (UTR) had no effect (Summerton et al. 1997). A recent study convincingly showed that NP oligonucleotides inhibited translation of a luciferase reporter construct in cell culture and in electroporated mouse muscle (Faria et al. 2001). The oligonucleotides were targeted to the 5′ UTR and caused downregulation of luciferase in a dose-dependant and sequence-specific manner.

Recent studies provide evidence that may reconcile the nuclear and cytoplasmic arguments. Iborra et al. (2001) showed the existence of discrete nuclear bodies where translation occurs along with RNA polymerase II transcription. While the fraction of overall translation in the nucleus is small, it may represent an initial quality control mechanism for transcripts before they enter the cytoplasm. Moreover, a pioneer round of translation, possibly taking place in the nucleus, seems to be responsible for the nonsense-mediated decay (NMD) pathway (Brogna 2001; Ishigaki et al. 2001). Thus, the binding of a foreign oligonucleotide could be recognized as a defect in the mRNA, leading to destruction of that message in the nucleus, by NMD-like mechanisms.

Another possible explanation of the nuclear/cytoplasmic paradox is that the oligonucleotide simply acts at both compartments. A recent study showed that phosphorothioate oligonucleotides can shuttle between the nucleus and the cytoplasm, and that this shuttling does not affect their antisense activity (Lorenz et al. 2000). The shuttling was apparently energy-dependent, being sensitive to both low temperature and ATP depletion. Furthermore, transport seemed to be mediated through the nuclear pore complex; it was not determined whether or not the shuttling oligonucleotide was bound by a target RNA.

6 Conclusions

Alternative splicing is emerging as a key mechanism responsible for the complexity of the human proteome. Selection of an alternative splice site is the culmination of a number of interrelated and competing interactions, and constitutes one of the most important molecular balancing acts in the cell. Splicing gone awry leads to a number of disease states, including cancer. Therefore, alternative splicing is a good target for manipulation by drugs, and antisense oligonucleotides represent the only class of such drugs with high sequence specificity. Furthermore, oligonucleotides provide diagnostic tools to identify important elements governing alternative splicing and splice sites that are susceptible to manipulation. As with any macromolecular drug, delivery to target tissues/cells is the major issue. This problem can be attacked through the use of quick readout splicing assays, such as the EGFP-654 system. In this way, oligonucleotide chemistries and delivery system combinations can be rapidly tested and developed, ensuring that antisense oligonucleotides will emerge as potent treatments for disorders caused by aberrant and alternative splicing.

References

Abeliovich D, Lavon IP, Lerer I, Cohen T, Springer C, Avital A, Cutting GR (1992) Screening for five mutations detects 97% of cystic fibrosis (CF) chromosomes and predicts a carrier frequency of 1:29 in the Jewish Ashkenazi population. Am J Hum Genet 51:951–956

Adachi T, Alam R (1998) The mechanism of IL-5 signal transduction. Am J Physiol 275:C623–C633

Akhtar S, Kole R, Juliano RL (1991) Stability of antisense DNA oligodeoxynucleotide analogs in cellular extracts and sera. Life Sci 49:1793–1801

Ambros V (2000) Control of developmental timing in *Caenorhabditis elegans*. Curr Opin Genet Dev 10:428–433

Balow G, Mohan V, Lesnik EA, Johnston JF, Monia BP, Acevedo OL (1998) Biophysical and antisense properties of oligodeoxynucleotides containing 7-propynyl-, 7-iodo- and 7-cyano-7-deaza-2-amino-2′-deoxyadenosines. Nucleic Acids Res 26:3350–3357

Bennett CF, Chiang MY, Chan H, Shoemaker JE, Mirabelli CK (1992) Cationic lipids enhance cellular uptake and activity of phosphorothioate antisense oligonucleotides. Mol Pharmacol 41:1023–1033

Berget SM (1995) Exon recognition in vertebrate splicing. J Biol Chem 270:2411–2414

Blencowe BJ (2000) Exonic splicing enhancers: mechanism of action, diversity and role in human genetic diseases. Trends Biochem Sci 25:106–110

Boise LH, Gonzalez-Garcia M, Postema CE, Ding L, Lindsten T, Turka LA, Mao X, Nunez G, Thompson CB (1993) bcl-x, a bcl-2-related gene that functions as a dominant regulator of apoptotic cell death. Cell 74:597–608

Brenner R, Yu JY, Srinivasan K, Brewer L, Larimer JL, Wilbur JL, Atkinson NS (2000) Complementation of physiological and behavioral defects by a slowpoke Ca(2+) -activated K(+) channel transgene. J Neurochem 75:1310–1319

Brogna S (2001) Pre-mRNA processing: insights from nonsense. Curr Biol 11:R838–R841

Bulfield G, Siller WG, Wight PA, Moore KJ (1984) X chromosome-linked muscular dystrophy (mdx) in the mouse. Proc Natl Acad Sci USA 81:1189–1192

Burge C, Tuschl T, Sharp P (1999) Splicing of precursors to mRNAs by the spliceosomes. In: Gesteland R, Cech T, Atkins J (eds) The RNA world. CSHL Press, Cold Spring Harbor, pp 525–560

Condon TP, Bennett CF (1996) Altered mRNA splicing and inhibition of human E-selectin expression by an antisense oligonucleotide in human umbilical vein endothelial cells. J Biol Chem 271:30398–30403

Consortium IHGS (2001) Initial sequencing and analysis of the human genome. Nature 409:860–921

Crooke S (ed) (2001) Antisense drug technology. Marcel Dekker, New York

DeDionisio L, Gryaznov SM (1995) Analysis of a ribonuclease H digestion of N3′→P5′ phosphoramidate-RNA duplexes by capillary gel electrophoresis. J Chromatogr B Biomed Appl 669:125–131

Dominski Z, Kole R (1993) Restoration of correct splicing in thalassemic pre-mRNA by antisense oligonucleotides. Proc Natl Acad Sci USA 90:8673–8677

Dominski Z, Kole R (1994) Identification and characterization by antisense oligonucleotides of exon and intron sequences required for splicing. Mol Cell Biol 14:7445–7454

Dunckley MG, Manoharan M, Villiet P, Eperon IC, Dickson G (1998) Modification of splicing in the dystrophin gene in cultured Mdx muscle cells by antisense oligoribonucleotides. Hum Mol Genet 7:1083–1090

Faria M, Wood CD, Perrouault L, Nelson JS, Winter A, White MR, Helene C, Giovannangeli C (2000) Targeted inhibition of transcription elongation in cells mediated by triplex-forming oligonucleotides. Proc Natl Acad Sci USA 97:3862–3867

Faria M, Spiller DG, Dubertret C, Nelson JS, White MR, Scherman D, Helene C, Giovannangeli C(2001) Phosphoramidate oligonucleotides as potent antisense molecules in cells and in vivo. Nat Biotechnol 19:40–44

Flanagan WM, Kothavale A, Wagner RW (1996) Effects of oligonucleotide length, mismatches and mRNA levels on C-5 propyne-modified antisense potency. Nucleic Acids Res 24:2936–2941

Flanagan WM, Wagner RW, Grant D, Lin KY, Matteucci MD (1999a) Cellular penetration and antisense activity by a phenoxazine-substituted heptanucleotide. Nat Biotechnol 17:48–52

Flanagan WM, Wolf JJ, Olson P, Grant D, Lin KY, Wagner RW, Matteucci MD (1999b) A cytosine analog that confers enhanced potency to antisense oligonucleotides. Proc Natl Acad Sci USA 96:3513–3518

Friedman KJ, Kole J, Cohn JA, Knowles MR, Silverman LM, Kole R (1999) Correction of aberrant splicing of the cystic fibrosis transmembrane conductance regulator (CFTR) gene by antisense oligonucleotides. J Biol Chem 274:36193–36199

Furdon PJ, Dominski Z, Kole R (1989) RNase H cleavage of RNA hybridized to oligonucleotides containing methylphosphonate, phosphorothioate and phosphodiester bonds. Nucleic Acids Res 17:9193–9204

Gemignani F, Landi S, DeMarini DM, Kole R (2001) Spontaneous and MNNG-induced reversion of an EGFP construct in HeLa cells: an assay for observing mutations in living cells by fluorescent microscopy. Hum Mutat 18:526–534

Giles RV, Spiller DG, Clark RE, Tidd DM (1999) Antisense morpholino oligonucleotide analog induces missplicing of C-myc mRNA. Antisense Nucleic Acid Drug Dev 9:213–220

Goedert M, Spillantini MG, Potier MC, Ulrich J, Crowther RA (1989) Cloning and sequencing of the cDNA encoding an isoform of microtubule-associated protein tau containing four tandem repeats: differential expression of tau protein mRNAs in human brain. EMBO J 8:393–399

Gorman L, Suter D, Emerick V, Schumperli D, Kole R (1998) Stable alteration of pre-mRNA splicing patterns by modified U7 small nuclear RNAs. Proc Natl Acad Sci USA 95:4929–4934

Griffey RH, Monia BP, Cummins LL, Freier S, Greig MJ, Guinosso CJ, Lesnik E, Manalili SM, Mohan V, Owens S, Ross BR, Sasmor H, Wancewicz E, Weiler K, Wheeler PD, Cook PD (1996) 2′-*O*-aminopropyl ribonucleotides: a zwitterionic modification that enhances the exonuclease resistance and biological activity of antisense oligonucleotides. J Med Chem 39:5100–5109

Gryaznov SM, Lloyd DH, Chen JK, Schultz RG, DeDionisio LA, Ratmeyer L, Wilson WD (1995) Oligonucleotide N3′→P5′ phosphoramidates. Proc Natl Acad Sci USA 92:5798–5802

Hanvey JC, Peffer NJ, Bisi JE, Thomson SA, Cadilla R, Josey JA, Ricca DJ, Hassman CF, Bonham MA, Au KG et al. (1992) Antisense and antigene properties of peptide nucleic acids. Science 258:1481–1485

Hastings ML, Krainer AR (2001) Pre-mRNA splicing in the new millennium. Curr Opin Cell Biol 13:302–309

Hastings ML, Ingle HA, Lazar MA, Munroe SH (2000) Post-transcriptional regulation of thyroid hormone receptor expression by *cis*-acting sequences and a naturally occurring antisense RNA. J Biol Chem 275:11507–11513

Heidenreich O, Gryaznov S, Nerenberg M (1997) RNase H-independent antisense activity of oligonucleotide N3′→P5′ phosphoramidates. Nucleic Acids Res 25:776–780

Herdewijn P (2000) Heterocyclic modifications of oligonucleotides and antisense technology. Antisense Nucleic Acid Drug Dev 10:297–310

Hertel KJ, Lynch KW, Maniatis T (1997) Common themes in the function of transcription and splicing enhancers. Curr Opin Cell Biol 9:350–357

Highsmith WE, Burch LH, Zhou Z, Olsen JC, Boat TE, Spock A, Gorvoy JD, Quittel L, Friedman KJ, Silverman LM et al. (1994) A novel mutation in the cystic fibrosis gene in patients with pulmonary disease but normal sweat chloride concentrations. N Engl J Med 331:974–980

Hutton M (2001) Missense and splice site mutations in tau associated with FTDP-17: multiple pathogenic mechanisms. Neurology 56:S21–S25

Iborra FJ, Jackson DA, Cook PR (2001) Coupled transcription and translation within nuclei of mammalian cells. Science 293:1139–1142

Imamura F, Takaki S, Akagi K, Ando M, Yamamura K, Takatsu K, Tominaga A (1994) The murine interleukin-5 receptor alpha-subunit gene: characterization of the gene structure and chromosome mapping. DNA Cell Biol 13:283–292

Ishigaki Y, Li X, Serin G, Maquat LE (2001) Evidence for a pioneer round of mRNA translation: mRNAs subject to nonsense-mediated decay in mammalian cells are bound by CBP80 and CBP20. Cell 106:607–617

Kalbfuss B, Mabon SA, Misteli T (2001) Correction of alternative splicing of tau in frontotemporal dementia and parkinsonism linked to chromosome 17. J Biol Chem 276:42986–42993

Kan Z, Rouchka EC, Gish WR, States DJ (2001) Gene structure prediction and alternative splicing analysis using genomically aligned ESTs. Genome Res 11:889–900

Kang SH, Cho MJ, Kole R (1998) Up-regulation of luciferase gene expression with antisense oligonucleotides: implications and applications in functional assay development. Biochemistry 37:6235–6239

Kang SH, Zirbes EL, Kole R (1999) Delivery of antisense oligonucleotides and plasmid DNA with various carrier agents. Antisense Nucleic Acid Drug Dev 9:497–505

Karras JG, McKay RA, Dean NM, Monia BP (2000) Deletion of individual exons and induction of soluble murine interleukin-5 receptor-alpha chain expression through antisense oligonucleotide-mediated redirection of pre-mRNA splicing. Mol Pharmacol 58:380–387

Kean JM, Kipp SA, Miller PS, Kulka M, Aurelian L (1995) Inhibition of herpes simplex virus replication by antisense oligo-2′-*O*-methylribonucleoside methylphosphonates. Biochemistry 34:14617–14620

Koenig M, Beggs AH, Moyer M, Scherpf S, Heindrich K, Bettecken T, Meng G, Muller CR, Lindlof M, Kaariainen H et al. (1989) The molecular basis for Duchenne versus Becker muscular dystrophy: correlation of severity with type of deletion. Am J Hum Genet 45:498–506

Kole R, Sazani P (2001) Antisense effects in the cell nucleus: modification of splicing. Curr Opin Mol Ther 3:229–234

Kole R, Shukla R, Akhtar S (1991) Pre-mRNA splicing as a target for antisense oligonucleotides. Adv Drug Delivery Rev 6:271–286

Kotsimbos AT, Hamid Q (1997) IL-5 and IL-5 receptor in asthma. Mem Inst Oswaldo Cruz 92:75–91

Krainer A (ed) (1997) Eukaryotic mRNA processing. Oxford University Press, New York

Krajewska M, Krajewski S, Epstein JI, Shabaik A, Sauvageot J, Song K, Kitada S, Reed JC (1996) Immunohistochemical analysis of bcl-2, bax, bcl-X, and mcl-1 expression in prostate cancers. Am J Pathol 148:1567–1576

Krawczak M, Reiss J, Cooper DN (1992) The mutational spectrum of single base-pair substitutions in mRNA splice junctions of human genes: causes and consequences. Hum Genet 90:41–54

Lacerra G, Sierakowska H, Carestia C, Fucharoen S, Summerton J, Weller D, Kole R (2000) Restoration of hemoglobin A synthesis in erythroid cells from peripheral blood of thalassemic patients. Proc Natl Acad Sci USA 97:9591–9596

Letsinger RL, Mungall WS (1970) Phosphoramidate analogs of oligonucleotides. J Org Chem 35:3800–3803

Lewis J, Yang B, Kim R, Sierakowska H, Kole R, Smithies O, Maeda N (1998) A common human beta globin splicing mutation modeled in mice. Blood 91:2152–2156

Lin KY, Matteucci MD (1998) A cytosine analog capable of clamp-like binding to a guanine in helical nucleic acids. J Am Chem Soc 120:8531–8532

Lin KY, Jones RJ, Matteucci M (1995) Tricyclic 2′-deoxycytidine analogs: synthesis and incorporation into oligonucleotides which have enhanced binding to complementary RNA. J Am Chem Soc 117:3873–3874

Liu HX, Chew SL, Cartegni L, Zhang MQ, Krainer AR (2000) Exonic splicing enhancer motif recognized by human SC35 under splicing conditions. Mol Cell Biol 20:1063–1071

Liu HX, Cartegni L, Zhang MQ, Krainer AR (2001) A mechanism for exon skipping caused by nonsense or missense mutations in BRCA1 and other genes. Nat Genet 27:55–58

Lorenz P, Misteli T, Baker BF, Bennett CF, Spector DL (2000) Nucleocytoplasmic shuttling: a novel in vivo property of antisense phosphorothioate oligodeoxynucleotides. Nucleic Acids Res 28:582–592

Mann CJ, Honeyman K, Cheng AJ, Ly T, Lloyd F, Fletcher S, Morgan JE, Partridge TA, Wilton SD (2001) Antisense-induced exon skipping and synthesis of dystrophin in the mdx mouse. Proc Natl Acad Sci USA 98:42–47

Manoharan M (1999) 2′-Carbohydrate modifications in antisense oligonucleotide therapy: importance of conformation, configuration and conjugation. Biochim Biophys Acta 1489: 117–130

Marcusson EG, Bhat B, Manoharan M, Bennett CF, Dean NM (1998) Phosphorothioate oligodeoxyribonucleotides dissociate from cationic lipids before entering the nucleus. Nucleic Acids Res 26:2016–2023

Mercatante DR, Bortner CD, Cidlowski JA, Kole R (2001) Modification of alternative splicing of Bcl-x pre-mRNA in prostate and breast cancer cells. Analysis of apoptosis and cell death. J Biol Chem 276:16411–16417

Micklefield J (2001) Backbone modification of nucleic acids: synthesis, structure and therapeutic applications. Curr Med Chem 8:1157–1179

Miller P (1998) Oligonucleotide methylphosphonates: synthesis and properties. In: Stein C, Krieg A (eds) Applied antisense oligonucleotide technology. Wiley-Liss, New York, pp 3–22

Miller PS, Cassidy RA, Hamma T, Kondo NS (2000) Studies on anti-human immunodeficiency virus oligonucleotides that have alternating methylphosphonate/phosphodiester linkages. Pharmacol Ther 85:159–163

Moulds C, Lewis JG, Froehler BC, Grant D, Huang T, Milligan JF, Matteucci MD, Wagner RW (1995) Site and mechanism of antisense inhibition by C-5 propyne oligonucleotides. Biochemistry 34:5044–5053

Nielsen PE, Egholm M, Berg RH, Buchardt O (1991) Sequence-selective recognition of DNA by strand displacement with a thymine-substituted polyamide. Science 254:1497–1500

Obika S, Uneda T, Sugimoto T, Nanbu D, Minami T, Doi T, Imanishi T (2001) 2′-O,4′-C-Methylene bridged nucleic acid (2′,4′-BNA): synthesis and triplex-forming properties. Bioorg Med Chem 9:1001–1011

Orum H, Wengel J (2001) Locked nucleic acids: a promising molecular family for gene-function analysis and antisense drug development. Curr Opin Mol Ther 3:239–243

Partridge M, Vincent A, Matthews P, Puma J, Stein D, Summerton J (1996) A simple method for delivering morpholino antisense oligos into the cytoplasm of cells. Antisense Nucleic Acid Drug Dev 6:169–175

Pertea M, Lin X, Salzberg SL (2001) GeneSplicer: a new computational method for splice site prediction. Nucleic Acids Res 29:1185–1190

Raviprakash K, Liu K, Matteucci M, Wagner R, Riffenburgh R, Carl M (1995) Inhibition of dengue virus by novel, modified antisense oligonucleotides. J Virol 69:69–74

Sazani P, Kang SH, Maier MA, Wei C, Dillman J, Summerton J, Manoharan M, Kole R (2001) Nuclear antisense effects of neutral, anionic and cationic oligonucleotide analogs. Nucleic Acids Res 29:3965–3974

Schmajuk G, Sierakowska H, Kole R (1999) Antisense oligonucleotides with different backbones. Modification of splicing pathways and efficacy of uptake. J Biol Chem 274:21783–21789

Schmucker D, Clemens JC, Shu H, Worby CA, Xiao J, Muda M, Dixon JE, Zipursky SL (2000) *Drosophila* Dscam is an axon guidance receptor exhibiting extraordinary molecular diversity. Cell 101:671–684

Schwartz E, Benz E (1995) Thalassemia syndromes._In: Hoffman R, Benz E, Shattil S, Furie B, Cohen H, Silberstein L, McGlave P (eds) Hematology: basic principles and practice. Churchill Livingston, New York

Sierakowska H, Sambade MJ, Agrawal S, Kole R (1996) Repair of thalassemic human beta-globin mRNA in mammalian cells by antisense oligonucleotides. Proc Natl Acad Sci USA 93: 12840–12844

Sierakowska H, Sambade MJ, Schumperli D, Kole R (1999) Sensitivity of splice sites to antisense oligonucleotides in vivo. RNA 5:369–377

Simons RW, Kleckner N (1988) Biological regulation by antisense RNA in prokaryotes. Annu Rev Genet 22:567–600

Spritz RA, Jagadeeswaran P, Choudary PV, Biro PA, Elder JT, deRiel JK, Manley JL, Gefter ML, Forget BG, Weissman SM (1981) Base substitution in an intervening sequence of a beta+-thalassemic human globin gene. Proc Natl Acad Sci USA 78:2455–2459

Stein D, Foster E, Huang SB, Weller D, Summerton J (1997) A specificity comparison of four antisense types: morpholino, 2′-*O*-methyl RNA, DNA, and phosphorothioate DNA. Antisense Nucleic Acid Drug Dev 7:151–157

Summerton J (1999) Morpholino antisense oligomers: the case for an RNase H-independent structural type. Biochim Biophys Acta 1489:141–158

Summerton J, Stein D, Huang SB, Matthews P, Weller D, Partridge M (1997) Morpholino and phosphorothioate antisense oligomers compared in cell-free and in-cell systems. Antisense Nucleic Acid Drug Dev 7:63–70

Taylor JK, Zhang QQ, Wyatt JR, Dean NM (1999) Induction of endogenous Bcl-xS through the control of Bcl-x pre-mRNA splicing by antisense oligonucleotides. Nat Biotechnol 17:1097–1100

Teplova M, Wallace ST, Tereshko V, Minasov G, Symons AM, Cook PD, Manoharan M, Egli M (1999) Structural origins of the exonuclease resistance of a zwitterionic RNA. Proc Natl Acad Sci USA 96:14240–14245

Thanaraj TA, Robinson AJ (2000) Prediction of exact boundaries of exons. Brief Bioinform 1:343–356

Torigoe H, Hari Y, Sekiguchi M, Obika S, Imanishi T (2001) 2′-O,4′-C-methylene bridged nucleic acid modification promotes pyrimidine motif triplex DNA formation at physiological pH. Thermodynamic and kinetic studies. J Biol Chem 276:2354–2360

Tsui LC (1992) The spectrum of cystic fibrosis mutations. Trends Genet 8:392–398

Valadkhan S, Manley JL (2001) Splicing-related catalysis by protein-free snRNAs. Nature 413:701–707

van Deutekom JC, Bremmer-Bout M, Janson AA, Ginjaar IB, Baas F, den Dunnen JT, van Ommen GJ (2001) Antisense-induced exon skipping restores dystrophin expression in DMD patient derived muscle cells. Hum Mol Genet 10:1547–1554

Wang L, Gryaznov S, Nerenberg M (1999) Inhibition of IL-6 in mice by anti-NF-kappaB oligodeoxyribonucleotide N3′→oligodeoxyribonnucleotide N3′→P5′ phosphoramidates. Inflammation 23:583–590

Wilton SD, Lloyd F, Carville K, Fletcher S, Honeyman K, Agrawal S, Kole R (1999) Specific removal of the nonsense mutation from the mdx dystrophin mRNA using antisense oligonucleotides. Neuromuscular Disord 9:330–338

Zamecnik PC, Stephenson ML (1978) Inhibition of Rous sarcoma virus replication and cell transformation by a specific oligodeoxynucleotide. Proc Natl Acad Sci USA 75:280–284

Zelphati O, Szoka FC Jr (1996) Mechanism of oligonucleotide release from cationic liposomes. Proc Natl Acad Sci USA 93:11493–11498

Zhang H, Cook J, Nickel J, Yu R, Stecker K, Myers K, Dean NM (2000) Reduction of liver Fas expression by an antisense oligonucleotide protects mice from fulminant hepatitis. Nat Biotechnol 18:862–867

Zhuang Y, Weiner AM (1989) A compensatory base change in human U2 snRNA can suppress a branch site mutation. Genes Dev 3:1545–1552

Subject Index

R

S

MIX
Papier aus verantwortungsvollen Quellen
Paper from responsible sources
FSC® C105338

If you have any concerns about our products,
you can contact us on
ProductSafety@springernature.com

In case Publisher is established outside the EU,
the EU authorized representative is:
Springer Nature Customer Service Center GmbH
Europaplatz 3, 69115 Heidelberg, Germany

Printed by Libri Plureos GmbH
in Hamburg, Germany